TM9-729

DEPARTMENT OF THE ARMY TECHNICAL MANUAL

M24 CHAFFEE LIGHT TANK TECHNICAL MANUAL

This manual is correct to 6 February 1951

By DEPARTMENT OF THE ARMY • MAY 1951

©2013 Periscope Film LLC
All Rights Reserved
ISBN#978-1-937684-33-4
www.PeriscopeFilm.com

DISCLAIMER:

This document is a reproduction of a text first published by the Department of the Army, Washington DC. All source material contained herein has been approved for public release and unlimited distribution by an agency of the US Government. Any US Government markings in this reproduction that indicate limited distribution or classified material have been superseded by downgrading instructions promulgated by an agency of the US government after the original publication of the document No US government agency is associated with the publication of this reproduction. This manual is sold for historic research purposes only, as an entertainment. It contains obsolete information and is not intended to be used as part of an actual training program. No book can substitute for proper training by an authorized instructor.

©2013 Periscope Film LLC
All Rights Reserved
ISBN#978-1-937684-33-4
www.PeriscopeFilm.com

DEPARTMENT OF THE ARMY TECHNICAL MANUAL
TM 9-729

This manual supersedes TM 9-729, 27 June 1944; TB 9-729-FE1, 5 October 1944; TB 9-729-FE2, 7 March 1945; TB 9-729-FE4, 15 May 1945; TB 9-729-FE5, 6 June 1945; TB 9-729-FE6, 28 June 1945;- TB 9-729-FE7, 21 July 1945; and those portions of TB 9-307-3, 17 June 1944; TB ORD 30, 3 February 1944; TB ORD 125, 22 July 1944; TB ORD 271, 29 March 1945; TB ORD 335, 4 March 1946; TB ORD 342, 8 July 1946; TB ORD 374, 16 February 1950; TB ORD 377, 27 February 1950; TB ORD 381, 23 March 1950; and TB 9-1729A-2, 20 December 1945 pertaining to the matériel covered herein.

LIGHT TANK

M24

DEPARTMENT OF THE ARMY • MAY 1951

United States Government Printing Office

Washington : 1951

Reproduced by, and sold by Portrayal Press,
P.O. Box 1190, Andover, N.J. 07821 USA

DEPARTMENT OF THE ARMY
WASHINGTON 25, D. C., *28 May 1951*

TM 9-729 is published for the information and guidance of all concerned.

[AG 470.8 (6 Mar 51)]

BY ORDER OF THE SECRETARY OF THE ARMY:

OFFICIAL:
Wm. E. BERGIN
Major General, USA
Acting The Adjutant General

J. LAWTON COLLINS
Chief of Staff, United States Army

DISTRIBUTION:
Tech Svs (2) except 9 (25); Arm & Svc Bd (2); AFF (2); AA Comd (2); OS Maj Comd (10); Base Comd (2); MDW (3); Log Comd (5); A (20); CHQ (2); D (2); R 9 (2), 17 (1); Bn 9 (2), 17 (1); C 9 (2), 17 (1); FC (2); Sch (5) except 9 (50); Gen Dep (2); Dep 9 (10); PE (Ord O) (5), OSD (2); PG 9 (10); Ars 9 (10); Dist 9 (10); One (1) copy to each of the following T/O & E's: 2-25; 6-10N; 6-160N; 7N; 7-25N; 7-26N; SPECIAL DISTRIBUTION.

For explanation of distribution formula, see SR 310-90-1.

CONTENTS

CHAPTER 1. INTRODUCTION   *Paragraphs*   *Page*

		Paragraphs	Page
Section I.	General	1–2	1
II.	Description and data	3–5	2

CHAPTER 2. OPERATING INSTRUCTIONS

Section I.	Service upon receipt of matériel	6–9	7
II.	Controls and instruments	10–42	12
III.	Operation under usual conditions	43–49	30
IV.	Operation of matériel used in conjunction with major item	50–51	37
V.	Operation under unusual conditions	52–57	38

CHAPTER 3. ORGANIZATIONAL MAINTENANCE INSTRUCTIONS

Section I.	Parts, special tools, and equipment for organizational maintenance	58–61	47
II.	Lubrication and painting	62–65	52
III.	Preventive maintenance services	66–69	53
IV.	Trouble shooting	70–84	87
V.	Engine description and maintenance in vehicle	85–95	107
VI.	Engine removal and installation	96–98	124
VII.	Ignition system	99–106	133
VIII.	Starting system	107–110	147
IX.	Fuel, air intake, and exhaust systems	111–116	151
X.	Cooling system	117–122	167
XI.	Generating and charging system	123–125	179
XII.	Batteries and lighting system	126–131	184
XIII.	Instrument panel, instruments, switches, and sending units	132–144	200
XIV.	Radio interference suppression	145–147	210
XV.	Hydramatic transmission	148–153	213
XVI.	Transfer unit	154–156	224
XVII.	Propeller shafts and support bearings	157–160	231
XVIII.	Controlled differential	161–166	238
XIX.	Final drives	167–169	249
XX.	Tracks and suspension	170–181	253
XXI.	Hull	182–192	282
XXII.	Turret	193–199	287
XXIII.	Fire extinguishers	200–201	299
XXIV.	Combination gun mount M64	202–203	301
XXV.	Machine gun mounts	204–206	302
XXVI.	Stabilizer	207–210	304
XXVII.	Maintenance under unusual conditions	211–214	317

III

CHAPTER 4. MATÉRIEL USED IN CONJUNCTION WITH MAJOR ITEM

	Paragraphs	Page
Section I. Armament	215–229	319
II. Sighting and fire control instruments	230–242	335
III. Ammunition	243–247	357
IV. Communication system	248–255	364

CHAPTER 5. SHIPMENT AND LIMITED STORAGE AND DESTRUCTION TO PREVENT ENEMY USE

	Paragraphs	Page
Section I. Shipment and limited storage	256–259	380
II. Destruction of matériel to prevent enemy use	260–264	391

		Page
APPENDIX. REFERENCES		395
INDEX		401

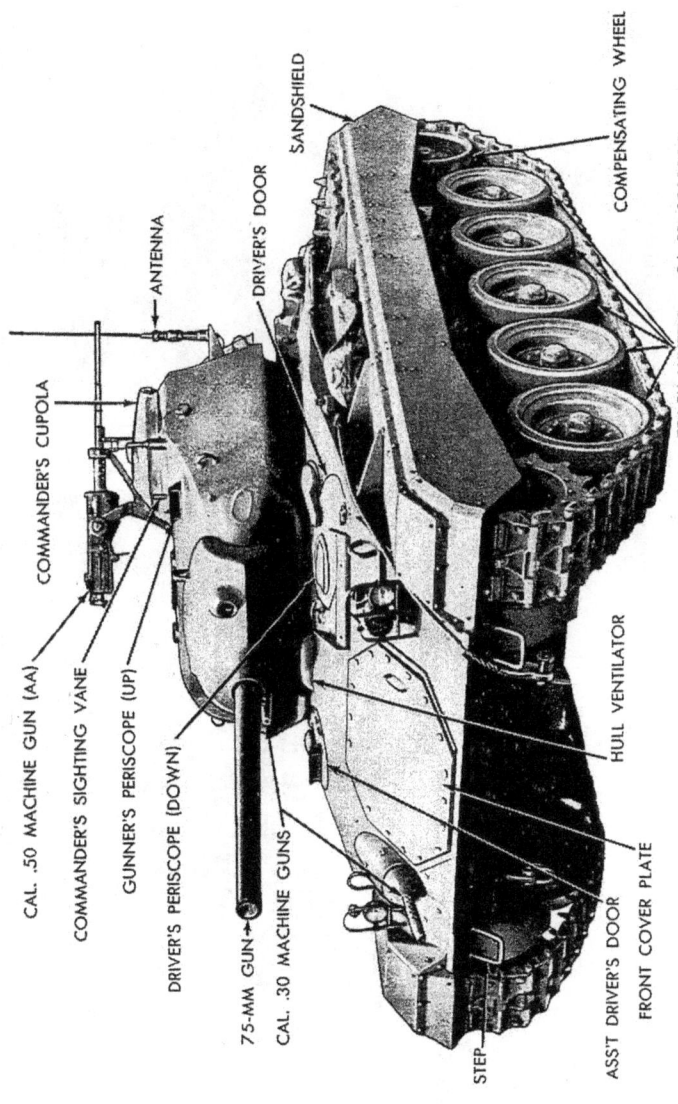

Figure 1. Light tank M24—left front view.

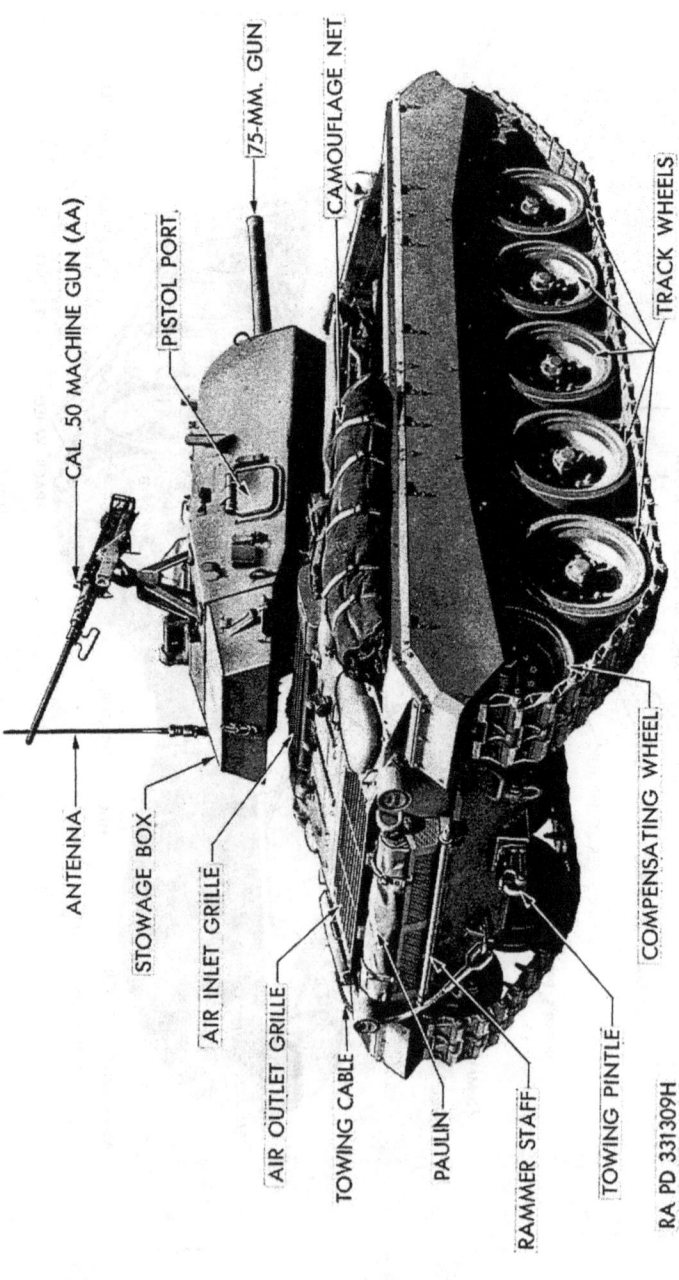

Figure 2. Light tank M24—right rear view.

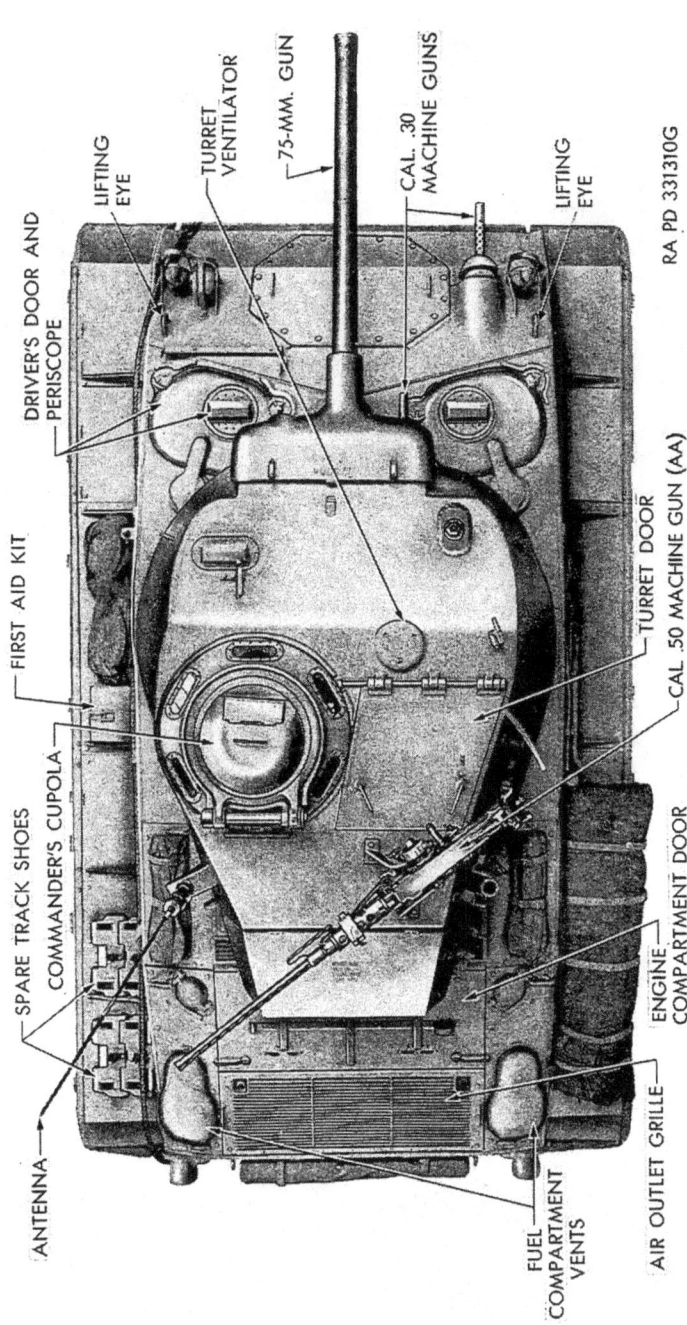

Figure 3. Light tank M24—top view.

VII

RESTRICTED

This manual supersedes TM 9-729, 27 June 1944; TB 9-729-FE1, 5 October 1944; TB 9-729-FE2, 7 March 1945; TB 9-729-FE4, 15 May 1945; TB 9-729-FE5, 6 June 1945; TB 9-729-FE6, 28 June 1945; TB 9-729-FE7, 21 July 1945; and those portions of TB 9-307-3, 17 June 1944; TB ORD 30, 3 February 1944; TB ORD 125, 22 July 1944; TB ORD 271, 29 March 1945; TB ORD 335, 4 March 1946; TB ORD 342, 8 July 1946; TB ORD 374, 16 February 1950; TB ORD 377, 27 February 1950; TB ORD 381, 23 March 1950; and TB 9-1729A-2, 20 December 1945 pertaining to the matériel covered herein.

CHAPTER 1

INTRODUCTION

Section I. GENERAL

1. Scope

a. These instructions are published for the information and guidance of the personnel to whom the matériel is issued. They contain information on the operation and organizational maintenance of the matériel as well as descriptions of major units and their functions in relation to other components of the matériel.

b. The appendix contains a list of current references, including supply catalogs, technical manuals, and other available publications applicable to the matériel.

c. This manual differs from TM 9-729, dated 27 June 1944, as follows:

(1) *Adds information on—*
    (*a*) 75-mm gun M6 and associated fire control equipment and ammunition.
    (*b*) Rubber track (T85E1).

(2) *Revises information on—*
    (*a*) Operation and maintenance under unusual conditions.
    (*b*) Destruction to prevent enemy use.
    (*c*) Organizational maintenance allowances.
    (*d*) Lubrication instructions.
    (*e*) Preventive maintenance schedules.
    (*f*) Trouble shooting.
    (*g*) Engine maintenance.
    (*h*) Ignition system.
    (*i*) Radio interference suppression.
    (*j*) Hydramatic transmission.
    (*k*) Communications equipment.
    (*l*) Stabilizer.

RESTRICTED

(3) *Deletes information on—*
   (a) 75-mm gun M5 and associated fire control equipment and ammunition.
   (b) 2-inch smoke mortar M3.

2. Forms, Records, and Reports

a. GENERAL. Forms, records, and reports are designed to serve necessary and useful purposes. Responsibility for the proper execution of these forms rests upon commanding officers of all units operating and maintaining vehicles. It is emphasized, however, that forms, records, and reports are merely aids. They are not a substitute for thorough practical work, physical inspection, and active supervision.

b. AUTHORIZED FORMS USED WITH THE VEHICLE. The forms generally applicable to units operating and maintaining these vehicles are listed in the appendix. No forms other than those approved for the Department of the Army will be used. Pending availability of all forms listed, old forms may be used. For a current and complete listing of all forms, refer to current SR 310–20–6.

c. FIELD REPORT OF ACCIDENTS. The reports necessary to comply with the requirements of the Army safety program are prescribed in detail in the SR 385–10–40 series of special regulations. These reports are required whenever accidents involving injury to personnel or damage to matériel occur. Whenever an accident or malfunction involving the use of ammunition occcurs, firing of the lot which malfunctions will be immediately discontinued. In addition to any applicable reports required above, details of the accident or malfunction will be reported as prescribed in SR 385–310–1.

d. REPORT OF UNSATISFACTORY EQUIPMENT OR MATERIALS. Any suggestions for improvement in design and maintenance of equipment, safety and efficiency of operation, or pertaining to the application of prescribed petroleum fuels, lubricants, and/or preserving materials, will be reported through technical channels, as prescribed in SR 700–45–5, to the Chief of Ordnance, Washington 25, D. C., ATTN: ORDFM, using DA AGO Form 468, Unsatisfactory Equipment Report. Such suggestions are encouraged in order that other organizations may benefit.

Section II. DESCRIPTION AND DATA

3. Description

a. GENERAL. The light tank M24 is an armored, full-tracklaying combat vehicle mounting a 75-mm gun. It carries a crew of four men; a commander, a gunner, a driver, and an assistant driver or five men

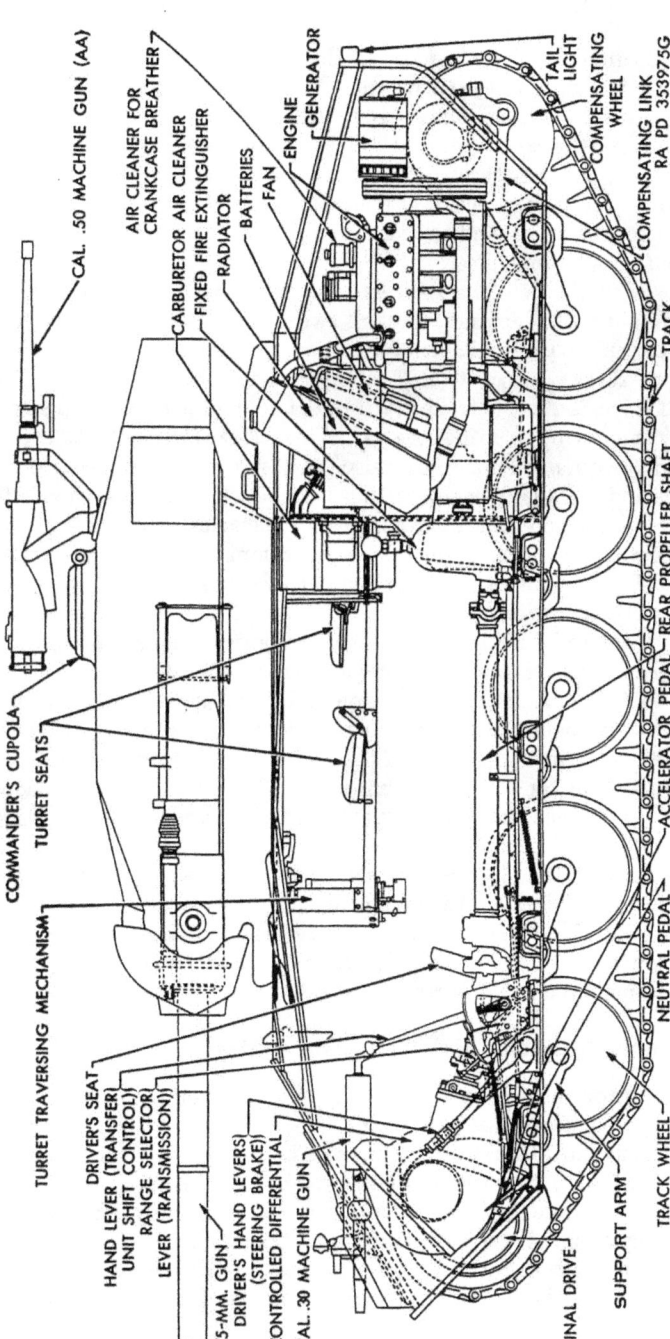

Figure 4. Light tank M24—longitudinal cross section.

when a cannoneer is included. In the four man crew, the assistant driver moves up into the turret and serves as loader when the vehicle is in combat. In the five man crew, the assistant driver operates the bow gun and the cannoneer loads. The vehicle is designed with a driving compartment at the front, a fighting compartment in the center, and an engine compartment at the rear. Dual controls are provided, one set for the driver and one for the emergency operation of the vehicle by the assistant driver.

b. POWER TRAIN. The vehicle is driven by two 8-cylinder, 90°, V-type liquid-cooled engines, through two hydramatic transmissions, a transfer unit with mechanically selected speed ranges (two forward and one reverse), a controlled differential for steering and braking which is located in the front of the hull, two final drives, and the necessary connecting propeller shafts. Wide rubber or steel block tracks provide traction. Torsion bar suspension is used for the dual track wheels and includes a compensating wheel at the rear on each side to keep track tension constant regardless of obstacles.

c. HULL. The hull of this vehicle is a completely welded structure, except for portions of the front, top, and floor, which are removable for service operations. The hull is divided into two compartments: the fighting and driving compartment at the front and an engine compartment at the rear. These compartments are separated by a bulkhead that extends from side to side and from the roof down to the bulkhead extensions, which in turn extend forward far enough to cover the transfer unit. The front of the hull slopes downward at the top and upward at the bottom to form a "V." The sides of the hull slope inward at the bottom. Lifting eyes are provided and should be used if the tank is to be lifted.

d. TURRET. The fighting compartment comprises a turret of approximately 60 inches inside diameter, mounted on a continuous ball bearing mounting, and having a 360° traverse by means of a power or hand traversing mechanism. The 75-mm gun and a coaxial machine gun are mounted in the turret.

4. Name, Caution, and Instruction Plates

a. INSTRUMENT PANEL CAUTION PLATE. This plate (figs. 6 and 10) gives engine starting information.

b. GEAR SHIFT INSTRUCTION PLATE. This plate (figs. 6 and 20) gives direction for shifting the transmission selector lever.

c. GENERATOR REGULATOR INSTRUCTION PLATE. This plate (left, fig. 5 and right, fig. 87) gives connection information.

d. PORTABLE FIRE EXTINGUISHER INSTRUCTION DECALCOMANIA. This decalcomania (fig. 23) gives instruction in the use of the fire extinguisher.

e. VENTILATING BLOWER INSTRUCTION PLATE. This plate (fig. 12) indicates switch handle positions for operation of the blower.

f. TURRET CONTROL SWITCH INSTRUCTION PLATE. This plate (fig. 19) identifies the switch buttons.

g. OIL FILTER INSTRUCTION DECALCOMANIA. This decalcomania (fig. 41) gives instructions on replacing filter element.

h. CHOKE THERMOSTAT INSTRUCTION. This instruction (fig. 69) gives instruction and warning in adjustment of choke thermostat.

i. AUXILIARY FUEL FILTER INSTRUCTION. This instruction (fig. 73) gives cleaning instructions.

j. GENERATOR INSTRUCTION PLATE. This plate (fig. 88) gives rotational direction information.

k. TURRET MOTOR NAME PLATE. This plate (fig. 162) includes manufacturer's model number, serial number and operating characteristics.

l. COMMUNICATION EQUIPMENT. Each item has a name plate for identification and indicating operating characteristics.

5. Tabulated Data

a. GENERAL DATA.

Crew	4 or 5 men
Armament:	
75-mm gun M6	1
Cal. .50 machine gun M2, HB	1
Cal. .30 machine gun M1919A4	2
Communication system	radio and interphone
Weight (fighting)	40,500 lb
Length (gun in traveling position)	16 ft 6 in.
Width	9 ft 8 in.
Height	9 ft 1 in.
Ground clearance	1 ft 6 in.
Ground pressure	11.08 psi
Engine	(2) Cadillac, 90° V-type, 8 cyl, Model 44T24
Electrical system	24 V
Number of batteries (6 V)	4
Hull armor:	
Type	welded steel plate
Front: upper 1 in. at 60°, lower	1 in. at 45°
Sides: front 1 in. at 78°, rear	¾ in. at 78°
Rear	¾ in.
Top	½ in.
Floor: front ½ in., rear	⅜ in.

Turret armor:
- Type ------ welded steel plate
- Sides ------ 1 in. at 68°
- Top ------ ½ in.
- Shield: upper ------ 1½ in. at 30°, lower ------ 1½ in. at 40°

Capacities:
- Fuel tank (2 tanks, 55 gal each) ------ 110 gal
- Cooling system (40 qt each independent system) ------ 80 qt
- Crankcase (refill) (9 qt each engine) ------ 18 qt
- Transmission (15 qt each transmission) ------ 30 qt
- Differential ------ 20 qt
- Transfer unit ------ 4½ qt
- Final drives (2 qt each) ------ 4 qt

b. PERFORMANCE.
- Allowable speed ------ 34 mph
- Cruising speed ------ 25 mph
- Grade ascending ability (max) ------ 60 percent
- Turning circle (diam) (right or left) (min) ------ 46 ft
- Fording depth (max) ------ 3 ft 4 in.
- Width of ditch vehicle will cross (max) ------ 8 ft
- Vertical obstacle vehicle will climb (max) ------ 3 ft
- Fuel consumption ------ 0.875 mpg
- Cruising range ------ 100 miles

c. DETAILED DATA REFERENCES. Additional detailed tabular data pertaining to individual components and systems are contained in the following paragraphs:

	Paragraph
Armament	217
Ammunition (authorized rounds)	244
Brake system	161
Communication system	248
Cooling system	117
Differential	161
Electrical system:	
Generator, and charging system	123
Ignition system	99
Lighting system and batteries	126
Starting system	107
Engine	85
Final drives	167
Fuel system	111
Propeller shafts and support bearings	157
Sighting and fire control instruments	231
Tracks and suspension	170
Transmission	148

CHAPTER 2

OPERATING INSTRUCTIONS

Section I. SERVICE UPON RECEIPT OF MATÉRIEL

6. Purpose

a. When a new or reconditioned vehicle is first received by the using organization, it is necessary for the organizational mechanics to determine whether the vehicle has been properly prepared for service by the supplying organization and is in condition to perform any mission to which it may be assigned when placed in service. For this purpose, inspect all assemblies, subassemblies, and accessories to be sure they are properly assembled, secure, clean, and correctly adjusted and/or lubricated. Check all tools and equipment with the listing in the ORD 7 supply catalog to be sure every item is present, in good condition, clean, and properly mounted or stowed.

b. In addition, perform a "run-in" of at least 50 miles on all new or reconditioned vehicles and a sufficient number of miles on used vehicles to completely check their operation according to procedures in paragraph 9 herein.

c. Whenever practicable, the vehicle crew will assist in the performance of these services.

7. Correction of Deficiencies

Deficiencies disclosed during the course of the "run-in" will be treated as follows:

a. Correct any deficiencies within the scope of the maintenance levels of the using organization before the vehicle is placed in service.

b. Refer deficiencies beyond the scope of organizational maintenance to ordnance maintenance personnel for correction.

c. Bring deficiencies of a serious nature to the attention of the supplying organization through proper channels.

8. Preliminary Service

a. FIRE EXTINGUISHERS. See that the portable and fixed cylinders are securely stowed. Inspect the operating valves. If the valves appear to have been opened or damaged, report to proper authorities for exchange or refill. Examine the lines, nozzles, and control cables of the fixed system to see that they are in good condition and that the nozzles are not corroded or clogged.

b. FUEL, OIL, AND WATER. Check the fuel tanks for quantity and fill if necessary.

Caution: Be sure master battery switch is in "OFF" position at all times while filling fuel tanks.

Check level of coolant in radiator and add as necessary to bring to correct level (par. 118). Allow room for expansion in fuel tanks and radiators. During freezing weather, test strength of antifreeze and add as necessary to protect system against freezing. Check level of oil in each engine.

Caution: If there is a tag in a conspicuous place in driver's compartment concerning engine oil in crankcases, follow instructions on tags before driving vehicle.

Note. Vehicles shipped from depots may have white unleaded gasoline in the tanks. This should be drained and 80-octane used for filling.

c. FUEL FILLER ELEMENT. Inspect filler element at carburetor inlet for leaks, damage, and accumulation of foreign matter (par. 111). Be sure fuel switches and shut-off valves operate properly. Drain accumulated dirt and water from bottom of the fuel tanks (par 114). Drain only until fuel runs clean. When draining dirt and water from fuel tanks, be certain that no fuel drains into or remains inside hull.

d. BATTERIES. Remove battery covers and clean any accumulated dust or dirt from tops of batteries. Make hydrometer and voltage test of batteries and if necessary, add distilled or clean water to bring electrolyte to $\frac{3}{8}$ inch above plates (par. 127).

e. AIR CLEANERS AND BREATHERS. Examine carburetor air cleaners (par. 115) and crankcase breather cleaners (par. 90) to see that they are in good condition and secure. Remove air cleaner elements, wash in dry-cleaning solvent or volatile mineral spirits paint thinner, and fill reservoirs with oil to the correct level as directed in lubrication order (par. 62). Install securely and make sure all connections are in good condition and ducts and air horn connections are tight.

f. ACCESSORIES AND BELTS. See that accessories such as carburetors, generators, regulators, distributors, starters, water pumps, and fans on both engines are securely mounted and that drive belts are in good condition and adjusted with $\frac{5}{8}$- to $\frac{3}{4}$-inch finger-pressure deflection (fig. 37).

g. ELECTRICAL WIRING. Examine all accessible wiring and conduits to see that they are in good condition, securely connected, and properly supported.

h. TRACKS. Examine track blocks for damage and loose or damaged nuts and connections. See that tracks are correctly assembled and mounted and that tension is adjusted so that there is a 2- to $2\frac{1}{4}$-

inch sag midway between the second and third support rollers (par. 171).

i. WHEEL AND FLANGE NUTS. See that sprockets, compensating wheel, track wheel and support roller assembly, mounting and flange nuts are present and secure.

j. FENDERS AND SHIELDS. See that fenders, brush guards, and sand shields (if tank is so equipped) are in good condition and secure and that shield hinges operate properly.

k. TOWING CONNECTIONS. Examine tow loops and pintle hook, for looseness and damage.

l. HULL. See that all hull attachments, hardware, lift loops, doors, hatches, and their release and locking mechanisms are in good condition, are secure, and operate properly. Examine entire hull and turret for looseness or damage.

m. VISION DEVICES. Inspect periscope prisms and windows (mounted or spares) and cupola vision block to see if they are in good condition and clean. See that mounted units are secure in holders and that holders are properly mounted. Test each periscope to be sure it will elevate, traverse, and depress through full range.

Caution: Clean periscopes with a soft clean cloth or brush. Periscope windows and eyepieces should be wiped with lens tissue only.

n. LUBRICATION. Perform a complete lubrication of the vehicle, covering all intervals according to instructions on lubrication order (par. 62) except gear cases and units lubricated or serviced in *a* through *m* above. Check all gear case oil levels and add oil as necessary to bring to correct levels; change oil only if condition of oil indicates the necessity or if oil is not of proper grade for existing atmospheric temperature.

Note. Perform *o* through *s* below during lubrication.

o. SUSPENSIONS. Examine all suspension brackets, arms, and shock absorber links and guides to see if they are in good condition, correctly assembled, secure, not leaking excessively at seals, or excessively worn. Inspect roller and wheel tires for damage or separation from rollers or wheels.

p. STEERING LINKAGE. Inspect all shafts, arms, rods, connections, levers, and grips to see if they are in good condition, correctly and securely assembled and mounted, and that they operate without excessive looseness or binding. Check brake linkage adjustment by pulling back evenly to make sure free travel is equal and does not exceed six notches and that brakes are fully applied within three additional notches (par. 162). Be sure parking brake mechanism locks the brake properly in the locking position.

q. PROPELLER SHAFTS. Remove all propeller shaft covers, and inspect shafts and universal joints to see if they are in good condition, correctly assembled and alined, secure, and not leaking excessively at seals.

r. VENTS. Examine breathers in each final drive to be sure they are in good condition, secure, and not clogged.

s. CHOKE. When starting engines (*t* below), observe if action of automatic choke is satisfactory. Adjust choke setting, if necessary, according to instructions (par. 112).

t. ENGINE WARM-UP. Start engines (par. 44), noting if there is any tendency toward difficult starting, inadequate starting speed, or any unusual noises. Set hand throttle to idle engines at 625 rmp during warm-up.

u. INSTRUMENTS (fig. 6).
  (1) *Warning signals.* Do not drive vehicle while engine or transmission warning signal lights are on.
  (2) *Ammeter.* Ammeter should show slight positive (+) charge at engine speeds above 1,100 rmp. High charge may be indicated until generator restores to battery current used in starting.
  (3) *Engine temperature gages.* Engine temperature should rise gradually during warm-up to normal operating range, 160° to 210° F. Maximum safe operating temperature is 240° F. at which point the red warning signal lights.
  (4) *Tachometers.* These instruments should register the engine speeds in revolutions per minute and record accumulating engine revolutions.

v. ENGINE CONTROLS. Observe if engines respond properly to controls and controls operate without excessive looseness or binding.

w. IGNITION TIMING. Check ignition timing and reset if necessary, as soon as engines are warm and before driving vehicle (par. 104).

x. SIREN. See if siren or horn is in good condition and secure. If tactical situation permits, test for proper operation and tone.

y. LAMPS (LIGHTS) AND REFLECTORS. Clean lenses and inspect all units for looseness and damage. If tactical situation permits, operate all light switches to see if lights respond properly.

z. LEAKS—GENERAL. Look under vehicle and within engine and fighting compartments for indications of fuel, oil, or coolant leaks. Trace any leaks found to their source and correct or report them to higher authority.

aa. TOOLS AND EQUIPMENT. Check tool and equipment lists (par. 61) to be sure all items are present and properly mounted or stowed. Inspect paulin on camouflage net to see that they are in good condition and securely stowed.

9. Run-In Procedures

For operation of the vehicle during "run-in," see instructions in section III. Perform the following procedures (*a* through *f* below) during the road test of the vehicle. Since these vehicles have been driven 50 miles or more before delivery to the using organization, confine the road test to the least mileage necessary to make observations listed below.

Caution: Avoid continuous operation of the vehicle at speeds approaching maximums indicated.

a. INSTRUMENTS AND GAGES. Do not move vehicle until engine temperature reaches 140° F. Maximum safe operating temperature is 240° F. at which point the red warning signal lights. Observe readings of ammeter, tachometers, temperature gages, and warning signal lights to be sure they are indicating the proper functioning of the units to which they apply. Also see that speedometer registers vehicle speed and that odometer records accumulating mileage.

b. BRAKES—STEERING AND PARKING. Test steering brakes to see if they will stop vehicle effectively with levers even and that they start to meet resistance at approximately six notches from released position. See that parking lock mechanism holds levers in applied position. Test levers separately to be sure vehicle steers properly with normal pull.

c. TRANSMISSION AND TRANSFER UNIT. Vehicle should start satisfactorily in either "DRIVE" or "LOW" range. Be alert for abnormal conditions in transmissions indicated by excessive slippage and engines racing when under load, severe shifting, or shifting in wrong speed range; also watch for unusual conditions in transfer unit, indicated by hard shifting or severe gear clash. Gears should operate without unusual noise and should not slip out of mesh during operation.

d. ENGINES. Be on the alert for any unusual engine operating characteristics or unusual noise, such as lack of pulling power or acceleration, backfiring, misfiring, stalling, overheating, or excessive exhaust smoke. Observe if engines respond properly to all controls and control synchronization appears to be satisfactory.

e. UNUSUAL NOISES. Be alert throughout road test for noise from hull or attachments, suspensions, running gear, or tracks that might indicate looseness, damage, excessive wear, or inadequate lubrication.

f. HALT VEHICLE AT 10-MILE INTERVALS OR LESS FOR SERVICE.

(1) *Temperatures.* Cautiously hand-feel each wheel, sprocket, idler, and support roller hub for abnormal temperatures. Examine transmissions, transfer unit, and controlled differential and final drives for indications of overheating or excessive leaks at seals, gaskets, or vents.

(2) *Leaks.* With engines running and fuel, oil, and cooling systems under pressure, look within engine and fighting compartments and under vehicle for leaks.

g. GUN ELEVATING AND TRAVERSING MECHANISMS. Inspect all mounted guns to see if they are in good condition, secure, and clean. Test elevating, traversing, and firing mechanism for proper operation. Place vehicle in position where it is tilted sidewise about 10°; traverse turret by manual and power controls to see if it operates through full 360° range and that turret gun can be elevated with hand control without binding, excessive lash, or erratic action.

h. TRACK TENSION. Check tracks to be sure they are adjusted to correct tension (par. 171).

i. VEHICLE PUBLICATIONS AND REPORTS.

(1) *Publications.* See that vehicle technical manuals, lubrication order, and Standard Form 91 (Operator's Report of Motor Vehicle Accident) are in the vehicle, legible, and properly stowed.

Note. USA registration number and vehicle nomenclature must be filled in on DA AGO Form 478 (MWO and Major Unit Assembly Replacement Record and Organizational File) for new vehicles. Forms 478 are carried in organization files. (Where winterization kits are issued, check for pertinent technical bulletin.)

(2) *Reports.* Upon completion of the "run-in," correct or report any deficiencies noted. Report general condition of the vehicle to designated individual in authority.

Section II. CONTROLS AND INSTRUMENTS

10. General

This section describes, locates, illustrates, and furnishes the crew sufficient information pertaining to the various controls and instruments provided for the proper operation of the vehicle. For the use of these controls and instruments in operation of the vehicle, refer to section III of this chapter. For the description and operation of controls and instruments pertaining to the armament, refer to chapter 4, section I.

11. Master Battery Switch

The master battery switch is located on the front of the bulkhead near the left side wall (fig. 5).

Note. Early production model vehicles did not have the remote control handle for the master battery switch, but have since been modified to include this handle.

Figure 5. Master battery switch.

12. Emergency Ignition Switch

An "emergency ignition" switch is located on the hull roof slightly to the rear of the driver (fig. 12).

Note. The wiring is so arranged (in series) that this "emergency ignition" switch must be kept in the "ON" position in order that the regular ignition switches on the instrument panel may function (fig. 6).

13. Ignition Switches

Two ignition switches, one for each engine, are located in the lower right corner of the instrument panel and are protected from accidental operation or damage by a sheet metal guard (fig. 6). These switches have two positions: "OFF," with the switch levers vertical and "ON," with the levers turned 90° clockwise. The right and left switches are for the right and left engines respectively. When the ignition switches are turned on, the warning lights light up.

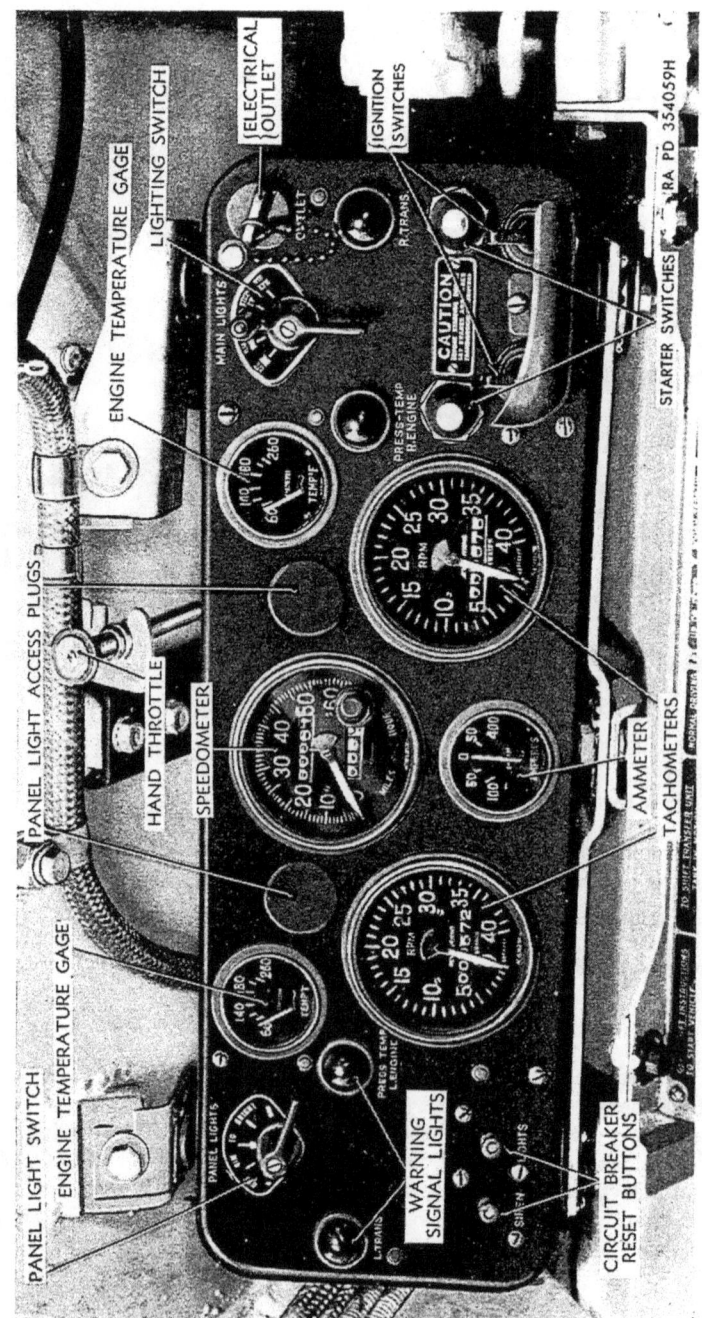

Figure 6. Instrument panel.

14. Fuel Pump Controls

The fuel pump controls consist of two levers mounted at the front center of the driving compartment (fig. 7). The left lever controls the fuel shut-off valve and the power feed to the fuel pump in the left fuel tank. The right lever controls the same devices for the right tank. When the levers are moved up, the valves are closed and the pumps are turned off. The two levers can be turned on or off independently.

Note. At least one of the ignition switches must be turned on to complete the feed circuit to the fuel pumps.

15. Starter Switches

The two starter switches are located in the lower right side of the instrument panel (fig. 6). Start both engines at the same time (except in extremely cold weather) by pressing both starter switch buttons.

16. Lighting Switches

The main lighting switch, located at the upper right of the instrument panel (fig. 6) is in the "OFF" position when the handle is vertical. Turning the switch counterclockwise to its first position turns on the blackout marker lights (labeled BO MK). Turning counterclockwise to the second position turns on the blackout driving light in addition (labeled BO DR), provided the blackout driving light is installed in place of the left headlight. Turning the switch to the first clockwise position from the "OFF" position provides for operation of the stop light for daytime driving (labeled STOP LT). Turning it to the second clockwise position turns on the headlights (labeled HD LTS). In order to move the switch to the clockwise positions, it is necessary first to depress the lock-out button at the top of the switch. The intensity of illumination of the instrument panel lights is controlled by a lever type switch handle at the upper left of the instrument panel (fig. 6). There are four interior lights. Two are located in the hull top, one above and forward of assistant driver and one between the driver and assistant driver. Two are located in the top of the turret. Each light has two lamps and two lenses, clear and red, which can be selected by means of the individual switch provided in each unit. The red light is normally used; to turn on the white light it is necessary first to release the stop.

17. Siren (or Horn) Switch

The siren (or horn) switch is located on the lower hull front plate and is operated by depressing the button located above the driver's left foot position (fig. 7).

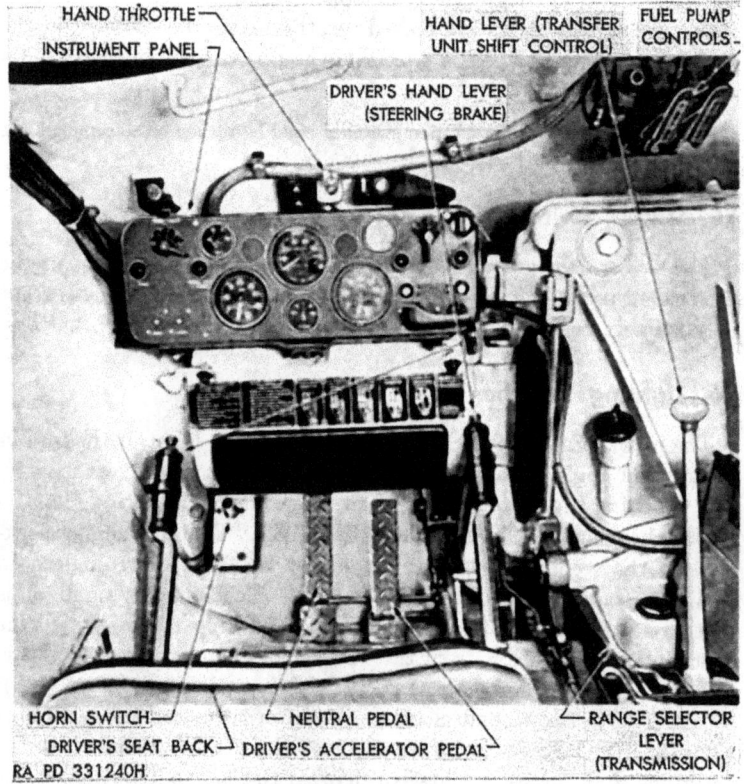

Figure 7. Driving controls.

18. Accelerator Pedal and Hand Throttle

Individual foot accelerator pedals are provided for the driver and assistant driver. Each accelerator pedal controls both engines. The accelerator pedal on the assistant driver's side should be disengaged when not in use by releasing the spring-loaded plunger provided on the assistant driver's side of the front throttle control relay (fig. 8). The hand throttle is on the driver's side, just above the instrument panel (fig. 7). The hand throttle is self-locking in any position and is released by pressing a spring button in the center of the control knob.

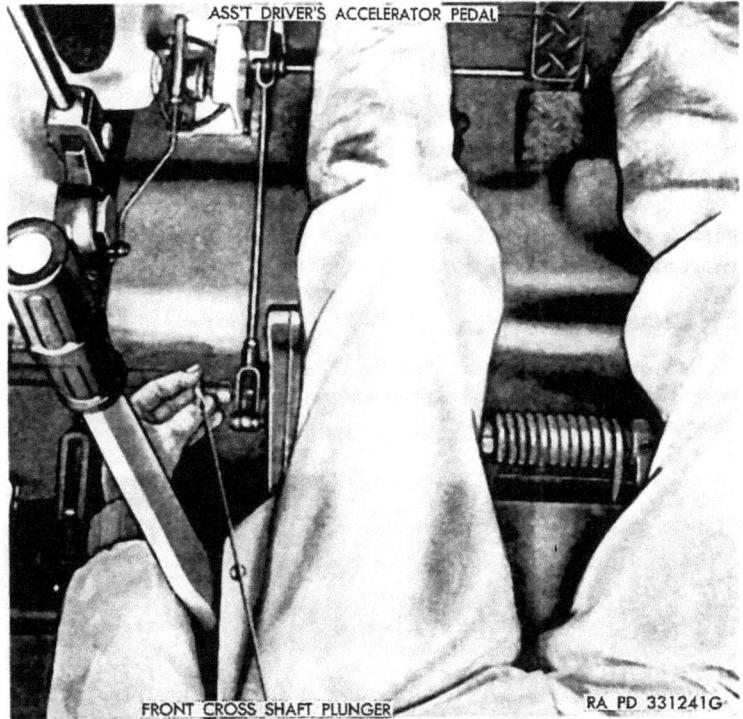

Figure 8. Disconnecting assistant driver's accelerator pedal.

19. Choke

An automatic choke mechanism is located on each carburetor to provide the correct fuel mixture for starting the engines. Fast idle with cold engines is obtained by using the hand throttle.

20. Spark Control

The spark control advance is provided by means of centrifugal weights in the distributor. It is fully automatic and governed by engine speed. It requires no attention from the driver.

21. Steering Brake Levers

Dual steering brake levers are mounted on the floor of the vehicle (fig. 7). One pair is provided for the driver and another for the assistant driver. To steer the vehicle, pull the lever on the side toward which it is desired to turn. Either set of levers may be swung forward when not in use.

22. Brakes

Pulling back simultaneously on both steering levers slows down or stops the vehicle, depending on the effort applied. The stop light is operated whenever both levers are pulled back, provided the main lighting switch is in any position other than "OFF." Parking brake controls are provided on the driver's steering brake levers. These levers may be locked in the applied position by depressing the small knob in the center of each steering lever (fig. 9). Parking brakes are released by pulling back on the steering levers.

23. Clutch

There is no clutch or clutch pedal. The use of the fluid coupling between the engines and the transmissions eliminates the need for a foot-operated clutch.

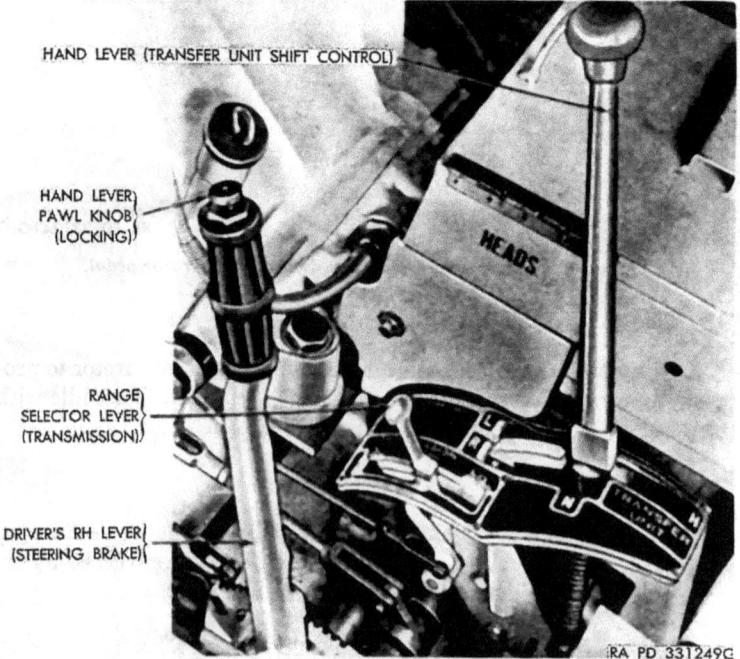

Figure 9. Transmission shifting controls.

24. Shifting Controls

Two shifting controls are provided; a range selector lever for the hydramatic transmissions and a manual shift control lever for the transfer unit (fig. 9). The transmission selector lever has three positions: "NEUTRAL," "DRIVE," and "LOW." This control does not shift any gears, but simply positions the valves in the transmission control mechanism. "NEUTRAL" prevents the units from transmitting the engine power; "DRIVE" permits shifting through all four speeds; and "LOW" limits the transmission to first and second speeds, in order to utilize the power of the engine through gear reduction for braking or low speed operation on very rough terrain. The transfer unit shift lever has three positions besides "NEUTRAL." Moving the lever to the right and rearward provides "HIGH" range; moving the lever forward provides "LOW" range; and moving the lever to the left and forward provides "REVERSE" (fig. 9).

25. Neutral Pedal

Although the hydramatic transmissions are automatic in operation, the transfer unit speed ranges are selected by manual shifting. In order to shift the transfer unit, it is necessary to have the hydramatic transmissions in "NEUTRAL" while the shift is being made. The transmissions can be shifted in "NEUTRAL" in one of two ways; either by moving the transmission selector lever to "NEUTRAL" (fig. 9) or by depressing the neutral pedal (fig. 7). The neutral pedal acts through the same linkage as the selector lever, but permits return to the selected lever position as soon as the pedal is released.

26. Spot Light

Tanks of early manufacture have a spot light which is located in the turret just forward of the commander's vision cupola (fig. 1). This light has a dual purpose. It can be operated for the purpose of illuminating the path selected by the driver or as a means of signaling. The switch is located at the operating handle for signaling. The spot light can be removed and attached to a special switch handle, the cord to which is carried on a self-winding reel. When spot lights already on tanks become inoperative or damaged, they are removed and no replacement is made.

27. Driver's Doors

The operation of each driver's door is identical.

a. To open, release both latches at front end of doors by turning latch handles approximately 90° (fig. 10). Pull down on driver's door hinge operating lever and push up on the opposite side to raise door; then rotate lever to swing door to open position, again pull down on lever, and push up on opposite side to lock in this position. The over-center action of the hinge operating lever keeps the door locked on the support.

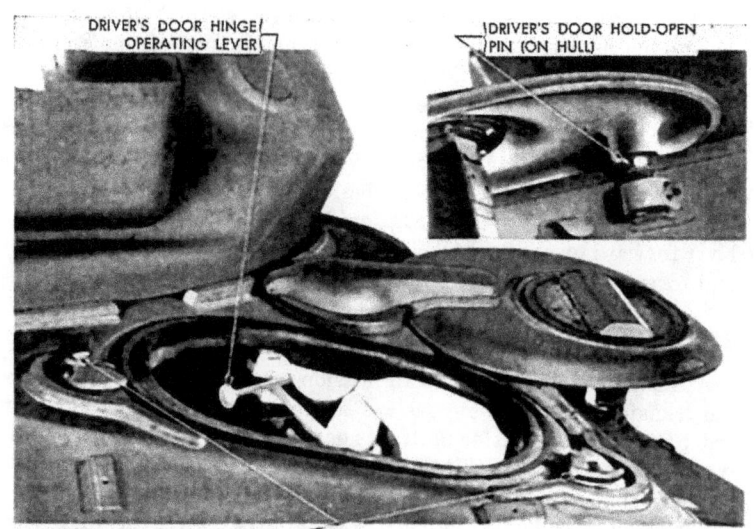

Figure 10. Driver's door controls.

b. To close driver's doors, pull down on door hinge operating lever and push up on opposite side, releasing door from support. After swinging door to closed position, pull down on lever and push up on opposite side to lock in closed position. Turn the two latches at the front of the doors 90°.

28. Seat Adjustment

Seats for the driver and assistant driver have two adjustment ranges. First—4-inch travel forward or backward, in 1-inch variations, is obtained by depressing the horizontal release lever on the left side of the seat (fig. 11) and sliding the seat forward or backward as desired. Second—there is a vertical adjustment range of 10 inches in 2-inch variations either up (to permit driving with doors open) or down (for driving when doors are closed). The lever for this adjustment is on the right side of each seat (fig. 11). To raise the seat, raise the vertical release lever and take weight off seat. It will move upward and forward into position by spring action. To lower the seat, raise the vertical release lever and body weight will force the seat down.

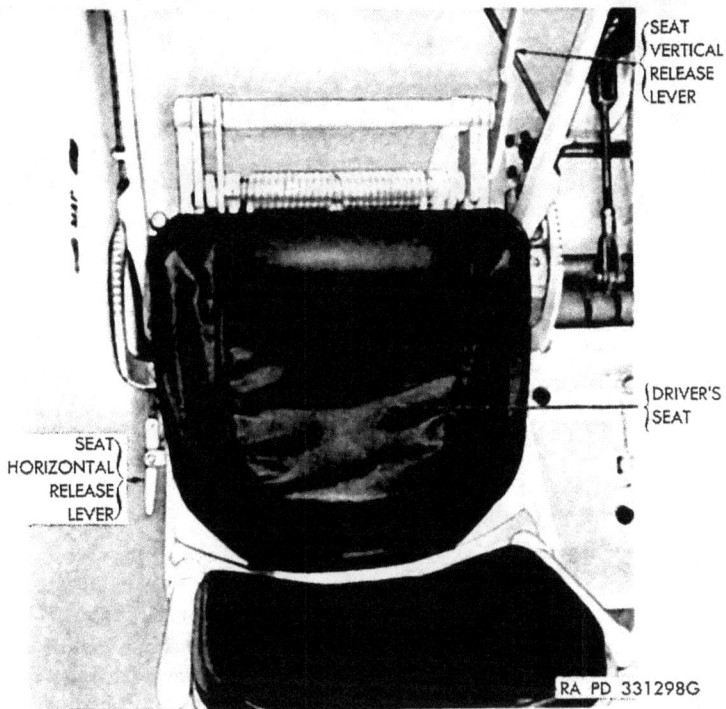

Figure 11. Driver's seat adjustment controls.

29. Ventilating Blower Controls

A two-way ventilating blower is located on the hull roof between the driver and assistant driver and is operated by a four-position switch (fig. 12). To draw air into the fighting compartment from the outside, the switch is turned to the left ("AIR-IN" position) and to the "FAST" or "SLOW" position as desired. To take foul air out of the fighting compartment, the switch is turned to the right ("AIR-OUT" position) and to the "FAST" or "SLOW" position as desired. A manual reset circuit breaker is located on the forward side of the unit and acts as a fuse in the circuit. A shut-off valve is provided in each air outlet elbow which will completely block the movement of air in or out of the vehicle at any time.

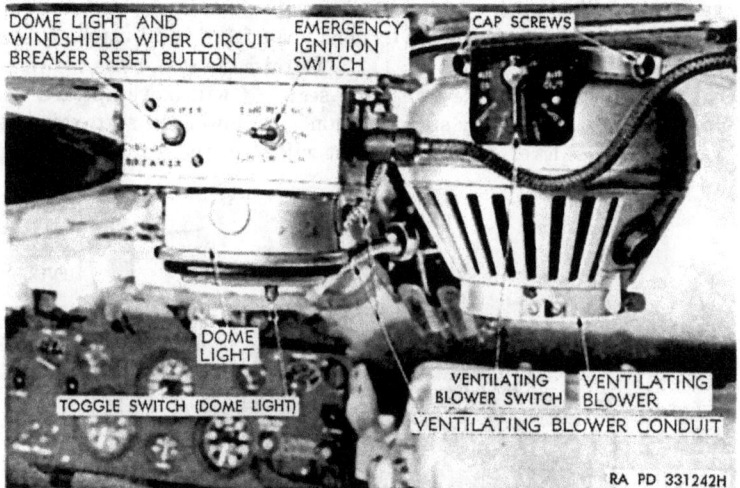

Figure 12. Ventilating blower controls and emergency ignition switch.

30. Ventilating Doors in Bulkhead

Additional ventilation for the fighting compartment can be secured by opening the ventilating doors in the bulkhead (fig. 156). This permits the engine fans to draw smoke-filled air from the fighting compartment when the guns are being fired.

31. Escape Door

a. OPENING. Remove subfloor over escape hatch by releasing subfloor lock located under floor at front. Subfloor section can then be slid forward and removed. Raise release lever on escape door (fig. 13) and let entire door (and 75-mm gun spare parts box) drop to the ground.

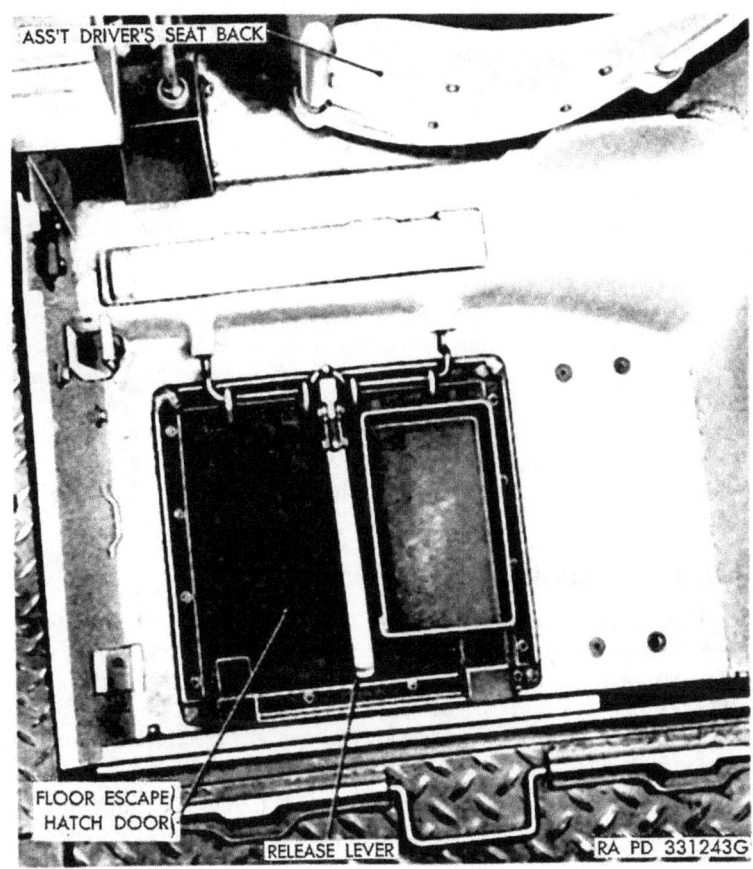

Figure 13. Escape door.

b. INSTALLING AND LOCKING. To install, raise rear end of door through opening in hull floor and hook clips over edge of opening, being sure door is pushed back as far as possible. Hold release lever in a vertical position and raise door into place. Swing lever downward towards door into its locking position until outer end of the release lever snaps snugly to the door.

32. Periscopes

a. The periscope M6 (fig. 14) or M13 (fig. 15) is used by the driver and assistant driver for observation from the interior of the tank. One periscope is secured in a holder in the driver's door and another in assistant driver's door by a latch mechanism and a knob. It may be clamped in either a viewing or retracted position. The periscope M6 is a hollow metal type with a head of plastic material that will shatter into small pieces if struck by a projectile. The periscope M13 is a solid plastic type which replaces the M6.

b. To operate, grasp the sides of the periscope and rotate, elevate, or depress the holder and periscope until the desired panorama is brought into the field of view.

c. To remove the periscope from the holder, open the latch on the front of the holder and loosen the knob on the front of the periscope. Pull the periscope straight out of the holder.

d. To remove the head from the periscope body, turn the two latches until the head is completely disengaged. Lift the head from the periscope body. When replacing the head, position it on the top of the periscope body with the window facing the front side of the periscope. Turn the latches until the reference arrow on each eccentric matches the corresponding arrow on the periscope body. See that the handles of the eccentric mechanism lie flat on the sides of the periscope body.

33. Circuit Breaker Reset Buttons

The circuit breaker reset button for the siren circuit is located at the extreme lower left of the instrument panel, while the button for the driving lights circuit is located just to the right of this first button. These circuit breakers are of the manual-reset type. The circuits can be closed, after the breakers have opened either circuit, simply by pressing the reset button.

Warning: Do not hold button down. If breaker continues to open each time button is pressed, investigate cause of excessive current draw (par. 77).

34. Ammeter

An ammeter is located in the lower center of the instrument panel. This ammeter indicates the amount of charge or discharge in the main battery circuit.

35. Warning Signal Lights

Four warning signal lights are located on the instrument panel. The two on the left are for the left engine and transmission, and the two on the right are for the right engine and transmission. In both locations the light to the left, when lit, indicates dangerously low transmission oil pressure (below 60 psi). The light to the right, when lit, indicates that the engine oil pressure has dropped below 10 psi or that the engine water temperature has risen above 240° F.

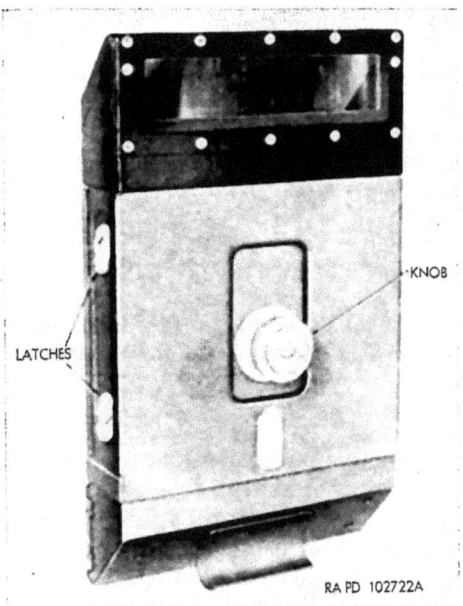

Figure 14. Periscope M6.

Figure 15. Periscope M13.

36. Engine Temperature Gages

Two temperature gages, one for each engine, are located in the upper center of the instrument panel. These gages are electrically connected to sending units mounted on the engines and register engine water temperature. Normal operating range is from 160° to 210° F. Maximum temperature is 240° F. at which temperature the red signal on the panel lights.

37. Speedometer

A speedometer in the upper center of the instrument panel indicates the speed of the vehicle as well as the total milage traveled.

38. Tachometers

Two tachometers, one for each engine, are located at the left and right of, and slightly below the speedometer, on the instrument panel. These instruments are connected to the distributor drive shafts through flexible drive shafts. They indicate the speed of each engine and the total revolutions each engine has been operated. Speed is indicated on the dial in hundreds, that is, 10 indicates 1,000 rpm and 25 indicates 2,500 rpm. Total revolutions are indicated in thousands.

39. Panel Light Access Plugs

Two access plugs, located one on each side on the speedometer, on the instrument panel cover the location of the two panel lights. These panel lights can be replaced by first removing the access plugs.

40. Electrical Outlet

An electrical outlet is located at the upper right-hand corner. Its purpose is to allow for connection of auxiliary electrical equipment such as a trouble light.

41. Turret Doors

a. DESCRIPTION (fig. 3). There are two doors in the top of the turret. One, on the right side of the turret top, is for the loader. Another, in the vision cupola to the left of the loader's door, is for the commander or gunner.

b. OPERATION. To open the commander's cupola door, the latch must first be released. This will allow it to spring partially open due to the tension of leaf springs. Continue to raise the door until it swings toward rear of vehicle where it rests on and is held by the door hold-open lock (fig. 16). To open the turret door, release the two latches

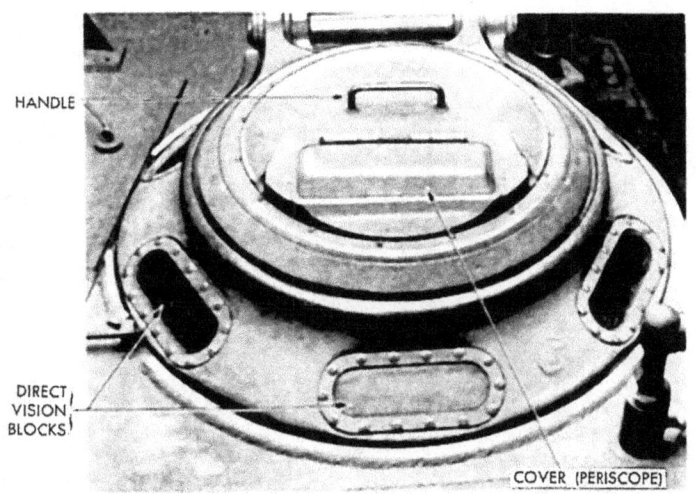

Figure 16. Commander's cupola door controls.

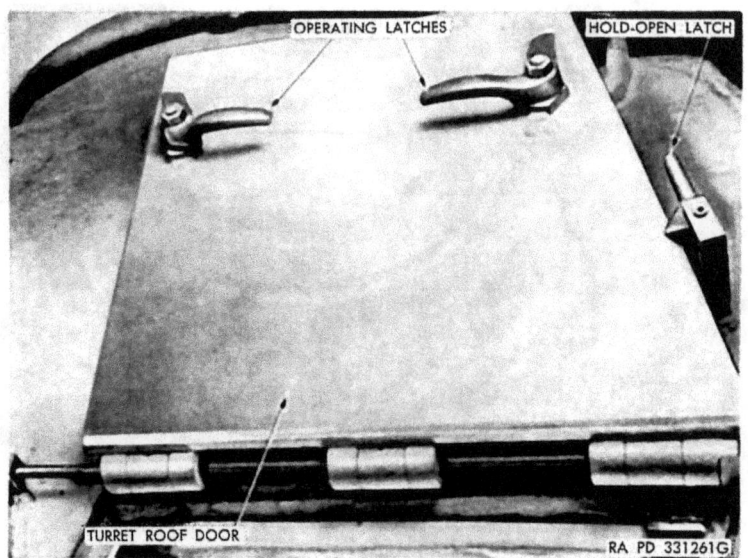

Figure 17. Loader's door controls.

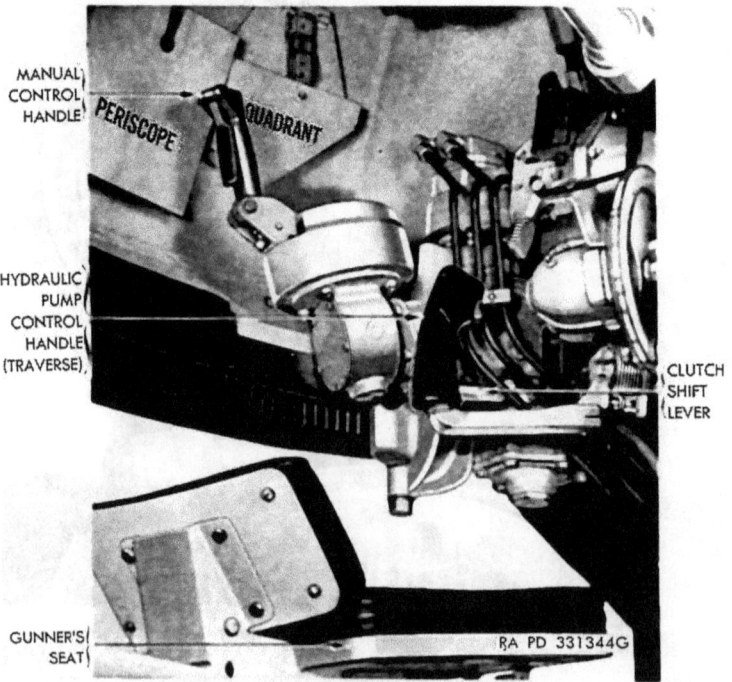

Figure 18. Turret traversing controls

either from inside or outside of turret (fig. 17) and raise door toward front of vehicle until it rests on hold-open latch on turret roof. Closing of both doors is accomplished by releasing hold-open locks or latches, pulling door back into closed position, and locking with latches.

42. Turret Traversing Controls

a. GENERAL. The turret rotates 360° on a continuous ball race mounting. This bearing is completely enclosed for protection from direct hits, lead splash, dirt, and water. The turret is traversed either by hand or by the hydraulic power traversing mechanism. The traversing method is selected by means of a clutch shift lever at the side of gear box housing to the left of the gunner (fig. 18). The turret can be locked in any position by means of the spring-loaded positive turret lock.

b. MANUAL TRAVERSING. To rotate the turret with the manual control handle, be sure that the turret travel lock is disengaged. To dis-

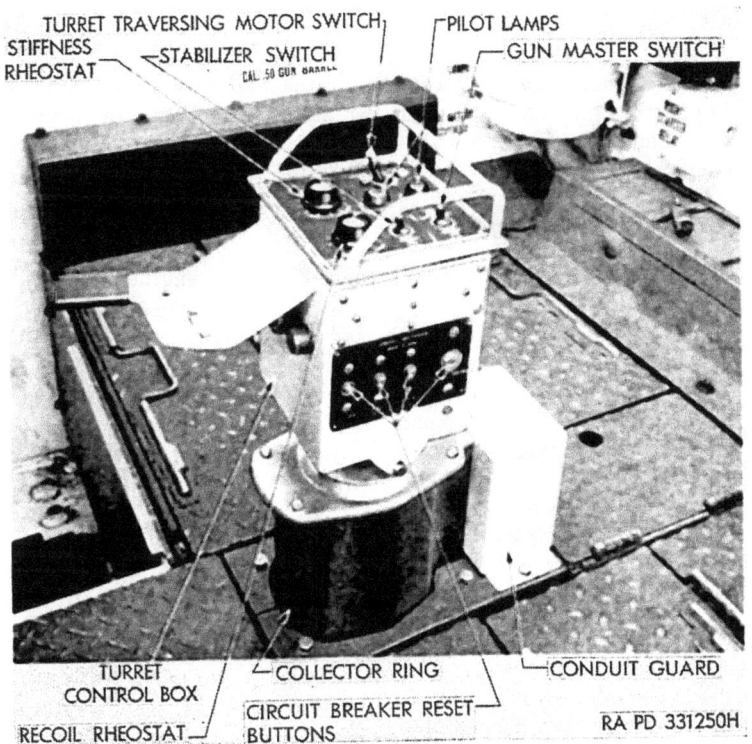

Figure 19. Turret electrical control switches.

engage the lock, rotate the handle clockwise against the spring pressure and pull handle lugs out of slot. Rotate handle counterclockwise to hold lock in the "OFF" position. Then move the clutch lever (fig. 18) down to engage the manual gears, so that the turret can be rotated by the manual control handle.

c. POWER TRAVERSING. To rotate the turret with the hydraulic power mechanism, be sure the turret lock is disengaged (*b* above). Turn the toggle for the turret motor switch to its "ON" position (fig. 19). Move the clutch lever up to disengage the manual gears. The turret is now ready for hydraulic power control. Turn the hydraulic traverse control handle (fig. 18) counterclockwise to rotate the turret to the left. Turn handle clockwise to rotate the turret to the right. The amount the handle is turned determines the speed of rotation.

Note. When power traverse is used, engines should be run at generator output speed to prevent batteries from being run down.

Section III. OPERATION UNDER USUAL CONDITIONS

43. General

This section contains instructions for the mechanical steps necessary to operate light tank M24 under conditions of moderate temperature and humidity. For operation under unusual conditions, refer to section V. For operation of auxiliary equipment, refer to section IV.

44. Starting the Engines

a. STARTING. Under normal conditions, the engines should be started as follows:

(1) Be sure that the before-operation services listed in chapter 3, section III have been performed.

(2) Set brakes. Place transmission selector in "NEUTRAL," and place transfer unit shift lever in the range in which the vehicle is to be driven. Place fuel tank control for right or left tank at "ON" position. Turn on ignition switches for both engines causing engine and transmission oil pressure warning signal lights to go on.

Caution: It is important to turn on electric fuel pumps before starting engines and equally important to turn them off when the engines are stopped. If the fuel, ignition, and master battery switches have been found in the "ON" position when the engines were not operating, check for hydrostatic lock (*f* below) by turning the engines over by hand before attempting to operate the starters. Do not allow pumps to operate in a dry tank.

(3) Press starter buttons for both engines until engines fire.

b. WARM-UP. After engines have started, pull hand throttle out to set idling speed at about 625 rpm to prevent stalling during warm-up. Check to see that oil pressure warning signal lights go out. If these signals remain lighted, shut off engines and investigate cause (par. 71). Check ammeter to see that generators are charging. Listen for unusual noises in power train or engines. The engines do not require extended warm-up periods in mild weather, except the time required to check the gages mentioned above. At temperatures below freezing, the warm-up periods given in paragraph 54 will be observed. Normal operating temperature is 160° to 210° F.

c. FLOODING. If the engines do not start readily in mild or warm weather, check for flooded condition. This can usually be corrected by depressing accelerator pedal all the way down and cranking flooded engine until it starts. The other engine can then be started with a minimum of "racing." Flooding may be caused by dirt on needle valve or by stuck choke shaft, which can be freed as described in paragraph 112.

d. STARTING ONE ENGINE WITH THE OTHER. If only one engine starts readily, it can be used to start the other engine by placing the transfer unit shift lever in "NEUTRAL," running the first engine at about 2,000 rpm and moving transmission selector lever to "DRIVE."

Caution: Inspect "dead" engine to be sure it turns over freely before attempting to start it by use of the other engine.

e. COLD-WEATHER STARTING. Refer to paragraph 54 for detailed instructions on cold-weather starting and warm-up.

f. HYDROSTATIC LOCK. If starter will not turn engine, check for hydrostatic lock (accumulation of fluid in the combustion chamber of cylinder, which prevents turning of engine). This condition can be caused by dirt under the carburetor float valve or by failing to turn off the fuel, ignition, and master battery switches when engine is not operating. Whenever an engine is affected by hydrostatic lock, the carburetor, carburetor float valve in particular, and fuel system in general must be inspected by ordnance personnel. Engines in this condition may be started before this inspection by the following procedure:

(1) Remove the spark plugs (par. 106).

Caution: Be sure ignition switch is "OFF."

(2) Lay cloths on the cylinder heads below the spark plug holes to absorb gasoline.

(3) Using a 1-inch socket with a 12-inch flexible handle on the crankshaft pulley nut, turn the engine six full revolutions.

Caution: Never turn over the engine with the starter, because arcing at the starting brushes can ignite the gasoline fumes and liquid gasoline forced out of the cylinders.

(4) Remove the cloths and mop up gasoline which was not absorbed.
(5) Install spark plugs (par. 106).
(6) Remove muffler of the engine affected by hydrostatic lock (par. 116); drain fuel from muffler and install (par. 116).
(7) Check the crankcase oil level (par. 62). If the gage shows more than one-half quart over full, gasoline dilution is excessive and the oil must be drained and crankcase refilled.
(8) Allow the vehicle to stand with all doors and engine compartment covers open for 20 to 30 minutes or until all traces of gasoline have evaporated.
(9) Start the engine which was not hydrostatically locked, to ventilate the hull and engine compartments further. While this is being done, station one man outside the vehicle near the fire extinguisher handle (fig. 24). Do not stand above the engine compartment.
(10) Start the engine which had the hydrostatic lock.

45. Placing Vehicle in Motion

a. DRIVING RANGES. Normally, the vehicle is driven with the transmission selector lever in "DRIVE" and the transfer unit shift lever either in "HIGH" range or in "LOW." Transfer unit "HIGH" range provides cruising gear ratios and is to be used for traveling on roads or on relatively smooth, level terrain. Transfer unit "LOW" range provides power gear ratios and is to be used for hill climbing or for rough, sandy, or muddy conditions.

b. FORWARD DRIVING. To start vehicle, shift transfer unit into "HIGH" or "LOW" as conditions warrant, move transmission lever to "DRIVE," release parking brake controls, and depress accelerator. Vehicle will start moving forward at a speed proportional with the distance the accelerator pedal is depressed. The vehicle will start out in first gear, and the transmission will automatically shift into second, third, and fourth gears as the vehicle speed increases and engine load decreases. As vehicle loses speed, either because accelerator pedal is released or due to upgrades, the transmissions will shift down automatically to a lower gear.

c. STEERING. Pull back on the right-hand steering lever to make right turn or on the left-hand lever for a left turn. The lever that is being used should be applied firmly and then released fully. This action should be repeated if necessary. The lever should not be held in a slightly applied position for long periods of time.

Caution: It is very important that the steering levers be held far enough forward to insure complete release of the steering brake bands

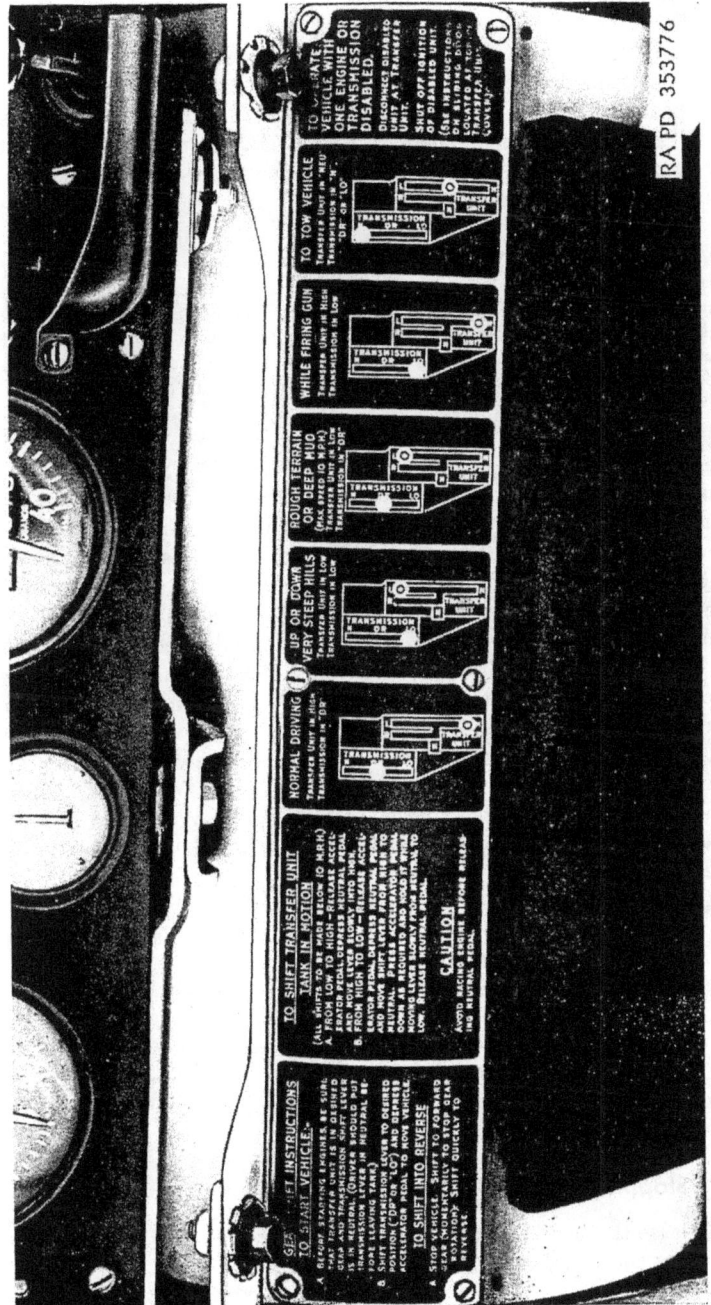

Figure 20. Driving gearshift instruction plate.

at all times, except when steering or stopping; otherwise, brake lining wear will be excessive and oil will become overheated.

d. REVERSE. Bring the vehicle to a complete stop, depress the neutral pedal to put the transmissions in "NEUTRAL," and move the transfer unit shift lever up to "NEUTRAL," over to the left, and forward into "REVERSE." Release the neutral pedal. Four speeds automatically selected in the hydramatic transmissions are available in reverse.

Note. Depressing the neutral pedal puts both transmissions in neutral by means of the same linkage as the transmission selector lever, but leaves the driver's right hand free to shift the transfer unit. Releasing the pedal puts the transmission back in the range for which the selector lever is set. Neutral pedal should be used only when throttles are closed (foot off accelerator pedal), except when shifting transfer unit from high to low while tank is in motion.

46. Driving Instructions

a. ROUGH TERRAIN. In negotiating rough or sandy terrain or heavy mud (par. 56), not only should the transfer unit be shifted to "LOW" range, but the transmission selector lever should also be moved to "LOW" range to hold the transmission in second gear, and thus obtain the benefit of a steady pull. When the transmissions are in "LOW" range, they will upshift only from first to second speeds and will not go into third and fourth, thus permitting a steadier pull through very bad terrain or providing more braking effect from engine compression and friction when descending hills.

b. DESCENDING HILLS. In descending moderately steep hills, the transfer unit should be shifted into "LOW" range. When descending long, very steep hills, the transmission selector lever should also be placed in "LOW" range and brakes should be used to keep engine speeds below 3,500 rpm.

Caution: Transfer unit shifts from "HIGH" to "LOW" must be made at speeds below 10 mph. The "NEUTRAL" pedal must always be depressed when making the shifts.

c. OPERATING VEHICLE WITH ONE ENGINE. When one engine or transmission fails and repair facilities are not available, the vehicle can be operated for a short time with one engine by disconnecting the dead engine at the transfer unit by means of the input clutch lever (fig. 114). Keep the transfer unit shift control lever in "LOW" range when driving with one engine.

47. Stopping the Vehicle

Release the accelerator pedal and pull back on both steering brake levers at the same time. There is no clutch to disengage. Both levers should be pulled back and engaged with heavy pressure and then released fully. If the stop is to be made from relatively high speed, it is advisable to use heavy pressure intermittently rather than a continuous but lighter pressure.

Caution: Do not press down on neutral pedal when stopping, unless the transfer unit is to be shifted. Do not use neutral pedal as a clutch when starting.

48. Stopping the Engines

Close the throttle until engines are idling at approximately 450 rpm. Run at this speed for 3 or 4 minutes and then turn off both ignition switches and shut off the fuel pump switch. If engines are extremely hot, idle at 2,000 rpm for a few minutes to cool down before reducing speed to normal idle. Finally, open master battery switch.

Caution: Always turn off ignition switches before opening master battery switch. Never operate engines with master switch off.

49. Towing the Vehicle

a. TOWING TO START VEHICLE. The engines can be started in an emergency by towing the vehicle, provided the following procedure is observed after checking to make sure the engines turn freely.
  (1) Release brakes.
  (2) Place transmission selector lever in "NEUTRAL" and transfer unit shift lever in "LOW."
  (3) When speed reaches approximately 4 mph, turn on both ignition switches and one fuel pump switch and move transmission selector lever to "DRIVE."
  (4) Continue towing until engines fire.

b. TOWING A DISABLED VEHICLE. When towing a disabled vehicle, several precautions must be taken:
  (1) Tow the vehicle with the transfer unit in "NEUTRAL," if possible. If transfer unit is damaged or disabled, disconnect the propeller shaft at the controlled differential pinion shaft yoke.
  (2) If controlled differential is disabled, disconnect the two short propeller shafts at the final drive yokes.
  (3) If final drive units are disabled, break tracks, and tow vehicle on track suspension wheels.

Caution: Tow bar must be used under these circumstances.

c. TOWING PROCEDURES. In towing a vehicle, changes in direction must be made by a series of slight turns so that the vehicle being towed is, as nearly as possible, directly behind the one doing the towing. This prevents the cable from contacting the track, which might damage both the cable and track blocks. If no operator is available to steer the disabled vehicle or if it is being towed with the tracks removed, the cable should be attached by the "short hitch," in which the cable is threaded through both eyes on the vehicle to be towed and then crossed and passed through both shackles of the towing vehicle.

d. RECOVERING DISABLED VEHICLES. As an emergency method, to be used in combat only, a disabled vehicle may be recovered by remote control when a tank recovery vehicle with a remote control towing hook is available. The disabled vehicle may be recovered without exposing crew members of either vehicle to enemy fire. To accomplish this, a tow bridle must be improvised for each vehicle from the standard 20 foot towing cable and wired in place when the vehicle is to be employed in combat. Typical examples are shown in figures 21 and 22.

Note. The standard towing cable will not be employed in routine training operations as the repeated sharp bend at the towing hook will fracture the strands of the cable and endanger the success of the recovery operation. For training requirements, make short pick-up loops locally, using unserviceable towing cables and attaching them between the lifting eyes of the vehicle.

(1) To recover a disabled vehicle, back the recovery vehicle up to the end of the disabled vehicle which is equipped with a bridle:

Note. The observer in the turret of the recovery vehicle uses the periscope and gives the driver instructions by interphone or other signals.

Lower the towing hook sufficiently for the point to engage the bridle, pull the control rope to open the safety latch, and

Figure 21. Improvised towing bridle on front of vehicle.

Figure 22. Improvised towing bridle on rear of vehicle.

slowly move the recovery vehicle away from the disabled vehicle while slowly lowering the hook until the bridle is completely engaged by the hook. Allow the latch to close. Raise the hook as high as possible and tow the disabled vehicle to a protected location. The hook is raised before towing the vehicle in order to improve the steering of the recovery vehicle, to make the towed vehicle track better, and to keep the towed vehicle from running over the cable on sharp turns.

(2) A bridle on the front of a recovery vehicle may be used for "hooking on" another recovery vehicle. This may be necessary if the recovery vehicle is itself disabled or if it requires assistance while towing through mud or negotiating a steep slope during a recovery operation.

Section IV. OPERATION OF MATÉRIEL USED IN CONJUNCTION WITH MAJOR ITEM

50. Fixed Fire Extinguishers

Two sizes of carbon dioxide fire extinguishers are carried in each vehicle. A fixed 10-pound unit is clamped in a vertical position on the left side of the bulkhead in the fighting compartment (fig. 24). This unit connects to lines leading to the engine compartment only and is used for extinguishing fires in the engine compartment. Its controls are accessible from the fighting compartment. In addition, a remote control pull handle (enclosed in a shield) on the outside of the hull, directly ahead of the engine compartment, and to the left (fig. 24) permits operation of the fire extinguisher from outside the vehicle.

51. Portable Fire Extinguishers

a. A 4-pound portable hand-operated extinguisher is strapped in a vertical position on the center post between the driver and assistant driver (fig. 23).

b. Because the pressure of carbon dioxide increases with increasing outside temperature, the fire extinguishers are equipped with safety valves which open when pressure increases beyond safe limits.

Section V. OPERATION UNDER UNUSUAL CONDITIONS

52. General

a. In addition to the operating procedures described in section III above for usual conditions, special instructions of a technical nature for operating and servicing this vehicle under unusual conditions are contained or referred to herein. In addition to the normal preventive maintenance service (ch. 3, sec. III) special care in cleaning and lubrication must be observed where extremes of temperature, humidity, and terrain conditions are present or anticipated. Proper cleaning, lubrication, and storage and handling of fuels and lubricants not only insure proper operation and functioning, but also guard against excessive wear of the working parts and deterioration of the matériel.

b. TM 21–301 contains very important instructions on driver selection, training, and supervision and TM 21–306 prescribes special driving instructions for operating full-track and tank-like vehicles under unusual conditions.

Caution: It is imperative that the approved practices and precautions be followed. A detailed study of these TM's is essential for use of this matériel under unusual conditions.

c. Refer to paragraph 64 for lubrication under unusual conditions, to table II, paragraph 68 and table III, paragraph 69 for preventive maintenance checks, and to chapter 3, section XXVII for maintenance procedures.

d. When chronic failure of matériel results from subjection to extreme conditions, report of the condition should be made on DA AGO Form 468 (par. 2).

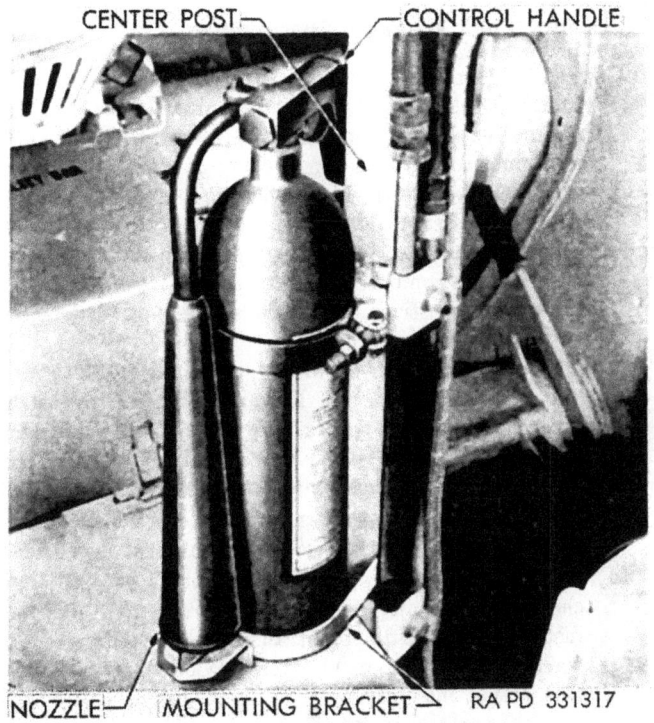

Figure 23. Portable fire extinguisher—installed.

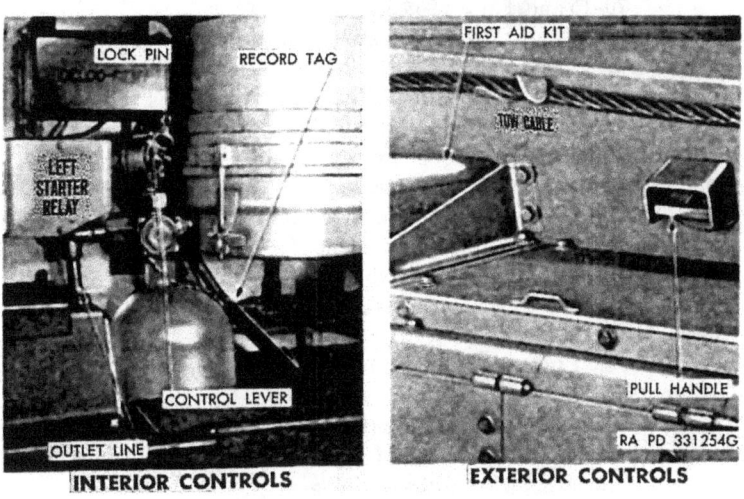

Figure 24. Fixed fire extinguisher controls.

53. Extreme Cold Weather Conditions

a. GENERAL PROBLEMS.

(1) Matériel scheduled for operation in extreme cold weather requires extensive preparation. Generally, extreme cold will cause lubricants to thicken or congeal, freeze batteries or prevent them from furnishing sufficient current for cold-weather starting, crack insulation and cause electrical short circuits, prevent fuels from vaporizing and properly combining with air to form a combustible mixture for starting, and will cause the various construction materials to become hard, brittle, and easily damaged or broken.

(2) The cooling system should be prepared and protected for temperatures below +32° F. in accordance with instructions given in TM 9-2855 on draining and cleaning the system and the selection, application, and checking of antifreeze compounds to suit the anticipated conditions.

(3) That manual also describes the method of connecting specific gravity readings for batteries exposed to extreme cold.

(4) Armament scheduled for extreme-cold operation should be checked for proper lubrication. Fire control and sighting instruments should not be transferred suddenly from cold to warm temperatures or vice versa. Condensation induced by this action may cause clouding of optics and rusting of internal parts. Strains may be set up in parts.

(5) For description of operations in extreme cold, refer to FM 70-15 and TM 9-2855, and TB ORD 193.

Caution: It is imperative that the approved practices and precautions be followed. TM 9-2855 contains general information which is specifically applicable to this vehicle as well as all other vehicles. It must be considered an essential part of this manual, not merely an explanatory supplement to it.

b. WINTERIZATION EQUIPMENT. Special equipment is provided for the vehicle when protection against extreme cold weather (0 to −65° F.) is required. This equipment is issued as specific kits. Each kit contains a technical bulletin which provides information on description, installation instructions, and methods of use. TM 9-2855 contains general information on winterization equipment and processing.

c. FUELS, LUBRICANTS, AND ANTIFREEZE COMPOUNDS (STORAGE, HANDLING, AND USE).

(1) The operation of equipment at arctic temperatures will depend to a great extent upon the condition of the fuels and lubricants used in the equipment.

(2) The manner in which the fuels and lubricants are stored,

handled, and used greatly affects the service the fuels and lubricants will give.
(3) In arctic operations, contamination with moisture is the source of many difficulties. Moisture can be the result of snow getting into the product, condensation due to "breathing" of a partially filled container, or moisture condensed from warm air in a partially filled container when a product is brought outdoors from room temperatures. Other impurities will also contaminate fuels and lubricants so that their usefulness is impaired.
(4) Immediate effects of careless handling of fuels and lubricants are not always apparent, but any deviation from proper handling of these products is likely to bring trouble at the least expected time. Instances of careless storage and handling include storing fuels or lubricants in unsealed or open containers, resulting in condensation of atmospheric moisture; neglect to filter stored fuel prior to pouring into vehicle tank, resulting in frozen fuel lines; adding water to arctic antifreeze compound, in spite of the fact that this material is intended for use in full strength throughout its range of use.
(5) Refer to TM 9-2855 for detailed instructions.

54. Extreme Cold Weather Operation

a. GENERAL.
(1) The driver and crew members must always be on the alert for indications of the effect of cold weather on the vehicle.
(2) The driver and crew members must be very cautious when placing the vehicle in motion after a shutdown. Congealed lubricants may cause failure of parts.
(3) Refer to TM 21-306 for special instructions on driving hazards in snow, ice, and unusual terrain encountered under extreme-cold conditions.

b. STARTING THE ENGINE.
(1) Before starting, see that everything is in readiness so that the engine will start on the first trial. Try to avoid having the engines fire a few times and then stop.
(2) Before cranking the engines, depress the accelerator pedal about one-quarter of the total travel and release. This will set the automatic choke correctly. No further choking is possible or necessary.

Caution: Do not pump or depress the accelerator pedal swiftly to the floor before cranking. This will force raw gasoline into the cylinders, causing flooding, decreasing oil film in the cylinders, and will hinder starting.

(3) After the engines have been started, set the hand throttle and allow the engines to run at 800 to 900 rpm for 4 or 5 minutes to warm up the oil before depressing the throttle further. This should be done with transmission lever in "NEUTRAL."

(4) One engine may be started with the other (par. 44), if battery capacity is low (par. 127).

c. OPERATING THE VEHICLE.

(1) After the engines have been started and the engine oil is warmed (b(4) above), shift the selector lever to "DR" and allow the engines to idle for several minutes more to warm up the oil in the transmissions.

Note. Do not move vehicle until engine has reached normal operating temperature of 160° to 210° F.

Do not drive the vehicle over 5 mph for at least 10 minutes after starting so that the oil in the transfer unit and differential will be warmed up before a higher speed is attempted. Be especially careful when first placing vehicle in motion, as congealed or thickened lubricant places added strain on driving mechanisms and failure may result.

(2) In extreme cold weather, gear clash will occur when shifting the transfer unit into "REVERSE" before the lubricant has warmed up sufficiently. If the vehicle is to be reversed immediately after making a cold start, shift the transfer case into "REVERSE" before starting engine. When it is necessary to shift into reverse from a forward speed position, the clash can be minimized in the following manner:

(a) Bring the vehicle to a complete stop.

(b) Depress the neutral pedal.

(c) Move lever into "REVERSE" position as rapidly as possible.

d. AT HALT OR PARKING.

(1) When halted for short shutdown periods the vehicle should be parked in a sheltered spot out of the wind. For long shutdown periods, if high dry ground is not available, effort should be made to prepare a footing of planks or brush. Chock in place if necessary.

(2) When preparing a vehicle for shutdown period, place control levers in the neutral position to prevent them from possible freezing in an engaged position. Freezing may occur when water is present due to condensation.

(3) Clean all parts of the vehicle of snow, ice, and mud as soon as possible after operation. Refer to table II, paragraph 68 for

detailed after-operation procedures. Be sure to protect all metal parts against entrance of loose, drifting snow during the halt. Snow flurries penetrating the engine compartment may enter the crankcase filler vent, etc. Cover and shield the vehicle but keep the ends of the canvas paulins off the ground to prevent them from freezing to the ground.

(4) If no power plant heating device is present, the battery should be removed and stored in a warm place.

(5) Refuel immediately in order to reduce condensation in the fuel tanks. Prior to refueling, open fuel tank drains and drain off any water collection.

(6) When the vehicle is equipped with a power plant heater as provided by the arctic winterization kit, start the heater and check to be sure that it is operating effectively. This heater should avoid the necessity of removing the battery to warm storage, and is designed to operate unattended during overnight stops. Instructions for operation of winterization equipment will be found in pamphlet packed with the kit.

Note. This heater is used only while the vehicle is halted.

(7) The use of any fluids (water, antifreeze compound, or ammudamp) in ammunition box cans has been discontinued. All ammudamp cans will be completely drained of ammudamp fluid. Where the ammunition racks can be removed easily, remove the rack, unscrew the filler and drain plugs from the cans, and pour out the fluid. In other locations where it may be impracticable to remove the racks, siphon the fluid from the cans by use of a suction type oil gun fitted with three nipples, straight tee, and rubber hose. Bring the free end of the rubber hose to the ammunition can, remove the filler and drain plug, and insert the hose to the lowest portion of the can. To start the fluid flowing, hold a finger tightly over the outlet nipple, pull the suction gun handle completely out, and then release the finger from the nipple. After all the fluid has been drained, thoroughly dry the ammunition boxes and ammudamp cans. After the cans have dried, replace the filler and drain plugs. Discard the ammudamp fluid drained from the cans. Remove from the ammunition boxes those ammudamp cans that have deteriorated to such an extent that they are no longer suitable for use as separators or barriers for the ammunition. Cut and shape well seasoned wooden blocks (procure wood locally) to the shape of the ammudamp cans that were removed. Fit the blocks into position in the ammunition boxes. Treat the wooden blocks with synthetic wood filler and apply two coats of paint before use. Paint

filler and drain plugs black to indicate that the ammudamp cans have been drained of all fluid.

e. ARMAMENT AND SIGHTING AND FIRE CONTROL MATÉRIEL. For information on extreme cold weather, refer to TB ORD 193.

55. Extreme Hot Weather Operation

a. GENERAL. Continuous operation of the vehicle at high speeds or long hard pulls in lower gear positions on steep grades or in soft terrain may cause the vehicle to register overheating. Avoid the continuous use of low gear ratios whenever possible. Continuously watch the temperature and halt the vehicle for a cooling-off period whenever necessary and tactical situation permits. Keep ventilating blower on during operation of vehicle. Make frequent inspections and servicing of cooling unit, oil filter, and air cleaner. If the engine oil temperature or transmission high oil temperature warning signal lights come on, look for dust, insects, or any other obstruction in radiator cores and remove.

b. AT HALT OR PARKING.
  (1) When practicable, park the vehicle under cover to protect it from sun, sand, and dust.
  (2) Cover inactive vehicles with paulins if no other suitable shelter is available. Where entire vehicle cannot be covered, protect the engine compartment against entry of sand or dust.
  (3) Vehicles inactive for long periods in hot humid weather are subjected to rapid rusting and accumulation of fungi growth. Remove and store batteries in a cool place. Make frequent inspections and clean and lubricate to prevent excessive deterioration.

c. SIGHTING AND FIRE CONTROL MATÉRIEL. This matériel should be shielded as much as possible from the direct rays of the sun.

56. Operation on Unusual Terrain

a. MUD. Select transmission low range to move vehicle steadily without digging in. When a convoy of vehicles is operating over soft ground in territory where circumstances permit, it is desirable to operate the vehicle so that the tracks overlap the trail of the preceding vehicle to avoid rutting. If the vehicle becomes mired, arrange to be towed out of the mud instead of digging in. When a drop to below freezing temperature is anticipated, make sure that the vehicle is parked on solid ground or footing to prevent the tracks from being frozen in the mud, and that accumulations of mud have been removed from track and wheel contacting surfaces.

b. Snow. It may be possible for the vehicle to ride heavily crusted snow with occasional breakthrough. To climb back onto the crust, reduce engine speed and shift into transmission low range to achieve very low track speed for forward movement without slippage. Avoid grades. Where grades must be taken, drive as nearly straight up and down as possible to equalize track load. Avoid sharp turns. For soft or fine snow, select the transmission low range shift position which gives good traction.

c. Ice. Skidding is the general hazard encountered on ice. Select the proper gear ratio to move the vehicle steadily, without imposing undue strain on engine. When skidding occurs, decelerate the engine and proceed with caution.

d. Sand. The main objective when driving in sand is to avoid spinning the tracks. Reduce speed and use a gear low enough to move the vehicle steadily. Do not allow the engine to labor.

57. Fording Operation

a. General. In fording, vehicles may be subjected to water varying in depth from only a few inches to depths sufficient to completely submerge the vehicle. Factors to be considered are spray-splashing precautions, normal fording capabilities, deep-water fording using fording kits, and accidental complete submersion. Optical instruments should not be submerged, except those periscopes used for driving.

b. Normal Fording. Fording of bodies of water up to maximum vehicle fording depth of 40 inches is based on the standard vehicle with special protection provided for critical units, but without deep-water fording kit. Observe the following precautions:

 (1) Do not exceed the known fording limits of the vehicle.
 (2) The engine must be operating at maximum efficiency before attempting to ford. Make sure that battery cell vent caps are snug.
 (3) Shift transmission into low range. Speed up engine to overcome the possibility of a "stall" when the cold water chills the engine. Enter the water slowly. Should the engine stall while submerged, it may be started in the usual manner.
 (4) All normal fording should be at speeds of 3 to 4 miles per hour to avoid forming a "bow wave."
 (5) If accidental complete submersion occurs, the vehicle will be salvaged, temporary preservation applied as outlined in paragraph 213 and then sent to the ordnance maintenance unit as soon as possible for necessary permanent maintenance.

c. Deep-Water Fording. Refer to TM 9-2853 for general information, descriptions, and methods of use of deep-water fording kits, and for general procedure for the operation of vehicles so equipped.

d. AFTER-FORDING OPERATION. Open all drain holes in body. Also, at the earliest opportunity check the engine oil level and check for presence of water in the crankcase. Heat generated by driving will evaporate or force out most water which has entered at various points. Refer to paragraph 213 for maintenance operations after fording. Optical instruments that have been wetted should be quickly wiped dry and examined for indication of leakage into the instrument. Such instruments should be turned in for reconditioning at the first opportunity.

CHAPTER 3

ORGANIZATIONAL MAINTENANCE INSTRUCTIONS

Section I. PARTS, SPECIAL TOOLS, AND EQUIPMENT FOR ORGANIZATIONAL MAINTENANCE

58. General

Tools, equipment, and spare parts are issued to the using organization for maintaining the matériel. Tools and equipment should not be used for purposes other than prescribed and, when not in use, should be properly stored in the bag provided for them.

59. Parts

Spare parts are supplied to the using organization for replacement of those parts most likely to become worn, broken, or otherwise unserviceable, providing such operations are within the scope of organization maintenance functions. Spare parts, tool, and equipment supplied for the light tank M24 are listed in Department of the Army Supply Catalog ORD 7 SNL G-200, which is the authority for requisitioning replacements.

60. Common Tools and Equipment

Standard and commonly used tools and equipment having general application to this matériel are listed in the ORD 7 catalog and by T/A and T/O&E.

61. Special Tools and Equipment

Certain tools and equipment specially designed for organizational maintenance, repair, and general use with the matériel are listed in table I for information only. This list is not to be used for requisitioning replacements.

Table I. Special Tools and Equipment for Organizational Maintenance

Item	Identifying number	References Fig.	References Par.	Use
ADAPTER, inner bearing race remover	41-A-12-550 7079201		179	Use with SCREW—41-S-1047-200. Remove inner bearing race from hub.
ADAPTER, puller, slide hammer, torsion bar and compensating link pin replacer.	41-A-18-245 7079311	25	174, 175, 177	Used with PULLER—41-P-2957-33.
ADAPTER, remover, track support outer bearing race roller.	41-A-18-296 7079455		179	Used with SCREW—41-S-1047-200. Remove outer bearing race from hub.
BOLT, eye, transmission lifting	41-B-1586-300 A266327	25	151	Used with hoist to disconnect transmission from engine.
CABLE, extn, ru covered, 2-conductor, stranded, w/female plugs at both ends, AWG No. 1, lgh 20 ft.	17-C-568 B257839	26		For charging batteries through bulkhead terminal box receptacle from an external source.
FIXTURE, track connecting and link pulling R and LH.* Composed of—	41-F-2997-86 D78191	26, 134, 135	172	Disconnect steel track.
FIXTURE, track connecting and link pulling, RH.	41-F-2997-388 7077673			
FIXTURE, track connecting and link pulling, LH.	41-F-2997-389 7077696			
GAGE, transmission oil pressure	41-G-446 B298875	26	150	Check transmission oil pump pressure.
HANDLE, remover and replacer, size of shank ¾ in, lgh 8⅜ in.	41-H-1397 A380406	25		Used with REMOVERS and REPLACERS.

LIFTER, track wheel, right and left, set	41-L-1400 7079701		176, 177	Raise track wheel (steel track).
PULLER, end, shock absorber	41-P-2907-196 7079316	155	180	Pull shock absorber off pin.
PULLER, slide hammer type, bogie gudgeon	41-P-2957-33 C73615	26	174, 175, 177	Pull link pin or torsion bar out of support arm.
REPLACER, bearing and seal, track and compensating wheel	41-R-2383-950 B296108	25	173, 176	Install oil seal in position on track wheel and compensating wheel hubs.
REPLACER, retainer, grease	41-R-2390-450 B296095	25	173, 176	Install grease retainer on track wheel and compensating wheel.
REPLACER, retainer, grease	41-R-2396-375 7080461		179	Install grease retainer on track support roller.
REPLACER, seal, track support roller	41-R-2397-875 A380369	25	179	Install oil seal in position on support roller bracket.
SCREW, remover, bearing, thd 1¼ in-12NF-2 lgh over-all 8¾ in.	41-S-1047-200 7079203		179	Used with ADAPTER—41-A-12-550 and —41-A-18-296.
SLING, engine lifting	41-S-3831-300 7079282	51	97, 98	Lift engine out of vehicle.
SLING, final drive lifting	41-S-3832-35 7079301	26, 30	169	Lift final drive out of vehicle.

*On vehicle tool.

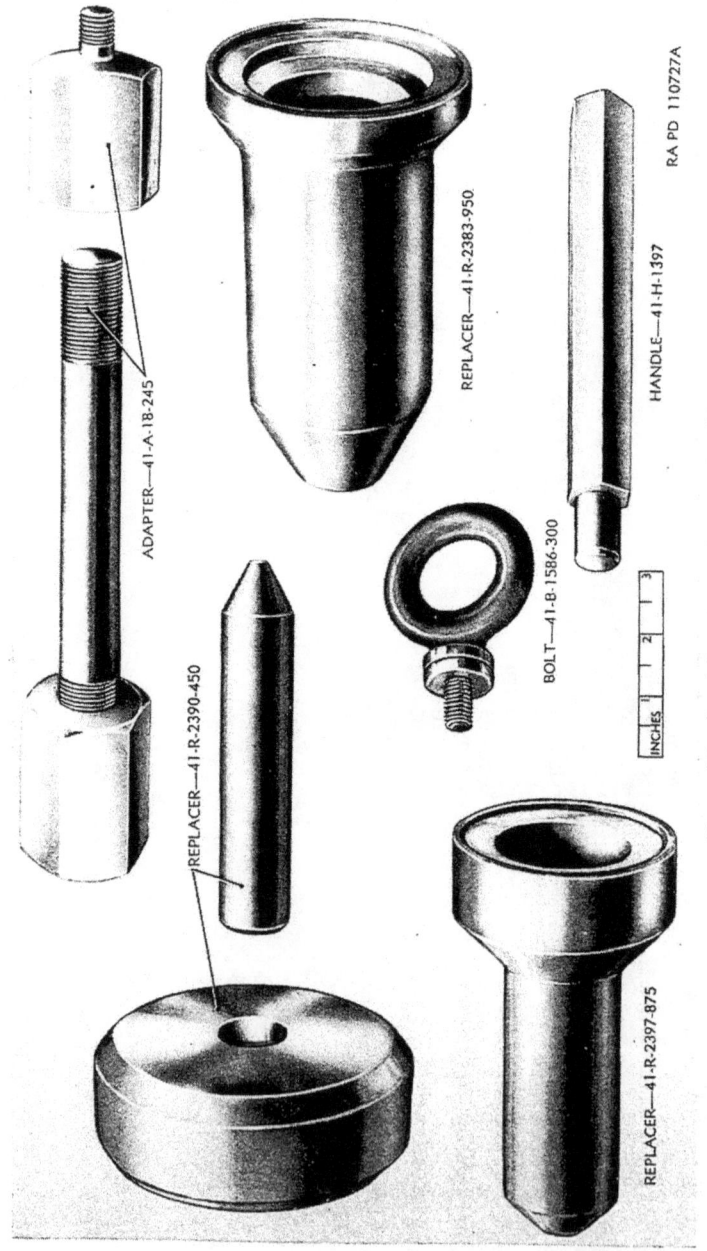

Figure 25. Special tools for light tank M24.

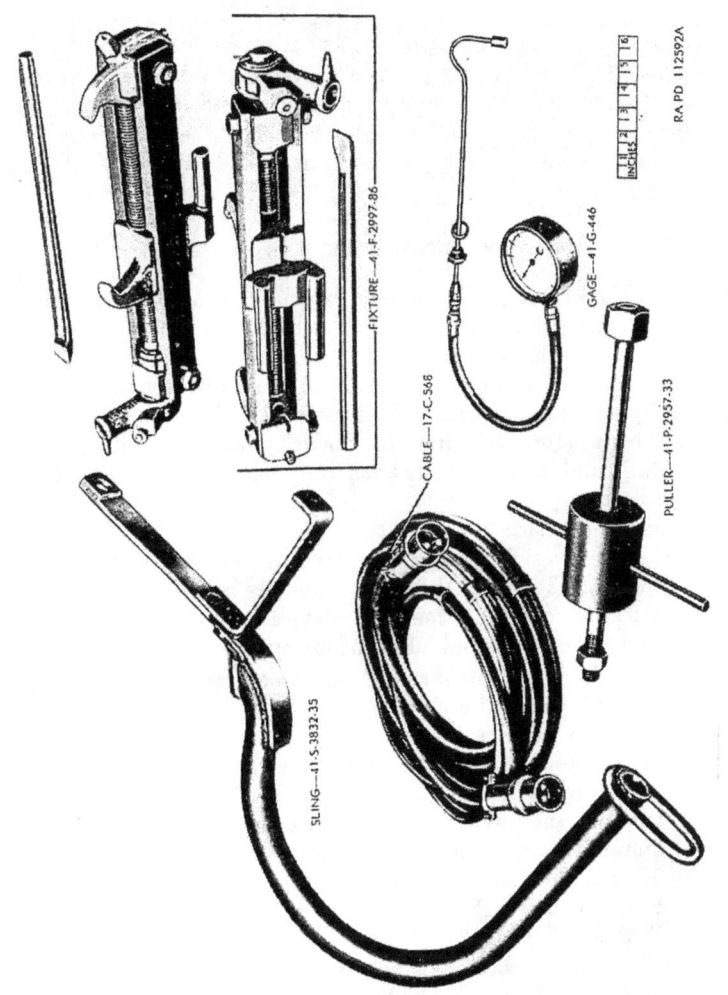

Figure 26. Special tools for light tank M24.

Section II. LUBRICATION AND PAINTING

62. Lubrication Order

Lubrication Order LO 9–729 (figs. 27 and 28) prescribes cleaning and lubricating procedures as to locations, intervals, and proper materials for this vehicle. This order is issued with each vehicle and is to be carried with it at all times. In the event the vehicle is received without a copy, the using organization shall immediately requisition one. See SR 310–20–4. Lubrication which is to be performed by ordnance maintenance personnel is listed in the **NOTES** on the lubrication order.

63. General Lubrication Instructions

a. Usual Conditions. Service intervals specified on the lubrication order are for normal operation and where moderate temperature, humidity, and atmospheric conditions prevail.

b. Lubrication Equipment. Each vehicle is supplied with lubrication equipment adequate for its maintenance. This equipment will be cleaned both before and after use. Lubricating guns will be operated carefully and in such a manner as to insure a proper distribution of the lubricant.

c. Points of Application.
 (1) Lubricating fittings, grease cups, oilers, and oilholes are shown in figures 29 through 34, and are referenced to the lubrication order. Wipe these devices and the surrounding surfaces clean before lubricant is applied.
 (2) A ¾-inch red circle should be painted around each lubricating fitting and each oilhole.

d. Reports and Records.
 (1) Report unsatisfactory performance of matériel or defects in the application or effect of prescribed petroleum fuels, lubricants, and preserving materials, using DA AGO Form 468, Unsatisfactory Equipment Report.
 (2) Maintain a record of lubrication of the vehicle on DA AGO Form 462, Work Sheet for Full-Track and Tank-Like Wheeled Vehicles.
 (3) Maintain a record of changes in grade of lubricant and recoil oil for the weapon in OO Form 5825, Artillery Gun Book.

64. Lubrication Under Unusual Conditions

a. Unusual Conditions. Reduce service intervals specified on the lubrication order to compensate for abnormal operation and extreme conditions, such as high or low temperature (refer also to TM 9–2855), prolonged periods of high-speed operation, continued operation in sand or dust, immersion in water, or exposure to moisture, any one of which may quickly destroy the protective qualities of the lubricant. Intervals may be extended during inactive periods. Refer to TM 9–2855 for lubrication for continued operation below 0° F.

b. Changing Grade of Lubricants. Lubricants are prescribed in "KEY" in accordance with four temperature ranges: above 32° F., from 32° to 0° F., from 0° to −40° F., and below −40° F. Change the grade of lubricants whenever weather forecast data indicate that air temperatures will be consistently in the next higher or lower temperature range or when sluggish starting caused by lubricant thickening occurs. Intervals may be extended during inactive periods commensurate with adequate preservation.

65. Painting

Instructions for the preparation of the matériel for painting, methods of painting, and materials to be used are contained in TM 9–2851. Instructions for camouflage painting are contained in FM 5–20B.

Section III. PREVENTIVE MAINTENANCE SERVICES

66. General Information

a. Responsibility and Intervals. Preventive maintenance services are the responsibility of the using organization. These services consist generally of before-operation, during-operation, at-the-halt, after-operation, and weekly services performed by the driver or crew and the scheduled services to be performed at designated intervals by organization mechanics of maintenance crews. Intervals are based on normal operations. Reduce intervals for abnormal or severe conditions. Intervals during inactive periods may be extended accordingly.

b. Definition of Terms. The general inspection of each item applies also to any supporting member or connection and is generally a check to see whether the item is in good condition, correctly assembled, secure, or excessively worn.

(1) Inspection for "good condition" is usually an external visual inspection to determine whether the unit is damaged beyond safe or serviceable limits. The term "good condition" is explained further by the following: not bent or twisted, not chafed or burred, not broken or cracked, not bare or frayed, not dented or collapsed, not torn or cut, not deteriorated.

(2) The inspection of a unit to see that it is "correctly assembled" is usually an external visual inspection to see whether it is in its normal assembled position in the tank.

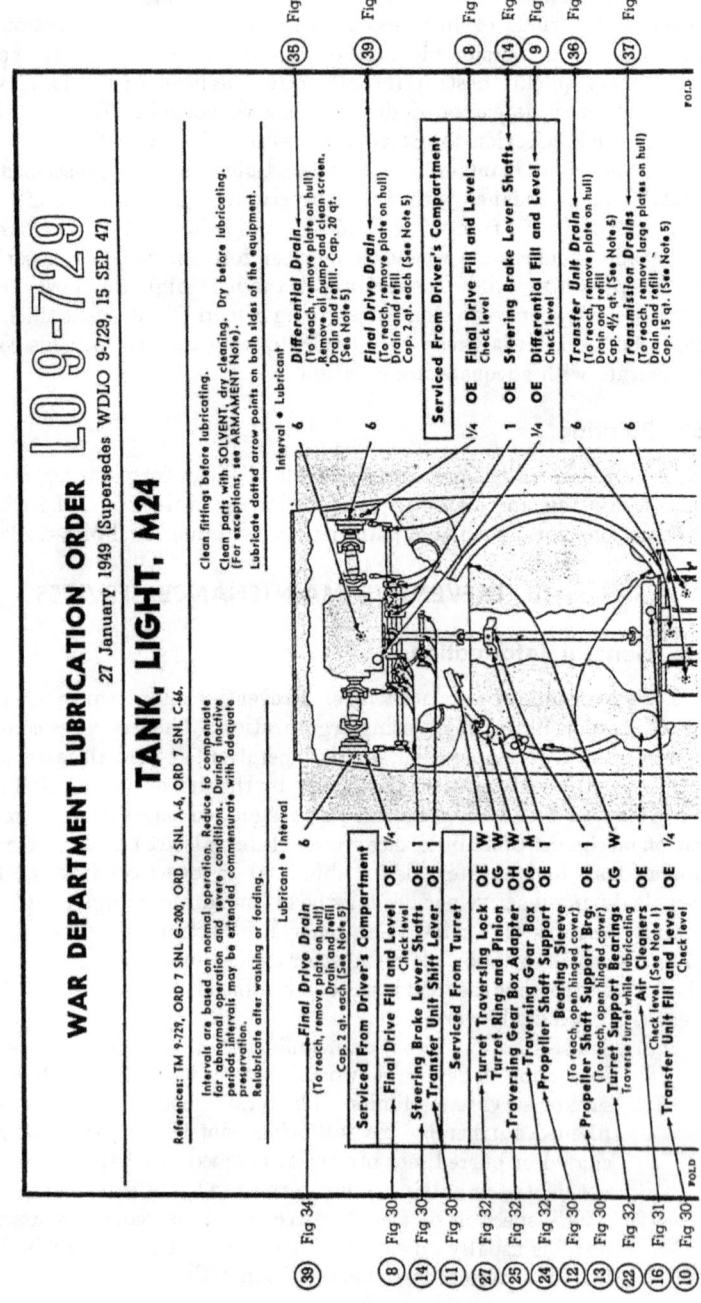

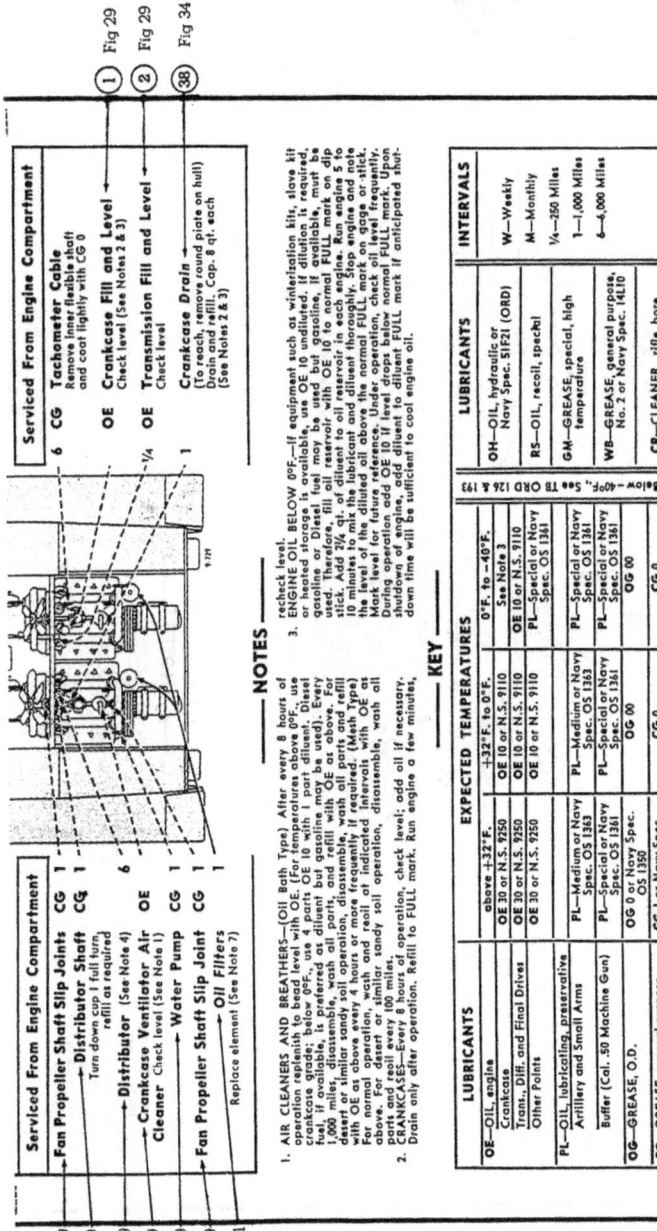

Figure 27. Front side of lubrication order—light tank M24.

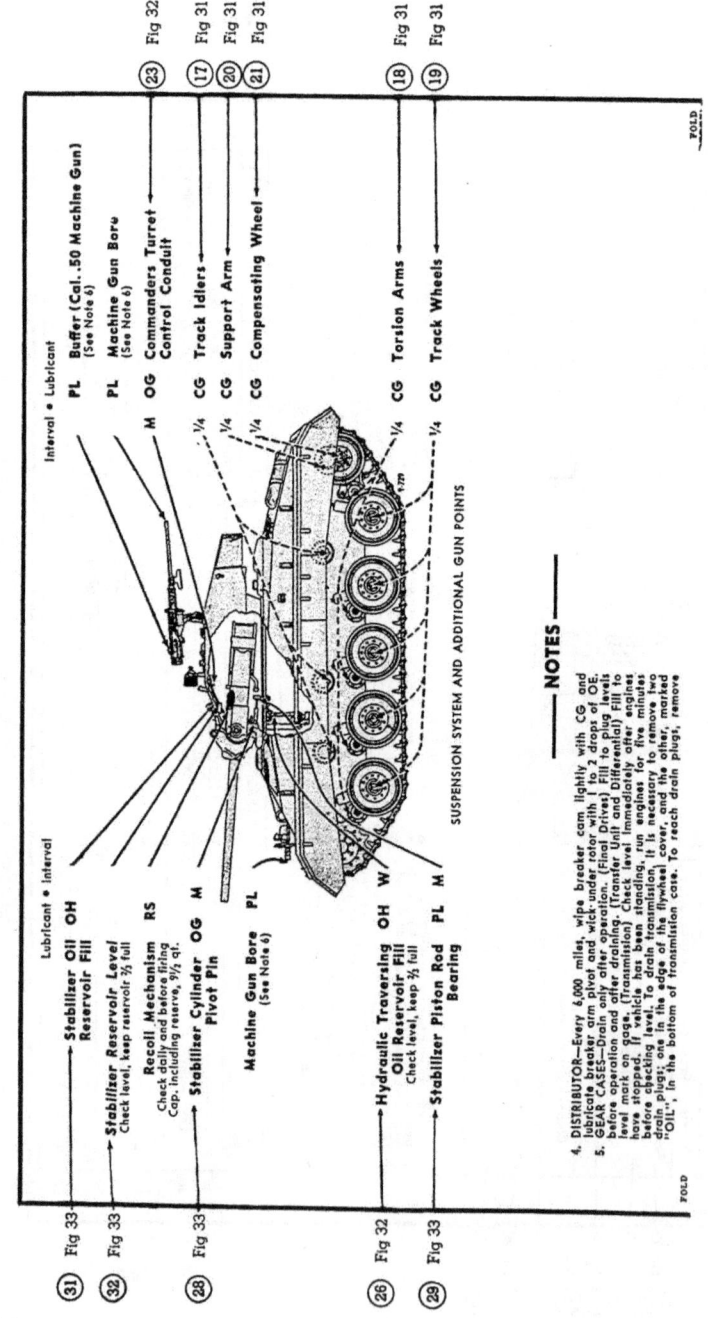

SUSPENSION SYSTEM AND ADDITIONAL GUN POINTS

NOTES

4. **DISTRIBUTOR**—Every 6,000 miles, wipe breaker cam lightly with CG and lubricate breaker arm pivot and wick under rotor with 1 to 2 drops of OE.
5. **GEAR CASES**—Drain only after operation. (Final Drives) Fill to plug levels before operation and after draining. (Transfer Unit and Differential) Fill to level mark on gage. (Transmission) Check level immediately after engines have stopped. If vehicle has been standing, run engines for five minutes before checking level. To drain transmission, it is necessary to remove two drain plugs; one in the edge of the flywheel cover, and the other, marked "OIL", in the bottom of transmission case. To reach drain plugs, remove

W	PL	**Gun Bore** (See Note 6)
	PL	**Mortar Bore** (See Note 6)
RS		**Recoil Mechanism Replenisher** — Fig 33 Check daily and before firing
M	PL	**Gyro Control Gear Box** — Fig 33
W	PL	**Mortar Firing Mechanism** (See Note 6)
W	PL	**Elevating Rack and Pinion**
M	OG	**Elevating Bevel Gear Case** — Fig 33
W	PL	**Breech and Firing Mechanism** Weekly and after firing, clean and oil

GUN, 75mm, M6
MOUNT, COMBINATION GUN, M64

6. ARMAMENT—After firing, and on 2 consecutive days thereafter, thoroughly clean the armament bores with CR, making sure that all surfaces including the rifling are well coated. Do not wipe dry. On the third day after firing, clean the bores with CR, wipe dry and lightly coat with PL. Weekly, thereafter, clean with CR, wipe dry, and recoil with PL. Clean all other exterior armament parts with CR before lubricating with PL. Buffer (Cal. .50 H.B.) Weekly, inject PL Special, then rotate buffer assembly and drain excess. In humid and salt air areas, use PL Medium daily in machine gun buffer and on external surfaces of armament.

7. OIL FILTERS—While crankcases are being drained, remove elements, clean inside of cases and install new elements.

8. OIL CAN POINTS—Weekly, lubricate Steering Lever Bearings and Braking Rod Clevises and Pins, Throttle Linkage, Pintle, Hinges and Latches, Handwheel Handle, Firing Button, Firing Lever Link, Machine Gun Mount Ball Socket, etc., with PL.

9. DO NOT LUBRICATE—(Tank) Generators, Commander's Cupola, Azimuth Indicator, (Gun Mount) Turret Traversing and Gun Stabilizer Motor Bearings.

10. DISASSEMBLED FOR LUBRICATION BY ORDNANCE PERSONNEL—Disassemble and lubricate at intervals with lubricants shown in parenthesis. Starters (Semiannually—GM), Propeller Shaft Bearing (Semiannually—WB), Final Drive Universal Joints (Semiannually—CG), Propeller Shaft and Fan Drive Universal Joints (Semiannually—CG), (Gun Mount) Cradle Trunnions—OG), Turret Traversing Gear Box (Semiannually—OG), Worm Gear Case (Semiannually—OG).

LO 9-729

Copy of this lubrication order will remain with the equipment at all times; instructions contained herein are mandatory and supersede all conflicting lubrication instructions dated prior to the date of this lubrication order.

BY ORDER OF THE SECRETARY OF THE ARMY:

OMAR N. BRADLEY
*Chief of Staff,
United States Army*

OFFICIAL:
EDWARD F. WITSELL
*Major General,
The Adjutant General.*

RA PD 355014J

Figure 28. Back side of lubrication order—light tank M24.

DISTRIBUTION: ARMY—Tech Sv (2); Arm & Sv Bd (1); AFF (2); OS Maj Comd (10); Base Comd (2); MDW (5); A (ZI) (18), (Overseas) (3); CHQ (2); D (2); R 9 (1); Bn 9 (1); C 9 (1); FC (1); USMA (2); Sch (5); Gen Dep (1); Dep 9 (2); Tng Cr (2); PE (Ord O) (5), (Overseas Sup Div) (1); PG 9 (3); Ars 9 (3); Dist 9 (3); T/O & E's: 17–17 (1); 44–75 (1); 44–77 (1); SPECIAL DISTRIBUTION. AIR FORCE—USAF (5).

For explanation of distribution formula, see TM 38-405.

Requisition additional Lubrication Orders in conformance with instructions in TM 38-405.

(3) Inspection of a unit to determine if it is "secure" is usually an external visual examination or a check by hand, wrench, or a pry-bar for looseness. Such an inspection must include any brackets, lock washers, lock nuts, locking wires, or cotter pins used.

(4) By "excessively worn" is meant worn beyond serviceable limits or to a point likely to result in failure if the unit is not replaced before the next scheduled inspection.

67. Cleaning

a. GENERAL. Any special cleaning instructions for specific mechanisms or parts are contained in the pertinent section. General instructions are as follows:

(1) Use dry-cleaning solvent or volatile mineral spirits paint thinner to clean or wash grease or oil from all parts of the tank. Refer to TM 9-313 for cleaning instructions pertaining to armament and sighting instruments.

(2) A solution of one part grease-cleaning compound to four parts of dry-cleaning solvent or volatile mineral spirits paint thinner may be used for dissolving grease and oil from engine blocks, chassis, and other parts. After cleaning, use cold water to rinse off any solution which remains.

(3) After the parts are cleaned, rinse and dry them thoroughly. Apply a light grade of oil to all steel parts having a polished surface to prevent rust.

(4) Before installing new parts, remove any preservative materials, such as rust-preventive compound, protective grease, etc.; prepare parts as required (oil seals, etc.); and for those parts requiring lubrication, apply the lubricant prescribed in the lubrication order.

b. GENERAL PRECAUTIONS IN CLEANING.

(1) Dry-cleaning solvent and volatile mineral spirits paint thinner are inflammable and should not be used near an open flame. Fire extinguishers should be provided when these materials are used. Use only in well-ventilated places.

(2) These cleaners evaporate quickly and have a drying effect on the skin. If used without gloves, they may cause cracks in the skin and, in the case of some individuals, a mild irritation or inflammation.

(3) Avoid getting petroleum products, such as dry-cleaning solvent, volatile mineral spirits paint thinner, engine fuels, or lubricants on rubber parts as they will deteriorate the rubber.

(4) The use of Diesel fuel oil, gasoline, or benzene (benzol) for cleaning is prohibited.

Figure 29. Localized lubrication points (points 1 through 7).

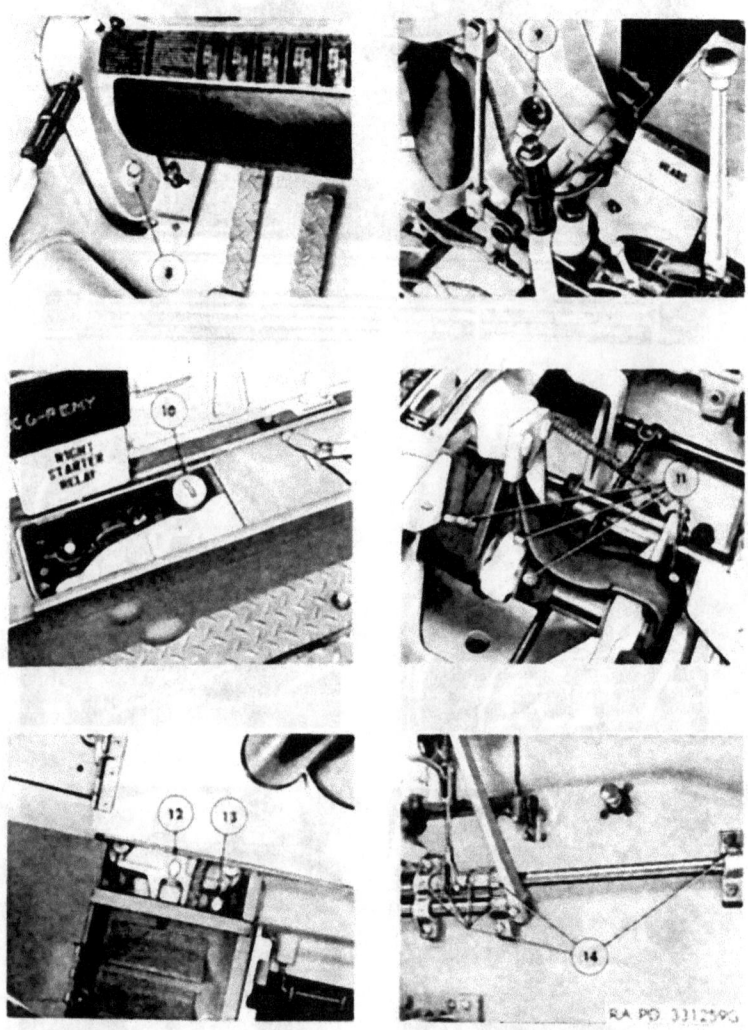

Figure 30. Localized lubrication points (points 8 through 14).

Figure 31. Localized lubrication points (points 15 through 21).

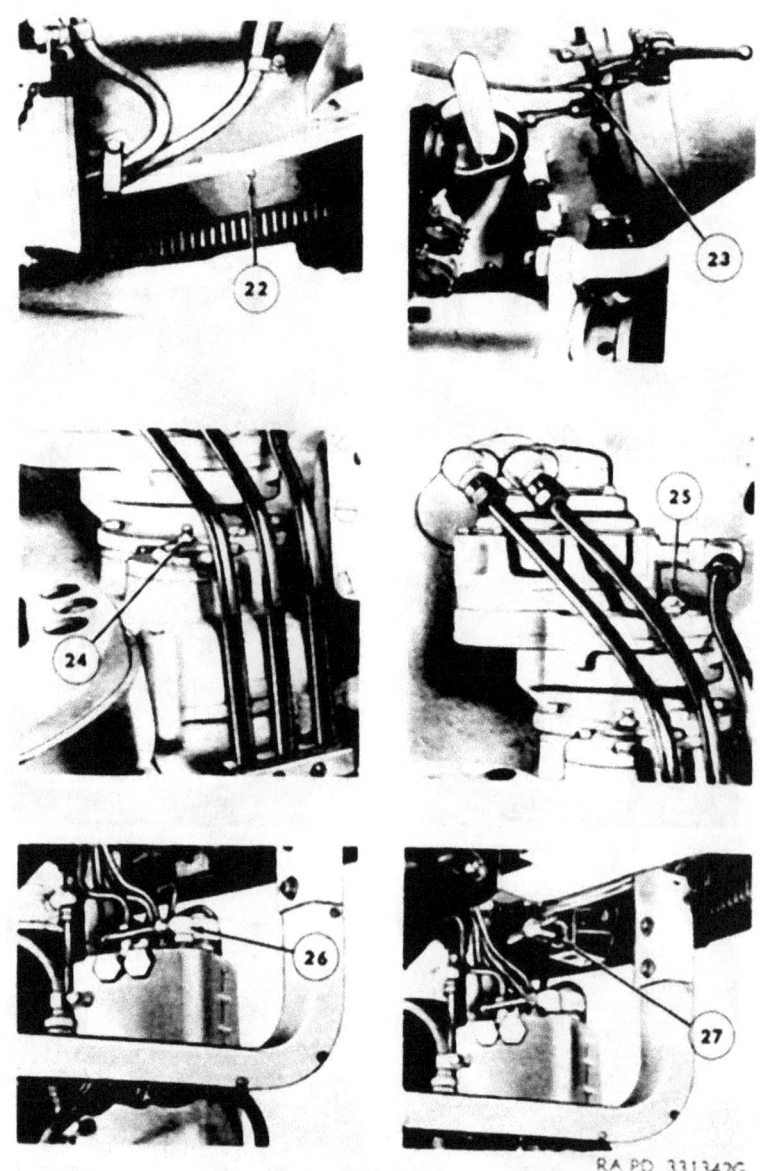

Figure 32. Localized lubrication points (points 22 through 27).

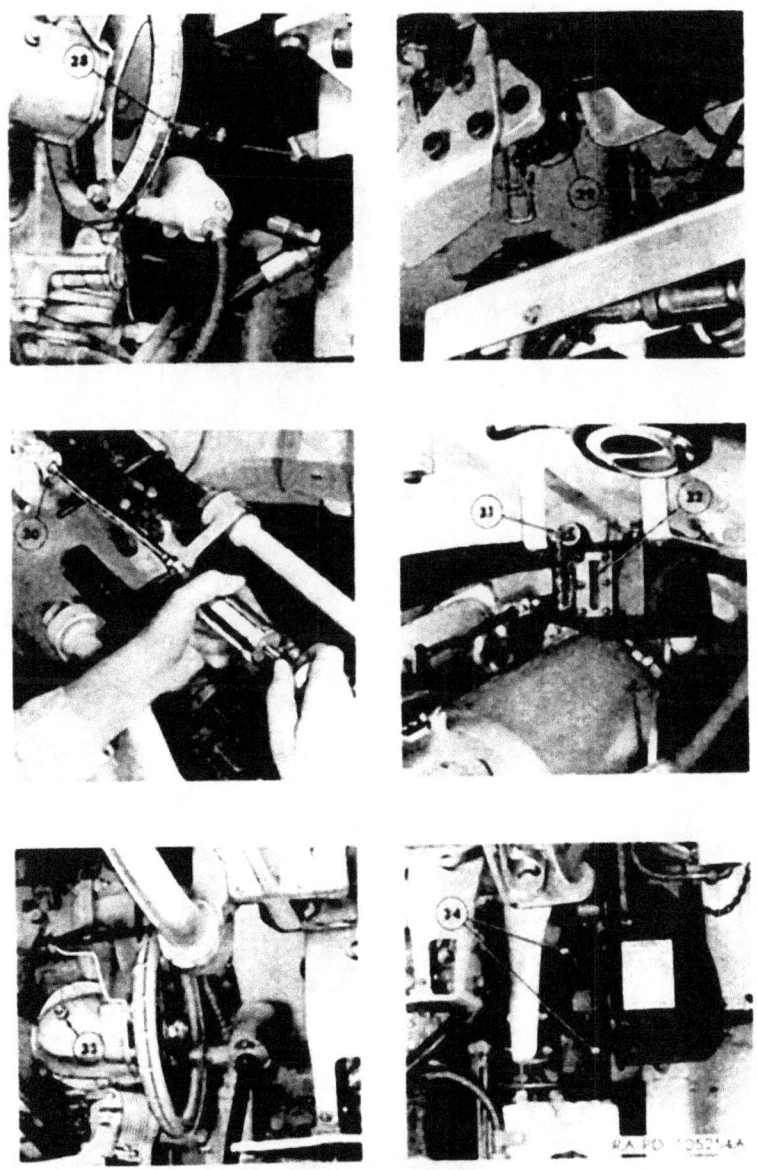

Figure 33. Localized lubrication points (points 28 through 34).

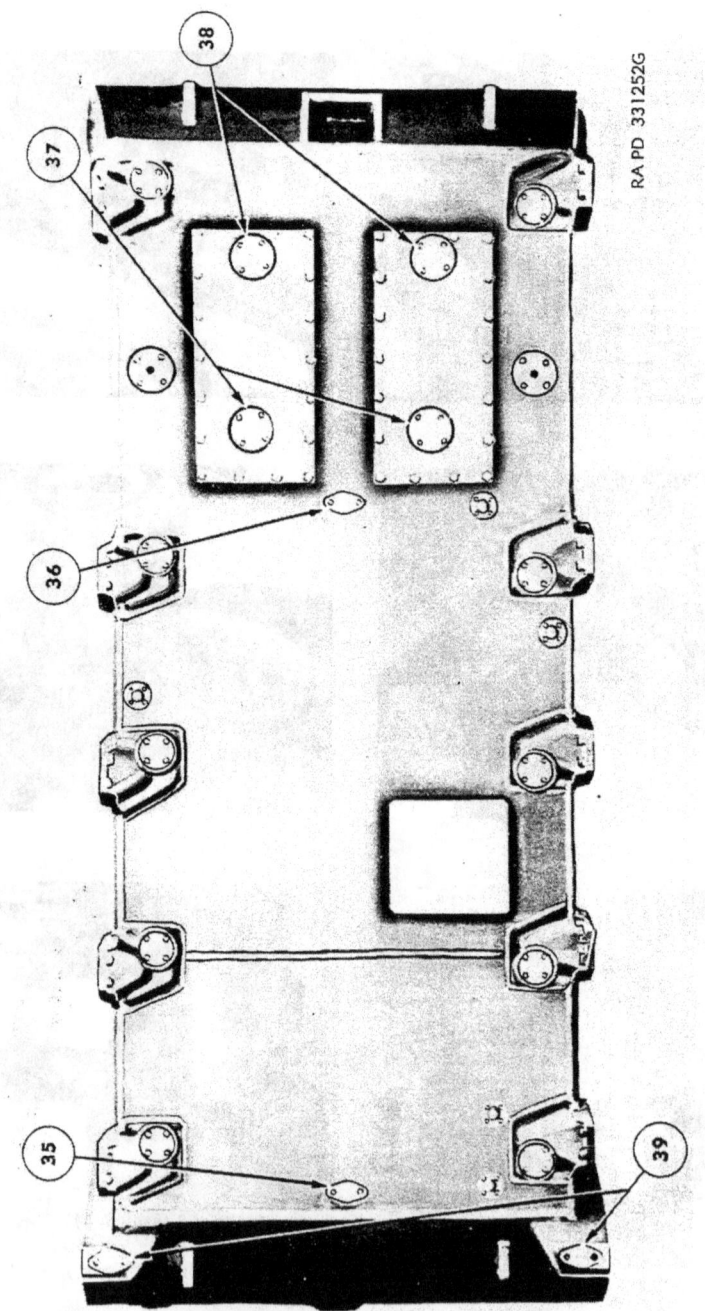

Figure 34. Localized lubrication points (points 35 through 39).

68. Preventive Maintenance by Driver or Operator

a. PURPOSE. To insure mechanical efficiency, it is necessary that the tank be systematically inspected at intervals each day it is operated and also weekly, so defects may be discovered and corrected before they result in serious damage or failure. Certain scheduled maintenance services will be performed at these designated intervals. Any defects or unsatisfactory operating characteristics beyond the scope of the driver or operator to correct must be reported at the earliest opportunity to the designated individual in authority. The services set forth in this paragraph are those performed by the driver or operator before-operation, during-operation, at-the-halt, after-operation, and weekly.

b. SERVICES. Driver and crew preventive maintenance services are listed in table II. Every organization must thoroughly school its personnel in performing the maintenance procedures for this tank as set forth in this manual.

TABLE *II. Driver's or Operator's Preventive Maintenance Services*

Before-operation	During-operation	At-the-halt	After-operation	Weekly	Procedure
					USUAL CONDITIONS
X	----	X	X	X	*Caution:* Place all tags describing condition of tank in the driver's compartment in a conspicuous location so that they will not be overlooked (fig. 7).
X	----	X	X	X	*Fuel, oil, water.* Check the amount of fuel in the tanks and inspect for indications of leaks. Check spare containers for contents. Add fuel if necessary (fig. 36). Check the engine and transmission oil levels and add oil if necessary (fig. 46, par. 151). Check the radiator. If water is added in cold weather, test solution with a hydrometer to determine if there is sufficient antifreeze (fig. 36, par. 118).
X	----	----	----	X	*Tracks.* Visually inspect tracks, idler wheels, and track drive sprockets for damage that may have developed since last operation (fig. 130, par. 171). Check track adjustment. Solid tires must be kept free of acids, oil and grease (fig. 131, par. 171).
				X	Tighten all wedge bolts securely. Inspect track links connections and wedges (par. 172).
		X	X	X	*Springs and suspensions.* Look at springs, suspensions, and shock absorbers, to see if they have been damaged (fig. 130 or 139, par. 69).

Table II. Driver's or Operator's Preventive Maintenance Services—Continued

Before operation	During operation	At-the-halt	After operation	Weekly	Procedure
					USUAL CONDITIONS—Continued
X	----	X	X	X	*Leaks, general.* Look under tank and in engine compartment for any indication of fuel, engine-oil, water, or brake-fluid leaks (fig. 36).
X	----	----	X	X	*Vehicle equipment.* Visually inspect fire extinguishers (fig. 5, par. 201).
				X	See that fire extinguishers are charged. Note seals (figs. 23, 24, par. 201).
X	----	----	X	X	Operate lights, horn (if tactical situation permits), and windshield wipers. Visually inspect reflectors, vision devices, etc. (fig. 7).
X	----	----	X	X	Visually inspect body or hull, armor, towing connections, traveling locks, doors, paulins, camouflage nets, tools, etc. Clean as required (fig. 2, pars. 58, 61).
X	----	----	----	X	Check for any tampering or damage that may have occurred since last inspection (figs. 1 and 2).
X	X	----	----	----	*Instruments.* Observe for normal reading during warm-up and during operation of tank (fig. 6, par. 8).
	X	X	----	----	*General operation.* Be alert for any unusual noises or improper operation of steering, clutch, brakes, or gear shifting. Investigate at next halt. Pull back on both steering levers and note whether both meet firm resistance at approximately 6 notches of lever travel. Stop the engine immediately if any warning signal lights come on (if tactical situation permits) and investigate cause (fig. 7).
X	X	----	X	X	*Armament.* Refer to TM 9-313.
			X	----	*Fuel filters.* Check all fuel filters for leaks (figs. 66 and 75).
				X	Remove the drain plug or sediment bowl and remove all water and sediment from the filter (figs. 73, 75, par. 112).
			X	X	*Lubrication.* Lubricate daily and weekly, items specified on lubrication orders (figs. 27, 28, par. 62).
			X	X	*Clean.* Clean glass, vision devices, and inside of tank. Wipe off exterior of tank (figs. 3, 16, par. 67).
				X	Wash tank, clean engine, and engine compartment (figs. 36 and 46).
			X	X	*Batteries.* Clean, check water level, and inspect terminals for tightness and coating of grease (figs. 90, 91, par. 127).

Table II. Driver's or Operator's Preventive Maintenance Services—Continued

Before-operation	During-operation	At-the-halt	After-operation	Weekly	Procedure
					USUAL CONDITIONS— Continued
				X	*Assemblies and belts.* Inspect assemblies such as carburetor, generator, heater, ventilator, compressor, starter, and water pump, for looseness of mountings or connection. Press drive belts to determine if tension is correct (figs. 35, 37, par. 88).
				X	*Electrical wiring.* Inspect visually all accessible electrical wiring, conduits, and shielding.
				X	*Towing connections.* Inspect towing shackles and pintle hook. Clean and lubricate as required. Make sure that latching mechanism closes completely and securely (fig. 2).
				X	*Publications.* Check to see that TM 9–313, TM 9–729, and ORD 7 SNL G–200 are on hand and in good order.
					UNUSUAL CONDITIONS *Extreme Cold*
		X	X	----	*Cooling and fuel systems.* Refuel and add denatured alcohol as required.
				X	Drain fuel tank and clean sump to remove condensation and sludge; refuel tank.
X	----	----	X	X	Check level. Check specific gravity of radiator coolant, using ethylene glycol and/or water if needed.
					Note. If using arctic antifreeze compound, make a warning tag and place it on or near the radiator filler neck. The tag should read: This cooling system is filled with arctic antifreeze compound. *Caution:* Do not add water or any other type of antifreeze.
				X	*Lubricants.* Check and, if necessary, change lubricants and special oils to conform with the LO.
				X	Check gear cases for collections of sludge and water clean out if necessary and refill.
					Note. It is necessary to have lubricant warm and fluid for draining and refilling.
		X	X	----	*Control levers.* Position levers in neutral position.
				X	*Batteries.* Check for proper charge and electrolyte level.
			X	----	Remove battery and store in warm place, if vehicle is not equipped with power plant heater.
X	----	----	X	X	*Clean.* Clean snow, ice, and mud from all parts of vehicle.

Table II. Driver's or Operator's Preventive Maintenance Services—Continued

Before operation	During operation	At-the-halt	After operation	Weekly	Procedure
					UNUSUAL CONDITIONS—Continued *Extreme Cold—Continued*
X					*Brakes.* Check for frozen brake shoes.
	X				*Operating characteristics.* Check for the feel of stiffness of lubricant in the final drives and suspension components. This will be indicated by unusual power demand when putting vehicle in motion. Listen for signs of malfunctions and inspect immediately to determine causes.
			X	X	*Winterization equipment.* Check all winterization equipment for secure installation and proper functioning.
					Extreme Heat
X		X		X	*Cooling and fuel systems.* Check air cleaner, fuel and oil filters, and radiator fins and clean as often as necessary to keep them in good condition.
X				X	*Batteries.* Check electrolyte level.
				X	Check for proper charge.
		X			*General operation.* Investigate when engine temperatures are consistently over 210° F.
X		X		X	*Ventilating blower.* Check for proper operation.
					Unusual Terrain
			X		*Lubrication.* Check for fouled lubricants and lubricate as necessary.
X				X	*Cooling and fuel systems.* Check air cleaner, fuel and oil filters, grilles and radiator fins to keep them in good condition. **WARNING:** Under extremely dusty conditions or blowing sand, it will be necessary to service the air cleaner several times daily during operation to prevent entry of dust or sand into the engine. Failure to do this may wear out engine parts in a short time.
					Fording Operations
X					*Fording limits.* Check fording depth (par. 57).
X					*Battery.* Check vent caps for tightness.
			X		Check for seepage of water into battery. Check charge and add electrolyte and charge as needed.

Table II. *Driver's or Operator's Preventive Maintenance Services*—Continued

Intervals					Procedure
Before-operation	During-operation	At-the-halt	After-operation	Weekly	
					UNUSUAL CONDITIONS—Continued *Fording Operations*—Continued
			X		*Drain hull.* Open drain valves to remove accumulated water.
			X		*Clean.* Remove water and sludge. If fording salt water, wash with fresh water.
			X		*Engine and transmission.* Check for evidence of water or grit and replace oil if necessary. Drain accumulation of water and sludge from fuel tank.
					Lubrication. Lubricate as specified in paragraph 64.
			X		*Leaks, general.* Check for entry of water into hull or engine compartment.
X			X		*Clean.* Check for tightness of vent caps on batteries.
			X		Check for seepage of water into batteries.
			X		*Lubrication.* See LO 9-729 and paragraph 57 for after-fording cleaning and lubrication instructions.

69. Preventive Maintenance by Organizational Maintenance Mechanics

a. INTERVALS. The indicated frequency of the prescribed preventive maintenance services is considered a minimum requirement for normal operation of vehicle. Under unusual operating conditions, such as extreme temperatures, dust or sand, or extremely wet terrain, it may be necessary to perform certain maintenance services more frequently.

b. DRIVER OR OPERATOR PARTICIPATION. The drivers or operators should accompany their vehicles and assist the mechanics while periodic organizational preventive maintenance services are performed. Ordinarily, the driver should present the vehicle for a scheduled preventive maintenance service in a reasonably clean condition.

c. SPECIAL SERVICES. These are indicated by the item numbers in the columns which show the interval at which the services are to be performed and show that the parts or assemblies are to receive certain mandatory services. For example, an item number in one or both columns opposite a *tighten* procedure means that the actual tightening of the object must be performed. The special services are as follows:

   (1) *Adjust.* Make all necessary adjustments in accordance with the pertinent section of this manual, or other current directives.

(2) *Clean.* Clean the unit as outlined in paragraph 67, to remove old lubricant, dirt, and other foreign material.

(3) *Special lubrication.* This applies either to lubrication operations that do not appear on the vehicle lubrication order or to items that do appear on such orders but which should be performed in connection with the maintenance operations if parts have to be disassembled for inspection or service.

(4) *Serve.* This usually consists of performing special operations, such as replenishing battery water, draining and refilling units with oil, and changing or cleaning the oil filter, air cleaner, or cartridges.

(5) *Tighten.* All tightening operations should be performed with sufficient wrench torque (force on the wrench handle) to tighten the unit according to good mechanical practice. Use a torque-indicating wrench where specified. Do not overtighten, as this may strip threads or cause distortion. Tightening will always be understood to include the correct installation of lock washers, lock nuts, lock wire, or cotter pins.

  d. SPECIAL CONDITIONS. When conditions make it difficult to perform the complete preventive maintenance procedures at one time, they can sometimes be handled in sections, planning to complete all operations within the week if possible. All available time at halts and in bivouac areas must be utilized, if necessary, to assure that maintenance operations are completed. When limited by the tactical situation, items with special service in the columns should be given first consideration.

  e. WORK SHEET. The numbers of the preventive maintenance procedures that follow are identical with those outlined on DA AGO Form 462, Work Sheet for Full-Track and Tank-Like Wheeled Vehicles. Certain items on the work sheet that do not apply to this vehicle are not included in the procedures in this manual. In general, the sequence of items on the work sheet is followed in the manual procedures, but in some instances there is deviation for conservation of the mechanic's time and effort.

  f. PROCEDURES. Table III lists the services to be performed by the organizational mechanic or maintenance crew at the designated intervals. Each page of table III has two columns at its left edge corresponding to quarterly and monthly maintenance respectively. Very often it will be found that a particular procedure does not apply to both scheduled maintenances. In order to determine which procedure to follow, look down the column corresponding to the maintenance procedure and wherever an item number appears, perform the operations indicated opposite the number.

Table III. Organizational Mechanic or Maintenance Crew Preventive Maintenance Services

Intervals		Procedure
Quarterly	Monthly	
		ROAD TEST
		Note. When the tactical situation does not permit a full road test, perform items 2, 3, 5, 9, 12, 13, 14, and 15 which require little or no movement of the vehicle. When a road test is possible, it should be for preferably 3 and not over 5 miles.
1	1	*Before-operation service.* Perform the before-operation service outlined in paragraph 68.
2	2	*Instruments and gages.* Observe readings of all instruments frequently during operation to see whether they are indicating properly. *Oil pressure warning signal light.* In case of lighting of oil pressure warning signal, vehicle should be stopped immediately and trouble corrected or reported to proper authority. *Engine temperature gages.* Engine temperature should rise gradually to normal range, 160° to 210° F. and gage should always indicate below 240° F. If gage indicates more than 240° F. or if warning signal lights, stop vehicle until correction can be made. *Ammeter.* The ammeter will register zero or slight discharge (−) with the engines idling or register zero or slight positive charge (+) with engines running at operating speeds. Any unusual drop or rise in reading for an extended period may indicate a dangerously low battery or a faulty generator regulator. *Tachometers.* Tachometers should indicate engine speed and accumulating revolutions without excessive noise or fluctuation. *Speedometer and odometer.* Speedometer should indicate vehicle speed. Odometer should register accumulating mileage without excessive noise or fluctuation. *Warning lights.* Transmission oil pressure warning lights should be off when engines are running faster than 450 rpm. Lights may flicker at slow idle speed.
3	3	*Windshield, windshield wipers, and siren.* Inspect windshield assembly and wipers (when in use) to see that they are in good condition and secure and whether wiper blades move through their full stroke and contact surface evenly. Inspect siren (or horn) for good condition and secure mounting and, if tactical situation permits, test for operation and tone.
5	5	*Steering brakes.* With vehicle stopped, pull back on steering brake levers simultaneously; if brakes are properly adjusted, levers should start to take hold at six notches of travel on ratchet. Accelerate vehicle to a moderate speed, release accelerator, apply both sterring brakes, and observe whether they stop the vehicle effectively. Apply steering brakes independently and see that they steer vehicle properly. Apply parking brake with vehicle on reasonable incline. It should hold vehicle effectively and locking device should hold levers in applied position.

Table III. Organizational Mechanic or Maintenance Crew Preventive Maintenance Services—Continued

Intervals		Procedure
Quarterly	Monthly	
		ROAD TEST—Continued
7	7	*Transmission (lever action, vibration, and noise).* Operate vehicle through each speed range of transmission. Observe whether selector lever operates properly and whether there are any unusual vibrations or noises in any speed range that might indicate damage, excessive wear, loose mountings, or improper operation.
9	9	*Engine (idle, acceleration, power, noise, smoke, and oil consumption):* *Idle.* With the vehicle stopped, observe if engines run smoothly at normal idling speed (450 rpm). Throughout road test, observe if there is any tendency for engines to stall when accelerator is released and hand throttle is closed. *Acceleration, power, vibration, and noise.* Test engines for normal acceleration and pulling power in each speed. While testing in driving range, accelerate from low speed to top speed and listen for unusual engine noise, knock, whine, or vibration that might indicate loose, damaged, excessively worn, or inadequately lubricated engine parts or accessories or loose mountings or drive belts. *Smoke.* During operation, observe if there is excessive smoke from exhaust. *Oil consumption.* Upon completion of road test, a check should be made to determine whether one or both engines have been consuming an excessive amount of oil.
10	10	*Unusual noises (propeller shafts and universal joints, differential and final drives, sprockets, wheels, support rollers, and tracks).* During road test, listen for any unusual noise from these units, indicating damaged, defective or loose parts, or inadequate lubrication.
11	11	*Temperatures (transmission, transfer unit, differential, and final drives, hubs, sprockets, wheels, and support rollers).* After operating, remove bottom inspection plate and examine transmissions for excessive heat. Check by hand-feel for any abnormal temperature in transfer unit, differential and final drives, hubs of sprockets, wheels, and support rollers. *Note.* If position on grade is selected for this check, time will be saved in performing item 12.
12	12	*Gun elevating and traversing mechanism.* Place vehicle in a position where it will be tilted (sidewise) about 10°. Traverse turret through its full 360° range, by both hand and power controls, checking for indication of binding. With gun pointed forward, elevate it through its entire range with hand controls; check for binding, excessive lash, or erratic action.
13	13	*Leaks.* Look in engine compartment and also beneath vehicle, for indications of fuel, oil, or coolant leaks.

Table III. *Organizational Mechanic or Maintenance Crew Preventive Maintenance Services—Continued*

Intervals		Procedure
Quarterly	Monthly	
		ROAD TEST—Continued
14	14	*Noise and vibrations (engine mountings, accessories, and drives).* While accelerating and decelerating the engines, listen for unusual noises in the engines, accessories, or accessory drivers paying particular attention to fan and generator drives. Notice whether there is any excessive vibration that might indicate loose engine mountings.
15	15	*Track tension after final road test.* Inspect for satisfactory track tension after final road test (par. 171).
		MAINTENANCE OPERATIONS
16	16	*Engine vacuum and fuel pump test:* *Vacuum test.* With the engine running at normal idling speed (450 rpm) and the engine compartment door open, attach vacuum gage to the intake manifold. The vacuum gage should read about 18–21 inches and the pointer should be steady. Accelerate the engine with full throttle momentarily. The gage indicator should drop to approximately 2 inches as the throttle is opened and recoil to at least 24 inches as the throttle is closed. Readings other than these should be reported to higher authority. *Note.* The above readings apply to sea level. There will be approximately a 1-inch drop for each 1,000 feet of altitude. *Fuel pump test.* Attach a fuel pump test gage in the carburetor inlet line and, with the engine running, the fuel pressure should not be less than 4½ psi. *Note.* The batteries must be fully charged, the wiring in good condition, and fuel tanks at least half full to perform this test satisfactorily. Check the fuel pump in each tank.
17	17	*Engine crankcase (leaks and level).* Stop engine and open battery switch. Observe both crankcases for indications of oil leaks, and inspect bayonet gages for oil level. Add oil as necessary. If oil change is due or condition of oil warrants a change, drain and fill crankcases to proper level with specified oil. Refer to lubrication order, paragraph 62. NOTE: If oil is changed, do not start engines until new oil filter elements have been installed as in item 54.
54	54	*Engine oil filters.* Inspect oil filters to see that they are in good condition, secure, and not leaking. Drain accumulated sediment from filter bowl.
54	------	*Clean and serve.* Remove oil filter elements, clean cases, and install new filter elements, using new gaskets and tightening covers securely. Refer to lubrication order, paragraph 62.

Table III. *Organizational Mechanic or Maintenance Crew Preventive Maintenance Services*—Continued

Intervals		Procedure
Quarterly	Monthly	

MAINTENANCE OPERATIONS—Continued

18	18	*Side armor (fenders, guards, paint and markings, shackles, and siren).* Inspect these items to see that they are in good condition, that armor, fenders, dust shields (if so equipped), guards, shackles, and siren are secure and that towing shackles are not excessively worn. Observe condition of paint for rust or polished surfaces that may cause reflections, and (unless covered for tactical reasons) see that all vehicle markings are legible.
19	19	*Bottom (armor, escape door, inspection plates, and drain plugs).* See that these items are in good condition and secure, that the bottom escape door hinges and latches operate properly and are adequately lubricated, and that bottom drain plugs are tight. Tighten all bottom inspection plates securely. Apply a few drops of oil to bottom escape door hinges and latches.
20	20	*Differential and final drives.* Inspect housings for good condition and leakage; check lubricant level. See that all assembly and mounting bolts are secure. If change of lubricant is due or condition of oil warrants a change, drain and refill with specified oil at this time. Refer to lubrication order, paragraph 62.
20	------	*Tighten.* Tighten all external assembly and mounting bolts securely.
21	21	*Tracks (shoes, pins, and bushings).* Inspect tracks to see that these items are in good condition, correctly assembled, secure, and not excessively worn. Tighten track screws.
21	------	*Note.* Whenever the tracks are disconnected and removed from the sprockets, support rollers, and wheels or at each quarterly maintenance service, the related items 22 and 25, marked on DA AGO Form No. 462 should be inspected as described below in the asterisk-marked (*) procedures. On the regular monthly and quarterly maintenance services, the tracks should not be removed unless repairs are needed. ***Caution:*** Whenever tracks are removed for repair or replacement, do not install tracks until the services followed by the asterisk (*) in items 22 and 25 have been completed.
22	22	*Compensating wheels, arms, adjustment, and lock nuts.* Inspect these items to see that they are in good condition, correctly assembled, and secure and that grease is not leaking excessively from wheel bearing seals. Be sure adjusting nuts and adjusting locks are secure. Tighten all assembly and mounting bolts and nuts securely. **Note.* In addition to the above, at each second quarterly maintenance service or whenever the tracks are removed, check the compensating wheel hub bearings for looseness or end play. Spin the wheels and listen for unusual noise that might indicate a damaged, excessively worn, or inadequately lubricated bearing.

Table III. Organizational Mechanic or Maintenance Crew Preventive Maintenance Services—Continued

Intervals		Procedure
Quarterly	Monthly	

MAINTENANCE OPERATIONS—Continued

23	23	*Track wheel arms, shock absorbers, links, and brackets.* Examine these items to see if they are in good condition, correctly assembled, securely mounted, and not excessively worn. Inspect shock absorbers for leaks. *Caution:* Shock absorbers should never be replaced merely because of external oil stains. As much as 6–8 ounces of oil can be lost before the operation of the shock absorber is affected. The most practical method of testing the shock absorber is by the temperature method. This test should be made immediately after a run of not less than 5 miles of high-speed operation or 4 miles of cross-country operation. Feel the small part of the shock absorber and then the hull near the shock absorber and note the difference in temperature. If the hull and the shock absorber are the same temperature, the shock absorber is not operating and should be replaced. The difference in temperature should be clearly evident but the reserve cylinder (small part of shock absorber) does not need to be extremely hot to indicate a satisfactory unit.
23	------	*Tighten.* Draw up all suspension unit assembly or mounting nuts or screws securely.
24	24	*Wheels (tires and support rollers).* Inspect wheels and rollers for good condition, correct assembly, and secure mounting. Pay particular attention to see that tire rubber has not separated from rim and that tires are not cut, torn, or excessively worn. Inspect for excessive lubricant leaks from bearings.
24	------	*Tighten.* Jack up track wheels and test bearings for looseness, roughness, and end play. Spin wheels and listen for any unusual noise. Tighten assembly and mounting bolts securely.
24	------	*Note.* Whenever the tracks are removed, the above operation should be performed before the tracks are installed.
25	25	*Sprockets (hubs, teeth, and nuts).* Inspect sprockets for good condition, correct assembly, and security of mounting bolts. Inspect sprocket teeth for excessive wear, and shaft flange gaskets or oil seals for excessive leaking of lubricant. If sprocket teeth are excessively worn, sprockets should be replaced or reversed (par. 168). Tighten assembly and mounting bolts securely. **Note.* In addition to the above, at each quarterly operation or whenever the track is disconnected and removed from the sprocket, check the sprocket teeth for excessive wear, see that sprockets are well secured to hubs, and that the hub-to-final-drive bolts are secure. Check the sprocket hub bearings for looseness and end play. After performing the above, install the tracks and connect them securely (par. 172).
26	26	*Track tension.* Adjust track to standard tension (par. 171).

Table III. Organizational Mechanic or Maintenance Crew Preventive Maintenance Services—Continued

Intervals		Procedure
Quarterly	Monthly	

MAINTENANCE OPERATIONS—Continued

27	27	*Top armor (turret, deck, paint and mrakings, grilles, doors, latches, and antenna mast).* Inspect these items to see that they are in good condition and secure. See that door hinges and latches operate properly, are not excessively worn, and are adequately lubricated. Be sure grilles are not obstructed. Examine paint for rust spots or polished surfaces that may cause reflections and, unless covered for tactical reasons, see that vehicle markings are legible.
28	28	*Caps and gaskets (fuel and radiator).* Inspect to see that fuel tank and radiator caps and gaskets are in good condition, secure, and not leaking.
30	------	*Engine removal (when required).* Remove engines on quarterly maintenance service only if inspections made in items 9, 10, and 13 and a check of oil consumption indicate a definite need.
30	30	*Clean.* Clean exterior of engine and dry thoroughly, taking care to keep dry-cleaning solvent away from electrical wiring, terminal boxes, and equipment.
		Note. In this and following services (items 32 through 60) the procedures should be followed as well as possible if engines are not removed from vehicle.
32	------	*Spark plugs (gaps and deposits).* Remove and clean spark plugs (par. 106). Inspect insulators for cracks or breaks and electrodes for excessive burning. Replace unserviceable plugs with new or reconditioned plugs. Adjust electrodes of all plugs to be installed to 0.030 inch by bending ground electrodes. Be sure to install new gaskets; do not overtighten plugs.
		Note. Perform item 33 before installing plugs. Inspect the spark plug resistor-suppresors for scorching and cracks and be sure all connections are secure.
33	------	*Compression test (record).* When spark plugs are removed for performance of item 32, test compression of each cylinder and record gage readings on DA AGO Form No 462. If there is more than 10 pounds variation between cylinders, report to designated authority.
34	34	*Generators and starters.* Inspect generators and starters for good condition, security of mounting, and secure wiring connections.
34	------	*Remove.* Remove commutator inspection covers and examine commutator for good condition. See that brushes are free in brush holders, clean, and not excessively worn, that brush connections are secure, and that wires are not broken or chafing.
34	------	*Clean.* Clean commutator end of units by blowing out with compressed air.
34	------	*Tighten.* Tighten mounting bolts securely. Inspect generator armature brush capacitors (condensers) to see that they are in good condition and securely connected.

Table III. *Organizational Mechanic or Maintenance Crew Preventive Maintenance Services*—Continued

Intervals		Procedure
Quarterly	Monthly	
		MAINTENANCE OPERATIONS—Continued
36	36	*Distributors.* Inspect distributor bodies, ground strap, external attachments, and radio noise suppressors for good condition and secure mounting. Inspect distributor ventilating lines for proper installation and good condition.
36	------	*Cap, rotor, and points.* Clean caps, rotors, and seals and inspect for good condition and correct assembly. Look for cracks or burns. Inspect breaker points for good condition, good alinement, and 0.015-inch gap. If the breaker-plate assemblies are unserviceably dirty, remove the distributor, clean in dry-cleaning solvent, dry with compressed air, lubricate the parts for the monthly maintenance service, and install the distributors in correct position for timing (par. 102). When cleaning distributors, remove wicks and lubrication cups, clean and dry while removed and install them only after the distributor assemblies are cleaned and blown dry with compressed air. If the breaker points are pitted, burned, or worn to an unserviceable condition, replace with a new set. If points are badly pitted, replace capacitor (condenser) also. If points are slightly pitted or burned, dress with point file or No. 2/0 flint paper and blow off filings with compressed air. *Shaft.* Check shafts by hand-feel for looseness to determine whether the bushings are worn. *Centrifugal advance.* Test range of movement of centrifugal advance mechanism by rotating shaft and rotor by finger force. Observe whether shaft returns to original position without binding.
36	------	*Special lubrication.* Lubricate cam surfaces, movable breaker-arm pin, camshaft, and the wick (par. 62). Take care to keep lubricant off the distributor points, not to apply more lubricant than is specified, and to wipe the cam clean before lubricating its surface.
36	------	*Adjust.* Adjust breaker-point to 0.015 inch at wide open position.
38	38	*Ignition wiring and conduits.* Inspect accessible ignition wiring and conduits for cleanliness, good condition, correct assembly, and secure mounting and see that they are not chafing against other engine parts.
39	39	*Coils.* Inspect ignition coils and radio noise suppressor capacitors (condenser) to be sure they are clean and that all mountings and connections are secure.
42	42	*Breather caps and ventilators.* Inspect engine crankcase ventilator air cleaner for good condition and correct assembly (par. 90). Clean oil bath air cleaner, fill with oil, and install securely. Refer to lubrication order, paragraph 62.

Table III. Organizational Mechanic or Maintenance Crew Preventive Maintenance Services—Continued

Intervals		Procedure
Quarterly	Monthly	

MAINTENANCE OPERATIONS—Continued

43	43	*Air cleaners (carburetors).* Disassemble, clean, fill with oil, and install carburetor air cleaners, following carefully the correct procedure outlined in paragraph 115 and on lubrication order, paragraph 62.
44	44	*Carburetors (automatic chokes, throttles, and linkage).* Inspect these items to see if they are in good condition, correctly assembled, and securely installed; that carburetors do not leak; and that control linkage, automatic choke, and throttle shafts are not excessively worn. Inspect automatic chokes for closed position when engine is cool. See that throttle valves open fully when accelerator is fully depressed.
45	45	*Manifolds (intake and exhaust).* Inspect accessible portions of manifolds for good condition, secure mountings, and possible leaks.
45	------	*Tighten.* Tighten manifold mounting and assembly nuts and cap screws to 25-30 foot-pounds torque.
46	46	*Cylinder (heads and gaskets).* Inspect cylinder heads for good condition and secure mounting. Note any indications of oil, coolant, or compression leakage around cap screws or gaskets. **Caution:** Cylinder head cap screws should not be tightened unless there is a definite indication of looseness or leaks. If tightening is necessary, a torque-indicating wrench must be used and screws tightened in proper sequence (fig. 38) to 70-75 foot-pounds.
49	49	*Water pumps, fans, and shrouds.* Observe whether these items are in good condition and securely mounted. Be sure fan blades do not interfere with shrouds or cores. See that the drive belts and pulleys are well alined; belts not excessively worn, frayed, oil-soaked, improperly adjusted, or bottoming in drive pulleys. Adjust belts to ⅝- to ¾-inch deflection (par. 88.)
51	51	*Engine compartment (bulkhead and control linkage).* See that engine compartment, including bulkhead, is in good condition, clean, and secure; and that control linkage in the engine compartment is in good condition, securely connected, and mounted.
51	------	*Clean.* Clean engine compartment as thoroughly as possible. Whenever engines are removed, clean out all fuel and oil drippings, dirt, and refuse; wipe out the engine compartment with cloths soaked in dry-cleaning solvent or volatile mineral spirits paint thinner and dry thoroughly.

Table III. *Organizational Mechanic or Maintenance Crew Preventive Maintenance Services*—Continued

Intervals		Procedure
Quarterly	Monthly	

MAINTENANCE OPERATIONS—Continued

53	53	*Fuel (tanks and lines).* Inspect accessible portion of these items for correct assembly, good condition, and secure mounting. Note whether there are indications of fuel leaks from tanks or lines. Drain water and sediment from each tank by removing drain plugs and allow fuel to drain briefly until it runs clean. Tighten plugs and fuel tank drain hole cover securely to prevent fuel and dust leaks. *Caution:* Catch drainings in container and use every precaution not to spill fuel. If fuel does spill, wipe up until dry before turning on the main battery switch.
53	------	*Tighten.* Tighten accessible fuel line support clips securely.
55	55	*Fuel filter.* Inspect auxiliary fuel filter for secure mounting and leaks at gaskets.
55	------	*Clean.* Remove element and wash in dry-cleaning solvent or volatile mineral spirits paint thinner until disks are clean and free (par. 112).
56	56	*Oil coolers (transmission and differential).* Examine differential oil cooler and lines and transmission coolant lines for good condition, secure attachment, and leaks. Inspect differential oil cooler to see that core is not clogged with dirt and trash; clean out foreign material.
57	57	*Exhaust pipes and mufflers.* Inspect these items for good condition, secure assembly, and mounting of exhaust pipes and mufflers and for indications of exhaust leaks.
57	------	*Tighten.* Tighten all accessible mounting bolts and connections securely.
58	58	*Engine mountings.* Inspect all engine mountings and radio bond straps to see that they are in good condition and secure.
58	------	*Tighten.* Tighten all mountings and brackets securely. When engines are removed, tighten the mount-to-engine and bracket-to-hull bolts securely.
60	60	*Fire extinguisher system (tank, valve, lines, nozzles, and mountings).* Inspect cylinder and valve of fire extinguisher system for good condition, secure mounting, and full charge. Full charge may be observed by presence of intact wire seal, but can be positively determined only by weighing. Examine control cables to see that they are in good condition and free to operate at a moment's notice. See that all lines and nozzles are in good condition, securely mounted, and connected, with nozzles properly aimed and not clogged. If dirty or corroded nozzles are found, disconnect main feed line and blow out cautiously with compressed air. *Caution:* Empty or partially empty cylinders should be reported for recharge or replacement immediately. Do not drop, strike, or roughly handle extinguisher cylinder. Do not expose to excessive heat.

Table III. *Organizational Mechanic or Maintenance Crew Preventive Maintenance Services*—Continued

Intervals		Procedure
Quarterly	Monthly	
		MAINTENANCE OPERATIONS—Continued
60	------	*Special lubrication.* Apply a few drops of oil to pulleys and guides through which control cables operate.
60	------	*Tighten.* Tighten all assembly and mounting bolts and screws.
61	------	*Engines (install mountings, lines and fittings, wiring control linkage, and oil supply).* Install removed engines according to instructions in paragraph 98. Take care to tighten mountings securely and to connect properly all fuel, oil, and coolant lines, wiring, and control linkage which were disconnected when engine was removed. *Caution:* Be sure to fill engine crankcase with specified oil (par. 62).
62	62	*Radiators (core, mountings, hose, antifreeze, and record).* Inspect these items for good condition, correct assembly, and leaks. See that radiator mountings and hose and tubing connections are secure. Clean insects and refuse from external air passages of cores. Remove if necessary. Test strength of antifreeze, if in use. Examine coolant to determine whether it is contaminated with rust, oil, or other foreign matter to the extent that the cooling system should be cleaned. If inspection indicates cleaning is necessary and always when antifreeze is installed or removed, proceed only according to paragraph 118 and current directives regarding proper procedure, cleaner, neutralizer, and inhibitor. Refill radiator with proper coolant. Do not fill to top, but allow room for expansion. On radiators which are removed, perform this inspection partly while removed from the vehicle and partly after replacement, as necessary to make the complete inspection properly.
62	62	*Serve.* Install removed radiators. Take care to tighten mountings and hose or tubing connections securely. Fill the cooling system afterward, adding antifreeze or inhibitor as required and check the cooling system for leaks.
63	63	*Battery (cables, hold-downs, battery compartment, and record of gravity and voltage.)* Inspect battery case for cracks and leaks. Clean top of battery. Inspect cables, terminals, bolts, posts, straps, and hold-downs for good condition. Test specific gravity and voltage and record on DA AGO Form No. 462. Specific gravity reading below 1.225 indicates battery should be recharged or replaced. Electrolyte level should be above top of plates and may extend three-eights of an inch.
63	------	*Serve.* Perform high rate discharge test according to instructions for "condition" test which accompany test instrument. Cell variation should not be more than 30 percent. Note: Specific gravity must be above 1.225 to make this test.

Table III. Organizational Mechanic or Maintenance Crew Preventive Maintenance Services—Continued

Intervals		Procedure
Quarterly	Monthly	

MAINTENANCE OPERATIONS—Continued

63	63	*Clean.* Clean entire exterior of battery and interior of battery box. Paint box if corroded. Add clean water to three-eights of an inch above cell plates. Clean battery terminals, terminal bolts and nuts, and battery posts. Inspect bolts for serviceability and grease lightly. Tighten terminals and hold-downs carefully to avoid damage to battery. *Note:* Turn main battery switch to "OFF" and close fuel shut-off valve at this time.
64	64	*Accelerators (pedals, linkage, and dual-throttle synchronization).* See that accelerator pedals and all of their connecting linkage are in good condition and securely connected. Press the accelerator pedals down fully to see whether or not the carburetor throttle valve plates open fully. Check synchronization of throttle linkage at each carburetor to see that throttle valve plates open and close together. After starting engines, test to see that throttles are so adjusted that, at idle speed (450 rpm), the variation between the two engines will be no more than 50 rpm and at 1,000 rpm there will be no more than 100 revolutions variation. Inspect throttle control linkage to see that it is properly adjusted.
64	------	*Adjust.* Adjust throttle control linkage (par. 112).
65	65	*Starters and instruments.* Start engines, observing all starting precautions (par. 44). Note whether the general action of the starters is satisfactory, particularly whether the drives engage properly without excessive noise and have adequate cranking speed. Observe whether each engine starts readily. As soon as the engines start, observe whether all instruments operate properly and particularly if the oil pressure signal and ammeter indications are satisfactory.
67	67	*Ignition timing.* With engine running, determine whether ignition timing of each engine is correct (par. 104). Also observe whether automatic controls advance the spark as engine is accelerated gradually. When necessary, adjust ignition timing to specifications (par. 104), taking care to see that distributor is well-secured when adjustment is complete. *Caution:* Be sure that distributor timing is properly adjusted for efficient engine performance with grade of fuel being used.
68	68	*Regulator unit (connections, voltage, current, and cut-out).* See that regulator units and the radio noise suppression capacitors (condensers) and bond straps are in good condition and that mountings and connections are secure.

Table III. *Organizational Mechanic or Maintenance Crew Preventive Maintenance Services*—Continued

Intervals		Procedure
Quarterly	Monthly	
		MAINTENANCE OPERATIONS—Continued
68	------	Connect low-voltage circuit tester to regulator and observe whether voltage regulator, current regulator, and circuit breaker (cut-out) control the generator output properly. Follow instructions which accompany test instrument. *Caution:* Make test only after regulator has reached normal operating temperature.
69	69	*Engine idle.* Adjust engine mixture and throttle stop to obtain smooth operation at 450 rpm. Refer to paragraph 112.
70	70	*Throttle synchronization.* After starting engines, see that throttles are so adjusted that, at idle speed (450 rpm), the variation between the two engines will be no more than 50 rpm and at 1,000 rpm there will be no more than 100 revolutions variation. Inspect throttle control linkage to see that it is properly adjusted (par. 112).
71	71	*Fighting compartment (paint, seats, stowage boxes, ammunition boxes, clips, and racks).* Inspect to see that these items are in good condition and securely assembled and mounted; that fighting compartment is clean and paint in satisfactory condition; that adjusting mechanisms of seats operate properly and are adequately lubricated. Pay particular attention to see that shell hangers and dividers are all present and properly installed in ammunition boxes. Inspect ammunition box cans to be sure they have been drained of all fluids. Use of fluids (water, antifreeze compound, or ammudamp) has been discontinued.
72	72	*Turret (locks).* Inspect locks to see if they are in good condition, securely assembled and mounted, that turret can be traversed easily when lock is released, and that locks operate properly.
73	73	*Periscopes.* Examine periscope prisms and windows to see that they are in good condition, clean, secure in holders, and that holders are securely mounted; that lever and locking devices operate freely and are not excessively worn; and that their traversing and elevating devices are free and not excessively worn. Examine spare periscopes and their stowage boxes to see that they are in good condition, clean, and secure.
75	75	*Brakes (steering levers, linkage, and shafts).* Inspect steering brake levers, linkage, and shafts to see that they are in good condition, securely mounted, properly connected, and not excessively worn. Apply steering brake levers and observe whether they begin to take hold at 6 notches of travel on ratchet.
75	------	*Tighten.* Tighten all assembly and mounting nuts and screws securely.

Table III. *Organizational Mechanic or Maintenance Crew Preventive Maintenance Services*—Continued

Intervals		Procedure
Quarterly	Monthly	

MAINTENANCE OPERATIONS—Continued

77	77	*Differential.* Examine accessible part of the differential case in driver's compartment to see that it is in good condition, that all mounting and assembly bolts and cap screws are secure, and that there are no oil leaks.
78	78	*Transmissions.* Inspect transmissions to see that they are in good condition, securely mounted, and not leaking. If condition or level of lubricant indicates necessity for oil change, drain and refill to correct level with specified oil. Refer to lubrication order, paragraph 62.
78	------	*Test and adjust.* Refer to paragraph 150.
79	79	*Transfer unit (seals and leaks).* Inspect the transfer unit to see that it is in good condition and securely assembled and mounted. Note whether oil is leaking from the case or seals. If condition or level of lubricant indicates necessity for oil change, drain and refill to correct level with specified oil. Refer to lubrication order, paragraph 62.
79	------	*Tighten.* Tighten all external assembly and mounting bolts and cap screws securely.
80	80	*Transmission and transfer unit (controls and linkage).* See that control levers and linkage for these units are in good condition, correctly assembled, and securely connected and mounted. See that joints are adequately lubricated and not excessively worn.
81	81	*Propeller shafts (universal joints, alinement and yokes, and support bearing).* Inspect propeller shafts for good condition and correct and secure assembly. See that universal joints and support bearing are not excessively worn. Inspect for adequate lubrication. Look for excessive leaks at seals of universal joints, at sliding joint, and at support bearing.
81	------	*Tighten.* Tighten all universal joint assembly and yoke cap screws securely.
84	84	*Compass (fluid and lamp).* Examine compass to see that it is in good condition and secure. Look for low level or indications of bubbles in fluid bowl; fill fluid bowl with ethyl alcohol if needed. See that compass light and switch operate properly.
85	85	*Lights and switches (head, tail, blackout, and internal).* If tactical situation permits, test switches and lights to see that they operate properly. Inspect all lights for good condition, secure mounting, broken lenses, and discolored reflectors.
86	86	*Wiring (junction and terminal blocks and boxes).* Inspect to see that all exposed electrical wiring and conduits, terminal blocks, and boxes are in good condition, well-supported, and securely connected. Be sure radio noise suppression capacitors (condensers) and/or wiring support clip bondings are in good condition, clean, and securely mounted or connected.

Table III. Organizational Mechanic or Maintenance Crew Preventive Maintenance Services—Continued

Intervals		Procedure
Quarterly	Monthly	

		MAINTENANCE OPERATIONS—Continued
87	87	*Collector ring (brushes, heads, cylinder, and cover).* With master battery switch off, remove collector ring cover and examine to see that above components are all in good condition and clean; that brushes contact cylinder properly under normal spring tension; and that leads are securely connected and not chafing. Install cover securely. Be sure radio bond strap is secure.
88	88	*Radio bonding (suppressors, filters, capacitors, and shielding).* See that all units not covered in the foregoing specific procedures are in good condition and securely mounted and connected. Be sure all additional noise suppression bond straps and shakeproof lock washers are inspected for looseness or damage and see that contact surfaces are clean. If objectionable radio noise from vehicle has been reported, make tests in accordance with paragraph 78. If cleaning and tightening of mountings and connections and replacement of defective radio noise suppression units do not eliminate the trouble, the radio operator will report the condition to the designated individual in authority.
		ARMAMENT
126	126	*Bore condition.* Examine breech recess, chamber, and rifling for signs of excessive wear, erosion, or other damage (refer to TM 9–313).
126	126	*Breech mechanism.* Operate breech mechanism, opening and closing the breech, noting indications of binding, excessive play or jamming. Action should be smooth (refer to TM 9–313).
126	126	*Firing devices.* With the breech empty, operate the hand firing knob (TM 9–313). After cocking the gun, press the gun master switch on the top of the control box to fire electrically. Check for smooth operation.
126	126	*Recoil mechanism.* Check recoil action. Normal recoil is 11½ inches and should not exceed 12½ inches. Check oil level, couplings, and packing (refer to TM 9–313).
126	126	*Recoil guard.* The recoil guard should be secure.
126	126	*Traversing mechanism.* Check the security of the mounting. Check oil level. Traverse the gun and turret completely, manually and under power, checking for binding, excessive backlash, or erratic motion.
126	126	*Elevating mechanism.* Elevate and depress the gun through entire range to check for binding, excessive backlash, or erratic motion.
127	127	*Stabilizer.* Check oil levels, controls, and operation.
72	72	*Locks.* The turret lock, travel lock, and cradle lock should all operate readily and positively.

Table III. *Organizational Mechanic or Maintenance Crew Preventive Maintenance Services*—Continued

Intervals		Procedure
Quarterly	Monthly	
		ARMAMENT—Continued
126	126	*Sight mounts.* Check periscope mounts and telescope mount for security of fastenings, ease of motion, and legibility of scales.
73	73	*Periscopes.* Check condition of attached heads and spare heads. Operate controls to determine ease of motion. Note condition of reticles and lighting devices. Boresight sighting periscope.
126	126	*Telescope.* Check for indication of damage. Note condition of reticle and lighting devices. Boresight.
126	126	*Elevation quadrant.* Check for security of mounting, condition of level vial and lighting device, and smooth motion in elevation.
126	126	*Azimuth indicator.* Dial should be legible and knobs should move readily.
71	71	*Hatches and seats.* All doors should securely and completely cover hatches. Check for quiet smooth motion and secure latching. Seats should be adjustable and in good working order. Padding should be in good condition.
131	131	*Ammunition stowage.* Check for complete supply and replenish if depleted. Rounds and covers should be secure.
137	137	*Lubrication.* Check lubrication order for lubrication to be performed.
71	71	*Paint.* Check paint, remove rust and touch up at first opportunity, taking care to avoid moving surfaces and glass surfaces.
86	86	*Electrical connections.* Cables to spark plugs should be secure and in good condition. Note flickering of lights or signals.
125	------	*Machine gun mounts.* Clamps, pins, and detents should function properly. Check bow mount for free movement. Spring should be active in operation.
		TOOLS AND EQUIPMENT
130	130	*Tools (vehicle kit and pioneer).* Check vehicle and pioneer tools against stowage lists to see that all items are present, in good condition, and properly stowed or mounted. Any tools mounted on outside of vehicle, having bright or polished surfaces, should be painted or otherwise treated to prevent glare or reflections. Tools with cutting edges should be sharp and edges should be protected.
131	131	*Equipment.* Check special equipment items against vehicle stowage lists to see if they are all present, in serviceable condition, and properly stowed or mounted.
132	132	*Spare track shoes.* Inspect to see if they are all present, in good condition, and properly stowed or mounted.
133	133	*Spare oil supply (recoil, hydraulic, and engine).* Check to see that supply of listed spare oil is present and properly stowed. This supply should be maintained at all times.

Table III. *Organizational Mechanic or Maintenance Crew Preventive Maintenance Services*—Continued

Intervals		Procedure
Quarterly	Monthly	
		TOOLS AND EQUIPMENT—Continued
134	134	*Decontaminator.* When use of decontaminator is directed in accordance with command directives, examine decontaminator to see that it is in good condition, secure, and fully charged. Make latter check by removing filler plug. *Note.* When in use, the solutio nmust be renewed every 3 months because it deteriorates.
135	135	*Fire extinguisher (portable).* Inspect extinguisher to see if it is fully charged, in good condition, and securely mounted and if seal on valve head is intact. Weigh cylinder to determine if it is fully charged.
136	136	*Publications and Standard Form No. 91.* Check to see whether TM 9-729, LO 9-729, and Standard Form 91 are present, legible, and properly stowed.
137	137	*Vehicle lubrication.* Perform a complete lubrication service on the vehicle according to instructions in lubrication order, paragraph 62, omitting only those items which have received attention in the foregoing specific procedures. Replace damaged or missing fittings, lines, or plugs.
138	138	*Modifications (MWO's completed).* Check DA AGO Form 478 to determine whether all modification work orders have been completed. A list of current modification work orders is contained in SR 310-20-4. Enter any modifications or major unit assembly replacements made during this service on DA AGO Form 478.
139	139	*Final road test.* Make a final road test, rechecking items 2 to 15 inclusive. Recheck transmission and differential to see that lubricant is at correct level and that there are no leaks. Confine this road test to the minimum distance necessary to make satisfactory observations. While testing vehicle, operate it in a normal manner. *Note.* Correct or report any deficiencies found during final road test to designated authority.
		UNUSUAL CONDITIONS
		Maintenance operations and road tests as prescribed under usual conditions will apply equally under unusual conditions for operations for all occasions except in extreme cold weather. Intervals are necessarily shortened in extreme cold weather servicing and maintenance. Vehicles subjected to salt-water immersion or complete submersion are evacuated to ordnance maintenance unit as soon as possible after the exposure. (See pars. 57 and 213.)

Section IV. TROUBLE SHOOTING

70. Scope

a. This section contains trouble shooting information and tests for locating and correcting some of the troubles which may develop in the vehicle. Trouble shooting is a systematic isolation of defective components by means of an analysis of vehicle trouble symptoms; testing to determine the defective component and applying the remedies. Each symptom of trouble given for an individual unit or system is followed by a list of probable causes of the trouble and suggested procedures to be followed.

b. This manual cannot cover all possible troubles and deficiencies that may occur under the many conditions of operation. If a specific trouble, test, and remedy therefor are not covered herein, proceed to isolate the system in which the trouble occurs and then locate the defective component. Use of all the senses to observe and to locate troubles. Do not neglect use of any test instruments such as voltmeter, ammeter, ohmeter, test lamp, hydrometer, and pressure and vacuum gages that are available (pars. 71 through 76). Standard automotive theories and principles of operation apply. Question vehicle crew to obtain maximum number of observed symptoms. The greater the number of symptoms of trouble that can be evaluated, the easier will be the isolation of the defective system and components thereof.

71. Engine

a. STARTER DOES NOT CRANK ENGINE.
   (1) *Open switch in starting circuit.* In order to complete the starting circuit, both ignition switches, as well as the master battery switches, must be "ON." Make sure that these switches are turned on.
   (2) *Battery ground circuit open.* Turn on lights. If lights do not light, battery ground circuit is open or master switch is inoperative. Check ground circuit and switch with test light or voltmeter.
   (3) *Loose connection in starting circuit.* If lights light properly, press starter button. If lights go out, there is a poor connection at one of the terminals, at the batteries, at the ground strap, at the master switch, or at the starter (fig. 63). These can be checked by a voltmeter between each successive terminal and ground. Clean and tighten loose connections, replace broken ground straps, or replace defective master switch.
   (4) *Battery run down.* If lights dim considerably when starter button is depressed, but still burn, batteries are low. Test

batteries with a hydrometer and a high-rate discharge tester, if available (par. 69 *f*, item 63). Replace any battery units that do not test satisfactorily.

(5) *No ground in starter relay circuit.* When the starter button is first pressed, the starter relay circuit is grounded through the transmission oil pressure signal switch (fig. 63). If lights do not dim at all and transmission warning light on the instrument panel is not lit, the signal switch is probably defective. A test to determine whether switch is defective can readily be made by grounding the signal switch cable. This will permit starting the engine.

(6) *Open starting circuit.* If lights do not dim at all when starter button is pressed, no current is flowing to the starter. If cause is not in ground circuit, use test light to determine if feed circuit is complete to relay terminal, then to solenoid coil, and finally to starter terminal (fig. 63). Replace relay lead, relay, solenoid, or cable as required.

(7) *Hydrostatic lock.* If previous checks indicate current to starter terminal and motor does not turn engine, check for hydrostatic lock (par. 44*f*).

(8) *Starter inoperative.* If previous tests indicate that there is current to starter terminal and that engine is not hydrostatically locked, yet starter does not turn, replace starter (par. 108).

b. ENGINE CRANKS BUT WILL NOT START.

(1) *Slow cranking speed.* Check for wrong grade of engine oil (par. 62). Check state of battery charge, connections, and ground strap (*a* above). Check for tight bearings or tight pistons and rings. Replace starter.

(2) *Fuel lines shut off.* Turn on fuel switch.

(3) *Fuel supply exhausted.* Replenish fuel supply.

(4) *Carburetor flooded.* A flooded condition may occur after repeated cranking. Usually one engine starts and the other does not. If flooded condition is suspected, hold accelerator pedal down while cranking engines.

(5) *Inoperative ignition system.* Open engine compartment, remove ignition cable from one spark plug and hold terminal one-quarter inch from cylinder head. Turn on ignition switch and crank engine. If strong, blue-white spark jumps gap, ignition is satifactory. If no spark appears or spark is weak or red, proceed with tests in paragraph 72.

(6) *Carburetor operating improperly.* Refer to paragraph 74.

c. ENGINE STOPS.

(1) *Break in ignition system.* Check ignition wires and connections for looseness or breaks. Check breaker points to see

that they are not stuck in the open position. Tighten, repair, or replace as necessary.

(2) *Inoperative fuel system.* Check all fuel lines and connections for breaks and clogging and reconnect, clean, or repair as required (par. 74). Clean fuel filters and screens.

d. ENGINE OPERATES UNEVENLY.

(1) *Ignition system operating improperly.* Check for inadequate ignition (par. 72).

(2) *Carburetor operating improperly.* Check carburetor adjustment and condition (par. 74).

(3) *Engines not synchronized.* Check tachometers for difference in engine speed, and tune engines (par. 86) or adjust throttle linkage as required (par. 112).

(4) *Leaking intake manifold or cylinder head gaskets.* Listen for air leaks and test for leaks by applying light engine oil to suspected areas. Tighten or replace affected gaskets.

(5) *Valves sticking, warped or burned.* Check compression pressure. The compression pressure at cranking speed should be 102 pounds. If the compression pressure at cranking speed falls below 95 pounds, notify ordnance maintenance personnel.

(6) *Pistons, rings, or cylinders worn.* Check compression pressure. If pressures vary considerably or if all cylinders are low, notify ordnance maintenance personnel.

e. ENGINE OVERHEATS.

(1) *Low level of coolant.* Check level of coolant and fill to correct level with water or antifreeze. Also check system thoroughly to determine cause of coolant loss (par. 75).

(2) *Belts loose or broken.* The belts drive the generator, water pump, and fan. Inspect belts and, if necessary, adjust to obtain $\frac{5}{8}$- to $\frac{3}{4}$-inch slack, measured as shown in figure 37.

Note. Belts are matched in sets. If one belt is broken, all three belts must be replaced at the same time.

(3) *Radiator air inlet plugged.* Open radiator clean-out doors on bulkhead (fig. 156). Check radiator core air passages for presence of dirt, leaves, or twigs. Clean out. See that inlet and outlet grilles on top of vehicle are not covered with paulin or equipment. Remove screen from inlet grille and clean.

(4) *Mechanical failure in cooling system.* Check thoroughly as outlined in paragraph 75.

(5) *Ignition timing late.* Check and reset ignition timing (par. 104).

(6) *Engine oil level low.* Check oil supply and add to the correct level (par. 62).

(7) *Radiator thermostat sticking.* Replace thermostat (par. 120).

f. ENGINE OPERATES NOISILY.
  (1) *Light knock or ping on acceleration.* Check grade of gasoline used; it should be 80 octane. Check or reset ignition timing (par. 104). Remove cylinder heads and clean out carbon (par. 89).
  (2) *Clicking noises synchronized with camshaft speed.* Check oil level; an overfilled crankcase will cause oil to foam and valve tappets to become noisy. Drain oil to correct level. Check condition of crankcase oil, and change oil if dirty or gritty (par. 62). Blow out oil lines to valve tappet assemblies. This may be accomplished by disconnecting the oil feed line from the valve tappet connection at the center of the "V" and applying a compressed air nozzle. If unsuccessful, notify higher authority.
  (3) *High-pitched squeals.* Inspect fan, generator, distributor, and water pump for underlubricated or frozen bearings. Lubricate or replace unit as required.
  (4) *Heavy knocks synchronized with crankshaft speed.* Crankshaft or connecting rod bearings burned out. Notify ordnance maintenance personnel.
  (5) *Fan universal joint noise.* Check if expansion plug on universal joint yoke has loosened and strikes journal during rotation. Reseat and stake in place.

g. ENGINE OIL PRESSURE LOW (WARNING LIGHT STAYS ON).
  (1) *Low oil level.* Check oil supply and add oil to full mark on gage.
  (2) *High oil level.* If oil level is high (above the full mark), it may be an indication that the fuel pump switch has been left on while the engine has not been running, and raw gasoline has passed through the carburetor and manifolds and into the crankcase, causing oil dilution and low oil pressure. Drain crankcase and install fresh oil (par. 62). Overfilling of crankcase should be avoided at all times.
  (3) *Oil worn out or grade incorrect.* Check oil for cleanliness and proper grade. Refill with correct seasonal grade of engine oil.
  (4) *Broken line to valve tappets.* Inspect oil feed line from side of cylinder block to valve tappet connection at center of "V" and replace if cracked or broken. Replenish oil.
  (5) *Oil pressure warning signal switch defective.* If no cause of low oil pressure can be found, the trouble may be in the oil pressure warning signal switch at the rear of the engine. Replace switch (par. 144).

(6) *Worn oil pump gears.* Remove oil filter inlet pipe at front of cylinder block and install pressure gage. If pressure is below 12 pounds, with engine idling, notify ordnance maintenance personnel.

h. WARNING SIGNALS DO NOT LIGHT WHEN IGNITION IS TURNED ON.
   (1) *Burned-out lamp.* If only one signal fails to light, the fault is probably in the lamp on the instrument panel. Check by replacing with new lamp (par. 129).
   (2) *Defective engine or transmission unit.* If only one signal fails to light, and the lamp is not at fault, disconnect cable to affected switch on engine or transmission and ground it. If lamp lights, replace switch.
   (3) *Open circuit.* If lamp fails to light, use test lamp to check back through affected circuit (fig. 103 or 104) and correct condition causing open circuit.

i. ENGINE DOES NOT DEVELOP FULL POWER.
   (1) *Preliminary instructions.* The many different factors that may cause loss of power make it advisable to perform the quarterly preventive maintenance service (par. 69) before proceeding further. If this service has been performed recently and the trouble was not eliminated, follow the procedure indicated below.
   (2) *Procedure.*
      (a) *Unsatisfactory spark.* Check as outlined in b (5) above.
      (b) *Inspect spark plugs.* Check as outlined in paragraph 72.
      (c) *Test compression.* Check as outlined in paragraph 69, item 33.
      (d) *Check ignition timing.* Check as outlined in paragraph 104.
      (e) *Check fuel mixture.* If the above inspections and tests do not locate the cause for lack of power, the trouble may be in a lean fuel mixture. Proceed as outlined in paragraph 112.
      (f) *Brakes dragging.* Steering brakes adjusted too tightly will cause sluggish vehicle action. Check as outlined in paragraph 162.

72. Ignition System

a. NO SPARK IN ONE CYLINDER.
   (1) *Spark plug wiring faulty.* Make visual inspection of wiring to determine if wet, disconnected, broken, or shorted. Dry off, reconnect, or replace as required. Check current to plug by cranking engine with cable disconnected and terminal

held one-quarter inch from cylinder head. If no spark jumps, cable is shorted or broken and must be replaced.

(2) *Spark plug faulty.* If strong spark jumps with above test, fault is in plug. Remove plug and inspect for cracks, broken electrodes, fouling, or incorrect gap. Clean, adjust, or replace as required.

b. NO SPARK TO ANY CYLINDER.

(1) *Distributor cap faulty.* Inspect distributor cap for loose mounting, moisture, dirt, or cracked or burned condition and dry off, clean, or replace as necessary.

(2) *Wiring defective.* Inspect high-tension ignition cables and low-tension cable from distributor to coil and back through feed circuit (fig. 56).

(3) *Contact points inoperative.* Remove distributor cap and inspect contact points for gap, burned condition, and spring tension. Service or replace in accordance with instructions (par. 102).

(4) *Defective coil or capacitor.* Check coil and capacitor (condenser) with instruments or by substitution and replace one or both units as required.

73. Starting System

a. STARTER WILL NOT OPERATE.

(1) *Ignition switches not turned on.* Turn switches on.

(2) *Emergency ignition not turned on.* Keep this switch turned on at all times.

(3) *Master battery switch not turned on.* Turn switch on.

(4) *Batteries run down.* Close master battery switch. Insert the voltmetal prods into master switch box receptacle outlet and observe reading. Depress starter buttons. If the voltage reading is below 12 volts, test the batteries with a hydrometer and a battery high rate discharge tester. If this test indicates that the batteries will not retain a charge, replace the batteries (par. 127). If batteries test satisfactorily, clean and tighten all battery connections.

(5) *Master battery switch inoperative.* Turn on all lights. Depress starter buttons. Under this test, the lights will normally dim somewhat, but if they dim excessively, connect the voltmeter positive cable to battery cable terminal on master battery switch and negative cable to the other terminal. Depress starter button. If the voltmeter reads more than 0.5 volt, indicating high resistance in the switch, and the batteries test satisfactorily ((4) above), replace the switch (par. 137).

(6) *Starter switch inoperative.* If the panel lights light, indicating that current is reaching the instrument panel, disconnect the connector on starter solenoid cable in the panel. Connect test voltmeter to the cable from the starter switch and depress the switch button. If no reading is obtained, replace the switch.

   Note. Check both switches.

(7) *Starter solenoid inoperative.* Tighten terminal connections on solenoid. Connect test voltmeter positive cable to the green wire terminal on solenoid and voltmeter negative cable to ground. Depress starter switch button.

  (*a*) If the voltmeter reads 24 volts, the circuit is complete to solenoid.

   Note. Make sure solenoid ground cable is making good contact.

  If solenoid still does not function, replace the solenoid (par. 109).

  (*b*) If no reading is obtained on the voltmeter, the circuit is open between the starter switch and the solenoid. If the terminals are secure and there is no obvious break in the wiring, notify higher authority.

   Note. In an emergency, current can be supplied to the solenoid by making connections with a short jumper. After engine is started, remove jumper.

(8) *Starter relay inoperative.* A click should be heard in the starter relay when the starter switch button is depressed. If this click is not heard and the starter switch and wiring are in good condition, the relay should be replaced (par. 110).

(9) *Starter inoperative.* Connect positive cable of test voltmeter to starter terminal and voltmeter negative cable to ground. Depress starter switch button. If a reading is obtained, starter should be replaced (par. 108).

(10) *Check wiring.* Make sure that wiring is in good condition. Check carefully for loose, broken, worn, or burned cables or wires. Replace any and all wiring found to be in bad condition.

  Caution: Broken cables or wires inside any type of covering are usually very difficult to locate. Particular care must be exercised in checking a cable or wire that may seem in good condition, but actually is broken.

b. SLOW CRANKING SPEED.

(1) *High electrical resistance.* This condition may be caused by loose or corroded terminals, wrong size cables or wire, faulty starter, or other starting items. Check as outlined in *a* above.

(2) *Engine oil too heavy.* Use proper grade of oil as indicated on the lubrication order.

(3) *Starter worn out or excessively noisy.* Replace starter (par. 108).

74. Fuel System

a. FUEL DOES NOT REACH CARBURETOR.

(1) *Fuel pump switch shut off.* Turn on switch.

(2) *Fuel supply exhausted.* Replenish fuel supply.

(3) *Carburetor strainer clogged.* Remove carburetor air inlet elbow (fig. 68), strainer plug, and screen; clean and replace.

(4) *Carburetor auxiliary fuel filter clogged.* Remove and clean as outlined in paragraph 112.

(5) *Fuel lines clogged or leaking.* Inspect lines and clean, tighten, or replace as necessary.

(6) *Fuel pump strainer clogged.* Remove pump support assembly, clean strainer screen, and install (par. 113).

(7) *Fuel pump defective.* Check by listening for pump operation with fuel switch turned on and neither engine running. If inoperative, replace pump support assembly or notify higher authority.

(8) *Fuel valves stuck.* Free or replace.

(9) *Cable to fuel valve disconnected or broken.* Connect or replace cable.

b. CARBURETOR NOT OPERATING PROPERLY.

(1) *Carburetor choke thermostat housing out of adjustment.* Set choke thermostat housing to indicator mark and tighten carefully (par. 112).

(2) *Carburetor choke sticking.* Free by working lever on end of choke shaft (fig. 68). Notify higher authority if condition persists.

(3) *Carburetor out of adjustment.* Adjust carburetor (par. 112). If these adjustments do not remedy condition, replace carburetor or notify higher authority.

(4) *Carburetor dirty, clogged, or worn.* Clean inlet screen. Replace carburetor (par. 112).

c. FUEL PUMPS (ONE OR BOTH) FAIL TO OPERATE.

(1) *Circuit breaker open.* Push circuit breaker reset button on fuel switch bracket.

Note. Open circuit breaker would make both pumps inoperative.

(2) *Defective switch.* Test fuel pump switch, using jumper. If pump operates with jumper in place, replace switch.

(3) *Open circuit to pump.* Check entire pump feed circuit (fig.

67), from ignition switches to connection at pump support. Locate and correct cause of open circuit.

(4) *Defective cable in fuel tank.* If circuit checks to terminal on pump support, remove support assembly from fuel tank (par. 113) and check cable to pump. Replace if defective.

(5) *Defective pump.* If cable to pump checks satisfactorily, pump is defective. Replace pump (par. 113).

75. Cooling System

 a. Loss of Coolant.
  (1) *Hose connection leak.* Tighten connections or replace connections and hose, if necessary.
  (2) *Leakage at radiator.* Remove radiator and repair or replace (par. 121).
  (3) *Leakage or loss of seal at radiator cap.* Replace cap or gasket.
  (4) *Leakage at thermostat housing.* Replace housing gasket and tighten securely.
  (5) *Leakage at water pump.* Replace pump gasket and tighten as required. If leakage is at packings, notify ordnance maintenance personnel.
  (6) *Leaks at transmission water pipe.* Tighten connections or replace.
  (7) *Leaks at cylinder heads.* Tighten head screws to 70–75 foot-pounds torque with torque wrench or replace head gasket and tighten (par. 89).
 b. Engine Overheats. Refer to paragraph 71e.
 c. No Water Circulation.
  (1) *Water level low.* Fill system and check for leaks.
  (2) *Belts loose or broken.* Adjust or replace (par. 88).
   Caution: Belts must be replaced in matched sets of three.
  (3) *Thermostat not operating.* Remove thermostat housing and check thermostat. Replace if defective.
  (4) *Cooling system badly clogged.* Clean system thoroughly (par. 118).
  (5) *Water pump not operating properly.* If water pump is found to be in poor working condition, or not working at all, replace water pump (par. 122).
 d. Engine Temperature Gage Does Not Register.
  (1) *Open feed circuit.* Connect jumper from engine temperature gage in instrument panel back to ignition switch (fig. 104). If gage operates, replace cable.
  (2) *Defective gage.* If gage does not operate when jumper is connected, gage is defective. Replace gage (par. 142).

e. TEMPERATURE GAGE READS HIGH AT ALL TIMES.
  (1) *Defective engine temperature sending unit.* Disconnect cable at engine sending unit. If gage drops to low side, engine temperature sending unit is defective. Replace unit (par. 144).
  (2) *Circuit to engine temperature sending unit grounded.* If gage remains high with cable disconnected, use test lamp to determine place where line between unit and gage (fig. 104) is grounded.
  (3) *Defective gage.* If no ground is found in circuit, replace gage on instrument panel (par. 142).
f. TEMPERATURE GAGE READS LOW AT ALL TIMES.
  (1) *Circuit to engine temperature sending unit open.* Connect test light to engine temperature sending unit terminal and to ground. If light does not burn, check back through circuit for loose or broken connections.
  (2) *Defective engine temperature sending unit.* If lamp lights, disconnect cable at engine temperature sending unit and watch gage. If gage has high reading, engine temperature sending unit is defective. Replace unit (par. 144).
  (3) *Defective temperature gage.* If gage does not show high reading under the above circumstances, it is defective. Replace gage (par. 142).

76. Generator and Charging System

a. GENERATOR CHARGES LITTLE OR NOT AT ALL.
  (1) *Belts excessively loose.* Adjust drive belts to obtain 5/8- to 3/4-inch slack (fig. 37).
    Note. Generator tests should be made on one generator at a time.
  (2) *Battery fully charged.* Test battery with hydrometer. If fully charged, regulator will prevent appreciable generator output.
  (3) *Defective ammeter.* Connect test ammeter in circuit. Replace vehicle ammeter if defective (par. 134).
  (4) *High resistance in charging circuit.* Check entire charging circuit (fig. 86) for loose or corroded connections, using voltmeter, and correct any conditions causing high resistance. High resistance makes regulator operate as if battery were fully charged.
  (5) *Defective generator.* Remove the field cable (1-201, fig. 86) from the feed terminal of the regulator. Start the engine and while it is running at idle speed, touch the free end of the field cable to the armature cable (2-202, fig. 86). Increase the engine speed and note the charging rate.

Caution: Do not increase charging rate above 50 amperes. The charging rate should increase as the engine is speeded up. If not, the generator is at fault and must be replaced (par. 124).

(6) *Defective regulator.* Start the engine and run it at approximately 1,100 rpm. Observe the ammeter. If no charging rate is indicated, connect the battery and armature terminals together with a jumper. If a reading is obtained, the circuit breaker relay and/or actuating relay of the regulator is at fault and the regulator must be replaced. If no reading is obtained, connect the battery and field terminals together with a jumper. If a reading is now obtained, the regulator is at fault and must be replaced (par. 125).

b. HIGH CHARGING RATE WHEN BATTERIES ARE FULLY CHARGED. A consistently high charging rate indicates that the voltage setting of the generator regulator is too high. Replace the regulator (par. 125).

Note. If the batteries gas freely and use water excessively, this is also an indication that the charging rate is too high.

77. Lighting System, Batteries, Siren, and Ventilator

a. NO LIGHTS BURN.

(1) *Circuit breaker open.* Set lighting system circuit breaker by pressing button on instrument panel (fig. 6).

Caution: If circuit breaker will not stay closed, do not hold in closed position. Check for short or ground.

(2) *No current in feed circuit.* Use test lamp or voltmeter to check power feed to lighting switch "BATTERY" terminal. If there is no current to this point, check back through the feed circuit (fig. 93, 94, or 95) until the "short" or "open" is located, and repair or replace wiring or conduit as required (par. 131).

(3) *Defective lighting switch.* If power feed to switch is satisfactory, use a jumper to connect "BATTERY" terminal with other terminals. If lights burn, replace lighting switch (par. 138).

b. ALL LIGHTS BURN DIMLY.

(1) *Battery voltage low.* Check battery condition with hydrometer and voltmeter. Charge or replace batteries as required (par. 127).

(2) *High resistance in feed circuit.* Check back through lighting switch feed circuit with voltmeter to locate and correct high-resistance condition.

(3) *High resistance in lighting switch.* If circuit checks satisfactorily and all connections are tight, replace lighting switch (par. 138).

c. ONE HEAD LIGHT, TAIL LIGHT, OR MARKER LIGHT DOES NOT BURN.
   (1) *Burned-out lamp.* Check lamp circuit by replacing with one known to be good.
   (2) *Dirt on lamp contact.* Clean contact.
   (3) *Open circuit to light.* If replacement of lamp does not correct the condition, check back through circuit to lighting switch, using test lamp or jumper, and correct "open" or "shorted" condition as required.
   (4) *Defective lighting switch.* If circuit checks back to lighting switch, connect a jumper between "BATTERY" terminal on switch and affected terminal. If light burns, replace lighting switch.

   Note. The foregoing procedure applies also to the instrument panel lights and to the dome lights, except that the circuits must be traced back through their respective switches and feeds (figs. 93 and 96).

d. STOP LIGHTS DO NOT OPERATE.
   (1) *Burned-out lamp.* If only one stop light fails to operate, check for burned-out lamp by replacement.
   (2) *Open circuit to one light.* If only one light fails to operate and replacement of the lamp does not correct the condition, check back through the circuit with jumper or test lamp until "open" or "short" is discovered and corrected.
   (3) *Defective stop light switch.* If neither service nor blackout stop light burns, check each of the stop light switches in turn by connecting the terminals with a jumper. Replace defective switch (par. 138).
   (4) *Open circuit to stop light switches.* If neither stop light switch proves defective, check circuit from main lighting switch through stop light switches and back; correct any breaks or loose connections.
   (5) *Defective main lighting switch.* If circuit through stop light switches checks correctly, other lights burn properly, but none of the stop lights operate, replace main lighting switch (par. 138).

e. NO CURRENT IN BATTERY CIRCUIT.
   (1) *Master switch open.* Close master battery switch.
   (2) *Loose cable connections.* Inspect all cable connections at battery, ammeter shunt, and terminal blocks (fig. 92) and clean and tighten if necessary.
   (3) *Battery ground circuit open.* Turn on lights. Connect a heavy jumper between negative terminal of left battery and ground (fig. 92). If lights burn, ground circuit is open or

master switch is defective. Check ground circuit and switches with test lamp.

(4) *Battery to shunt circuit open.* If lights do not burn, run a heavy jumper from positive terminal of right-hand battery to left terminal of ammeter shunt (fig. 92). If lights burn, circuit is open. Check with test light or jumper from terminal to terminal.

(5) *Battery connecting cable loose or broken.* If lights still do not burn, test cable that connects batteries, using a jumper. Replace cable if broken or defective (par. 131).

(6) *Battery discharge (one or more cells dead).* Check each cell of each battery with hydrometer. If battery is low, charge with outside source or engine generators or replace as tactical situation permits. If one or more cells are dead, replace units containing defective cells (par. 127).

(7) *Generator circuit breaker not operating.* Set generator circuit breaker on master switch box.

(8) *Switches left on when not in use.* Turn switches off when not in use.

(9) *Generator regulator circuit breaker faulty.* If the ammeter on the instrument panel shows a heavy discharge when the generator is not operating and all switches are off except the battery master switch, disconnect the battery cable in the regulator terminal box. If the condition is corrected, the regulator circuit breaker contact points are stuck. If the ammeter does not show charge until generator is operating at high speed, the regulator circuit breaker is adjusted to operate at too high a voltage. In either of these cases, replace the regulator (par. 125).

(10) *Generator regulator inoperative.* Start the engine and note reading on the ammeter. If no charging rate is indicated, connect the battery and armature terminals together in the regulator terminal box using a short piece of insulated wire. Hold this wire across the two terminals and watch the ammeter. If a reading is obtained, the regulator is not connecting the generator to the battery circuit. If, however, this test does not reveal the trouble, connect the battery and field terminals together with a jumper. If a reading is obtained, the regulator is not allowing current to reach the generator field coils, preventing charging. If excessive charge is experienced and the batteries and circuits test properly, the trouble is caused by improper regulator adjustment. Replace the regulator (par. 125).

(11) *Generator inoperative.* First check to see if ground strap is loose or broken. Tighten or replace. Connect or test volt-

meter between armature terminal in the regulator terminal box and ground. This test will show if the generator is charging. If no voltage reading is shown, leave the voltmeter connected and connect the battery and field terminals together with a jumper. A flash will be seen and the test voltmeter will show a reading when this jumper is connected if the circuit is complete. Check the ammeter on the instrument panel. If a charge is now shown, the trouble has been corrected by flashing the field coils which has increased the magnetism, or properly polarized the field coil shoes. If no reading is obtained on the voltmeter, inspect the terminals at the generator for loose, or broken connections. If no trouble is observed in the connections or leads, the generator should be replaced (par. 124).

f. BATTERIES DO NOT STAY CHARGED.
 (1) *Excessive use of electrical equipment.* Keep one or both engines running at 1,500 rpm while using electrical equipment in accordance with the load.
 (2) *Generator charging rate inadequate.* Check and service charging circuit (par. 76).
 (3) *Batteries not capable of holding a charge.* Test each battery with hydrometer and high-discharge tester. Replace substandard battery units. Check for batteries connected with reversed polarity.

g. SIREN (OR HORN) OPERATES WEAKLY.
 (1) *Battery voltage low.* Check with voltmeter. Charge or replace affected battery units (par. 127).
 (2) *High resistance in siren circuit.* Check back through siren circuit (fig. 95) with voltmeter. Locate and correct high resistance condition.
 (3) *Defective siren.* If circuit checks satisfactorily, siren is defective. Replace siren (par. 140).

h. SIREN (OR HORN) FAILS TO OPERATE.
 (1) *Circuit breaker open.* Push reset button on instrument panel.
 (2) *Siren switch defective.* Test switch with jumper. If siren operates, replace switch (par. 141).
 (3) *Open circuit to siren.* Check back through siren circuit (fig. 95) with test lamp or voltmeter. Locate and correct "open" or "short" condition.
 (4) *Defective siren.* If circuit checks satisfactorily, siren is defective. Replace siren (par. 140).

i. VENTILATOR FAILS TO OPERATE.
 (1) *Circuit breaker open.* Push circuit breaker reset button on left front of ventilator assembly.

(2) *Open feed circuit.* Check feed circuit with test lamp at connector. If there is no current, check back through circuit (fig. 96) until open circuit is located and corrected.

(3) *Defective ventilator.* If feed circuit checks, ventilator circuit has internal short, broken switch, or other defect. Replace complete ventilator circuit breaker assembly (par. 130).

78. Radio Interference Suppression

a. Preliminary Instructions. When checking radio interference test the vehicle in a location free from high-tension lines, other vehicles, machinery, and electrical equipment which could be a source of interference. Stop engine, and turn off all vehicle electrical equipment. Turn on radio and check noise level. Listen carefully to type of noise present under these conditions. This registering of noise sounds is done so that when vehicle equipment is checked the presence of a new noise or interference can be immediately detected. Checking noise level is also done so that noises already present will not be attributed to lack of vehicle suppression. If noise level, with all vehicle equipment shut off, is too high due to atmospheric conditions or other outside causes, delay further checking, if tactical situation permits, until such time as moderate noise levels prevail. Disconnect the radio terminal box capacitor (condenser) (terminal end). If radio interference does not increase, replace capacitor. If interference increases, as it should, leave terminal box capacitor disconnected for the remainder of the tests or until any defective equipment has been located, replaced, and tested. Examine all shielded conduits and cables to make sure couplings are tight and conduits and cables are clamped or bonded to hull at least every 2 feet.

b. Radio Interference With Engine Running (Vehicle Stopped).

(1) *Ignition system.* If a rhythmic "popping" noise is heard in the radio receiver with changes in frequency as the speed of the engine is increased or decreased, the trouble is probably caused by the ignition system (burned distributor points, faulty capacitor, loose contacts, broken wiring, or improperly set spark plugs). Inspect all shielding, bonding, and wiring for damage or looseness. Replace damaged conduits, tighten loose connections, and correct any faulty bonding as necessary. If interference is still present, test all spark plugs for cracked spark plug shielding and eroded or improperly set points.

(2) *Generator.* If a "whining" or "squealing" sound is detected in the radio receiver while the engine is running, the trouble is probably in the generator. Stop the engine. If the inter-

ference decreases as the generator slows down, inspect to see that the generator is properly grounded. If the ground is satisfactory, remove the shield from generator terminals and inspect the connections. If the connections are tight and there is no obvious damage to the conduit containing the generator cables, disconnect the field cable (1–201) (fig. 86) from the generator and start the engine. The noise will disappear while the field cable is disconnected because there will be no output from the generator. Consequently, the regulator will not function as a field control, eliminating the action of relay contacts and the generator brush noise. Connect the field and armature terminals of the generator together and allow the generator to build up.

Caution: Do not run the engine above fast idle speed or allow generator to charge above 50 amperes. If the noise is again evident and increases in amplitude as the generator builds up, capacitors (condensers) in the generator are faulty. Replace generator (par. 124).

(3) *Generator regulator.* If an intermittent "clicking" sound is heard in the receivers, the trouble is probably due to a faulty generator regulator. Tighten bonding strap connections between regulator mounting legs and bulkhead. If noise continues, replace regulator (par. 125.).

c. NOISY GUN MOUNT TRAVERSING AND ELEVATING ELECTRICAL SYSTEM. If radio noise is believed due to faults within the gun mount traversing or elevating system, it can be determined by traversing the mount at least one complete revolution or elevating gun with vehicle engine stopped. If radio interference is caused when mount is traversed or gun elevated, the fault lies within one or more of the units designated in (1) and (2) below.

(1) *Faulty conduits, wiring, or bonding.* Tighten all coupling nuts and bonding clips. Replace all defective conduits and/or wiring. Tighten bonding strap and bonding cable under collector ring assembly. If either strap or cable is damaged, refer to higher authority. Tighten all conduit fittings in junction and control boxes and tighten switches and junction blocks in switch boxes. Replace the capacitor in switch box. If noise continues and trouble is not remedied, refer to (2) below.

(2) *Faulty pump motors or collector ring.* If cause of radio interference is not traceable to causes given in (1) above, it is probably due to a faulty motor or a faulty collector ring assembly; if so, refer to higher authority.

d. RADIO INTERFERENCE WITH ENGINE NOT RUNNING (VEHICLE STOPPED). The most probable cause for interference under this condition is that units do not have a good ground contact with the hull. To test, use a long jumper with a handle containing a prod which will penetrate painted surfaces. Attach the other end of the jumper to the negative terminal of the battery and contact the prod on all units of the vehicle, all conduits, and terminal boxes. If improvement is obtained in radio reception by making contact with the jumper at any of these points, the unit or part must be grounded to the part to which it is attached. These parts are grounded by bonding straps and/or internal-external tooth washers.

e. NOISE CAUSED BY MISCELLANEOUS SYSTEMS.

(1) *Causes of radio noise.* Noise may be caused by loose or damaged conduits or cables of the lighting systems and by various instruments, including their respective sending units, such as fuel tank gages, engine temperature gage and switches, oil pressure gage and switches, and speedometer and tachometer. Various instruments on the instrument panel and in control boxes may become loose. Battery terminals and ground cables of the batteries may work loose or become corroded or otherwise damaged. The lights of the vehicle and their switches, as well as the capacitors in windshield wipers, are also sources of radio noise.

(2) *Location and remedy of miscellaneous causes or radio noise.* The same general rule for maintenance of various systems, instruments, and devices must be followed, as given previously, concerning correction of radio interference. A process of elimination can be applied to locate trouble within any system.

f. RADIO INTERFERENCE NOT TRACEABLE TO ANY ELECTRICAL SYSTEM. If radio interference persists in spite of remedies set forth in *b* through *e* above, the trouble is probably due to defective radio apparatus or to faults within the structure of the vehicle. Refer to higher authority.

79. Hydramatic Transmission

a. LUBRICATION LEAKS. Check transmission for oil leaks at the drain plug, oil pan, and side cover. Tighten drain plug or the cover attaching screws to stop leak. If leak will not stop, replace gaskets or drain plug. If fluid coupling is found to be leaking, do not attempt to repair; notify higher authority.

b. TRANSMISSION FAILS TO TRANSMIT POWER.

(1) *Manual control linkage improperly adjusted.* Check adjustment of manual linkage and adjust if necessary (par. 149).

(2) *Transmission front band loose.* Check and adjust transmission front band (par. 150).

> *Note.* Rear band has automatic adjustment and does not require service.

(3) *Low oil level.* Fill transmission to proper level (par. 62). Check for leaks.

c. Transmission Shift Speeds Are Abnormal.
   (1) *All shifts occur at excessively high engine speeds.* Check vacuum line between transmission and intake manifold for leaks at fittings or cracks in line. Tighten or replace if necessary.
   (2) *Engine speeds up in first and third, while being driven forward.* Front band out of adjustment. Adjust front band (par. 150).
   (3) *Engine speeds up in front and second.* Rear band out of adjustment. Notify higher authority.
   (4) *All shifts above second occur at excessive engine speeds.* Manual control linkage improperly adjusted. Adjust manual linkage (par. 149).

d. Vehicle Creeps.
   (1) *Engine idling speed excessive.* Adjust carburetor to bring engine idling speed down to 450 rpm.
   (2) *Throttle linkage sticking open.* Eliminate binds in throttle linkage. Replace missing or weak throttle return springs.

80. Transfer Unit

a. Transfer Unit Does Not Respond to Control Lever.
   (1) *Control rods out of adjustment.* Adjust rods (par. 155).
   (2) *Shifter shaft extensions loose on shifter shafts.* Tighten shifter shaft extensions and adjust rods (par. 155).
   (3) *Yokes loose on shifter shaft.* Replace transfer unit or notify higher authority.

b. Transfer Unit Overheats.
   (1) *Low oil levels.* Check and add oil if necessary.
   (2) *High oil level.* High oil level will cause a foaming condition and overheating. Check oil level and drain or syphon until oil is at proper level on gage.
   (3) *Oil pump not operating.* Open sliding door over transfer unit and remove one of the $\frac{1}{8}$-inch pipe plugs from top of case. Start engines, place transfer unit shift lever in neutral position and transmission lever in "DRIVE." If oil pump is operating properly, oil will flow out of the plug opening. If no oil flows from plug opening, notify higher authority.

c. Gears Clash When Shifting.
   (1) *Cold or heavy lubricant.* In cold weather the gears may clash when shifting, if the transfer unit lubricants have not been warmed up sufficiently. Start engines and place transmission selector lever in "DRIVE" and transfer unit shift lever

in "NEUTRAL." Allow engines to run until the transfer unit warms up. Check for proper seasonal grade of lubricant.

(2) *Improper use of neutral pedal.* Review driving instructions (par. 45).

(3) *Shifting linkage out of adjustment.* Adjust linkage (par. 155).

(4) *Synchronization clutches inoperative.* If fault is not in (1), (2), or (3), above notify higher authority.

81. Propeller Shafts

a. Vibration at Universal Joints. Replace all propeller shaft joints having worn needle bearings. Check shafts for run-out and replace all defective shafts (pars. 158, 159, and 160).

b. Heavy Thumps on Rough Terrain. Check transfer unit universal joint slip yokes to make sure yokes are free on shaft splines. Replace frozen slip joints.

c. Oil Leakage at Universal Joints. Check oil seals at universal joints and replace defective propeller shaft joints (par. 158).

82. Controlled Differential

a. Vehicle Hard To Steer or Stop. If vehicle is hard to steer or stop, check steering brake adjusting and adjust as necessary (par. 162). If brake shoes are highly glazed or worn, replace shoes (par. 163).

b. Brakes Require Frequent Adjustment. If vehicle brakes require adjustments frequently, driver should observe driving instructions (par. 45) to insure proper handling of vehicle. If brakes continue to require frequent adjustments, check brake rims for roughened condition. If found to be rough or damaged, notify higher authority.

83. Tracks and Suspension

a. Suspension Wheel Tire Wear. If suspension wheel tires show excessive wear, check track shoes for bent guide or other damage which would have tendency to cut or wear tires. Repair or replace damaged shoes or track as necessary (par. 172).

b. Thrown Tracks. If track was thrown, check complete suspension to determine the cause. Improper track tension, damaged compensating wheels, foreign matter between track and wheels, or a broken front torsion bar are the most likely causes for the track to be thrown. Adjust track tension (par. 171), replace damaged compensating wheels (par. 173), clean out foreign matter, or replace broken torsion bar (par. 177).

c. Inoperative Track Support Roller. Check roller to make certain no foreign matter is caught between track roller and track or between track roller and support. If roller is free of foreign matter, check bearing adjustment. If bearing has seized, replace roller assembly (par. 179).

d. INOPERATIVE COMPENSATING WHEEL. Check compensating wheel to make certain no foreign matter is caught between wheel and track or compensating wheel support. If wheel is free of foreign matter, check wheel bearing adjustment. If bearing has seized, replace bearing (par. 173).

84. Turret Traversing Mechanism

a. TURRET MOTOR FAILS TO OPERATE.
  (1) *Circuit breaker open.* Push circuit breaker reset button on stabilizer control box (fig. 19).
  (2) *Feed circuit open.* Using test lamp or voltmeter, check feed circuit back from motor terminal to collector ring (fig. 96) or, if necessary, back to shunt box (fig. 56). Repair loose connections or replace defective wiring.
  (3) *Motor defective.* If current reaches motor terminal, motor is defective. Replace motor (par. 197).
  (4) *Collector ring defective.* If current reaches collector ring input terminal but will not pass through collector ring, replace collector ring and stabilizer control box as an assembly.

b. TRAVERSE PUMP FAILS TO OPERATE (MOTOR COUPLING SHEARED). Replace coupling (par. 197).

c. OIL LEAKING FROM SYSTEM.
  (1) *Shaft oil seals worn.* Report to higher authority.
  (2) *Gaskets worn.* Tighten or replace defective gaskets.
  (3) *Loose fittings or connections.* Tighten or replace.

d. UNSTEADY TURRET OPERATION.
  (1) *Low oil level.* Fill reservoir.
  (2) *Loose gear pump suction tube connections.* Tighten connections.

e. PUMP RUNS BUT TURRET WILL NOT TURN.
  (1) *Turret lock engaged.* Release lock.
  (2) *Shifter lever in intermediate position.* Set to power traverse position.
  (3) *Low oil level.* Fill reservoir.

f. TRAVERSE SPEED LOW IN EITHER DIRECTION. Fill reservoir.

g. TURRET CREEPS IN ONE DIRECTION WHILE VEHICLE IS HORIZONTAL. Adjust pump control shaft to neutral position.

h. TURRET DRIFTS WHEN VEHICLE IS NOT HORIZONTAL.
  (1) *Low oil level.* Fill reservoir.
  (2) *Loose gear pump suction tube connections.* Tighten connections.
  (3) *Cracked gear pump housing.* Replace pump (par. 197).

Section V. ENGINE DESCRIPTION AND MAINTENANCE IN VEHICLE

85. Description and Data

a. DESCRIPTION.

(1) This vehicle is powered with two 90° V-type, 8-cylinder engines (fig. 35), mounted side by side in the engine compartment in the rear of the hull. The engine cylinder blocks and crankcase, made in one casting of grey iron, support the crankshaft and camshaft and enclose the reciprocating parts. Cylinder heads are of cast iron also. The cylinders are arranged in two banks of four each, located 90° apart. Opposite connecting rod assemblies operate side by side from the same crankshaft journal.

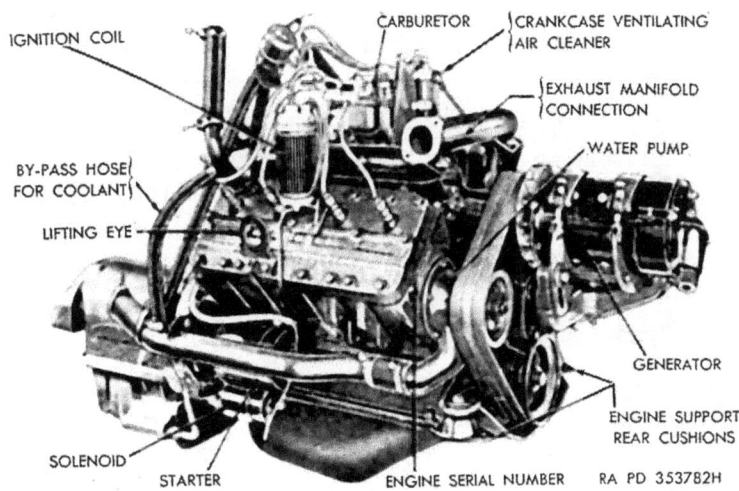

Figure 35. Engine.

(2) The distributor, ignition coil and wiring, carburetor, intake and exhaust manifolds, cylinder heads, water pump, generator, and oil filter are accessible through the engine compartment roof door. The starter and the engine oil pan are accessible through an inspection opening with removable cover in the hull floor.

(3) The generator end of these engines is designated the front end, even though it faces the rear of the vehicle; the flywheel end is designated the rear end. The right and left sides of each engine (and transmission) are determined by standing

107

at the transmission and looking toward the generator end. Right- and left-hand engines in the vehicle, however, are designated according to their relation to a man sitting in the driver's seat. The engine on the right of the hull is the right-hand engine. Engine supports, cushions, and attaching brackets and bolts are considered with reference to their position in the hull rather than their relation to the engines. Thus, the support under the transmission is designated as the engine front support, although it attaches to the rear of the engine, and the two supports at the generator end are the engine rear supports.

 b. DATA.

  Make_____ Cadillac, Model 44T24
  Number of engines_____ 2
  Type_____ V–8, 4-cycle
  Number of cylinders (each engine)_____ 8
  Piston displacement (each engine)_____ 346 cu.-in.
  Bore_____ 3½ in.
  Stroke_____ 4½ in.
  Compression ratio_____ 7.06 to 1
  Brake horsepower_____ 110 at 3,400 rpm
  Torque_____ 240 ft-lb at 1,200 rpm
  Firing order_____ 1, 8, 7, 3, 6, 5, 4, 2 (fig. 63)
  Valve arrangement_____ L-head
  Weight with accessories (each engine)_____ 1,045 lb
  Type of ignition_____ distributor
  Engine supports_____ 4

86. Engine Tune-Up

The operations required for a complete engine tune-up are as follows:

 a. Clean and adjust or replace spark plugs (par. 106).
 b. Clean and adjust or replace distributor contact points (par. 102).
 c. Check ignition timing (par. 104).
 d. Clean fuel filters (par. 112).
 e. Service carburetor air cleaners (par. 115).
 f. Adjust carburetors (par. 112).

87. Operations Performed with Engines in Vehicle

The following operations can be performed without removing the engines from the vehicle.

 a. Belts, adjust or replace (par. 88).
 b. Carburetor, adjust or replace (par. 112).
 c. Cooling system, service (par. 118).

d. Distributor, adjust or replace (par. 102).
e. Distributor contact points, adjust or replace (par. 102).
f. Fuel filter, service (par. 112).
g. Generator, replace (par. 124).
h. Hose connections, replace (par. 119).
i. Ignition coil, replace (par. 103).
j. Ignition capacitor (condenser), replace (par. 102).
k. Ignition timing, adjust (par. 104).
l. Ignition wiring, replace (par. 105).
m. Manifold gaskets, replace (par. 94).
n. Manifolds, replace (par. 94).
o. Oil filter, service or replace (par. 91).
p. Oil filter connections, replace (par. 91).
q. Oil pan, service and replace gaskets (par. 92).
r. Oil pump strainer, replace (par. 93).
s. Signal sending units, replace (par. 144).
t. Spark plugs, service, adjust, or replace (par. 106).
u. Starter, replace (par. 108).
v. Water pump, replace (par. 122).

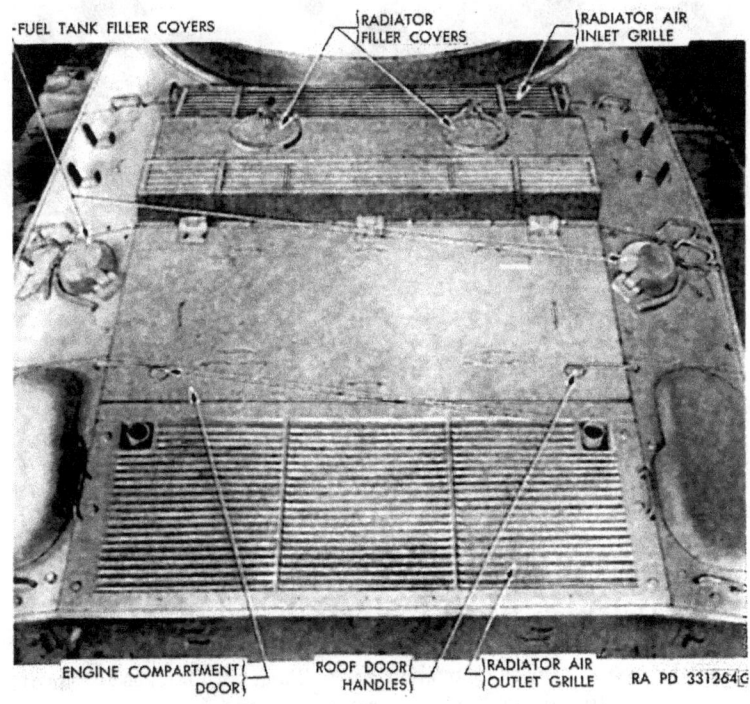

Figure 36. Engine compartment hinged door.

88. Belt Adjustment and Replacement

a. GENERAL. A triple-grooved pulley, mounted on the front of the crankshaft, carries three "V" belts which drive the water pump and generator (fig. 35). A bracket on the engine front cover provides for mounting the generator, and a movable cradle assembly permits belt adjustment.

b. BELT TENSION. The tension of the three water pump and generator belts must be adjusted if belts can be compressed more than three-quarters of an inch between pulleys (fig. 37).

Figure 37. Belt tension.

c. BELT ADJUSTMENT.
   (1) *Open master battery switch.* Master battery switch is located on the front of the bulkhead near the left side wall (fig. 5).
   (2) *Open engine compartment door.* Turn both latches at rear of door to unlocked position, raise door, and fasten to chain on rear of turret.

(3) *Remove radiator air outlet grille.* Remove the six attaching screws on the ends of the radiator air outlet grille (fig. 36). Remove the grille from the vehicle.

Caution: Care must be exercised in the removal of this grille because of its weight.

(4) *Adjust belts.* Loosen adjusting nut and pivot nut for generator cradle. Insert large screwdriver or utility bar between front cover and cradle and raise generator and cradle until slack in belts between water pump pulley and generator pulley is five-eighths to three-quarters of an inch, as measured with a straightedge pressed firmly on belts (fig. 37). After belts have been adjusted properly, tighten adjusting nut and pivot nut.

(5) *Install radiator air outlet grille.* Install the radiator air outlet grille using the six screws.

Caution: Care must be exercised in the installation of this grille because of its weight.

(6) *Close engine compartment door.* Unfasten door from chain on the turret, lower door into closed position, and fasten the two latches at rear of door.

d. REMOVAL. Open engine compartment door as directed in *c* above. Remove radiator air outlet grille. Loosen generator cradle lock nut and pivot nut and allow generator to drop to its lowest position. Disconnect generator cables. Remove all three belts from pulleys.

Caution: Do not use a screw driver or sharp instrument to pry belts off pulleys, as this may cause nicks or burs on pulleys and shorten life of belts. Remove and install belts by hand.

e. INSTALLATION. Install three new belts over pulleys, being careful not to make any nicks or burs on pulley. Belts are tied in matched sets of three and must not be separated until ready for actual installation.

Caution: Never install one new belt with two old belts except in an emergency; belts should always be replaced in sets of three.

Connect generator cables. Adjust belts. Install radiator air outlet grille. Close engine compartment door (*c* above).

89. Cylinder Head Gaskets

a. REMOVAL.

(1) *Open engine compartment door.* Turn both latches at rear of door to unlocked position, raise door, and fasten to chain on rear of turret.

(2) *Drain cooling system.* Remove drain cover under transmission by removing four screws. Remove plug at bottom of transmission, marked "WATER" (fig. 108). Remove radiator cap to vent system.

(3) *Remove spark plugs and suppressors.* Disconnect spark plug ignition cables from suppressors and remove suppressors from spark plugs. Using a deep socket wrench, remove spark plugs and spark plug gaskets.

(4) *Remove engine signal sending unit conduit (left cylinder head).* Remove two screws holding engine signal sending unit conduit to left cylinder head (fig. 99). Disconnect cables from units by pulling straight out of units.

(5) *Remove carburetor air intake intermediate elbow (late model engine).* Loosen clamp and remove air intake hose from upper end of air intake intermediate elbow. Loosen clamp and remove air intake hose from carburetor. Remove air intake pipe ground strap. Disconnect choke heater line and distributor ventilating line at air intake pipe. Remove two screws securing intermediate elbow to engine compartment cross support and two screws securing intermediate elbow to engine compartment side wall. Lift out intermediate elbow and air intake pipe as a unit.

(6) *Remove ignition coil and mounting bracket (late model engine—right cylinder head).* Disconnect cable from resistor terminal at bottom of coil by removing nut and washer.

Note. Make sure master battery switch is in "OFF" position.

Remove two cylinder head screws which hold coil mounting bracket to cylinder head. Position coil and support bracket out of way to provide access to head screws.

Caution: Do not kink coil or spark plug conduits; disconnect if necessary (par. 106).

(7) *Remove water outlet elbow.* Remove two screws holding water outlet elbow to cylinder head. Leave outlet elbow attached to hose. On left cylinder head, leave coil attached to outlet elbow. On right cylinder head, leave bypass hose attached to elbow.

(8) *Remove cylinder head.* Remove all remaining cylinder head screws except the lower row on outside bank of cylinders and remove cylinder head and gasket. Screws in lower row on outside bank of cylinders have to be unscrewed and removed with the cylinder head.

Caution: Support cylinder head while removing last screw to prevent head from sliding down and damaging valves.

b. INSTALLATION.

(1) *Install cylinder head gasket.* Coat new cylinder head gasket with gasket cement on both sides and position on cylinder block. Be sure surfaces of block and head are clean and free of carbon or other particles.

(2) *Install cylinder head.* Position cylinder head over gasket, being careful not to damage gasket or valves. Due to the slight clearance between the hull side wall and cylinder head, it is necessary to place the lower row of cylinder head screws and the engine lifting ring on the outside cylinder head before head is lowered into final position. Be sure engine lifting ring support is in proper location on cylinder head (fig. 38). Install (but do not tighten) all screws ($2\%_{16}$ in. long) with the exception of the two longer screws ($2^{11}/_{16}$ in. long) which hold the water outlet elbow and the two which attach the ignition coil mounting bracket to the cylinder head.

Caution: The two screws which hold the outlet elbow to cylinder head have special heads. They are longer than the other screws and must not be used in any other position or cylinder block may be damaged.

(3) *Install water outlet elbow.* Coat a new water outlet elbow gasket with gasket cement and position on cylinder head. Place outlet elbow over gasket and install the two long screws (fig. 38).

(4) *Install carburetor air intake intermediate elbow (late model engine).* Position air intake elbow and secure with two screws to engine compartment side wall and with two screws to engine compartment cross support. Secure ground strap to engine compartment side wall. Slide air intake hose over air intake elbow and tighten hose clamp. Install and tighten hose clamp on air intake hose at carburetor. Connect distributor ventilating line and choke heater line to air intake pipe at carburetor.

(5) *Install ignition coil and mounting bracket (late model engine).* Position coil and mounting bracket over the two center holes on cylinder head (right cylinder head only, and install two attaching screws (fig. 60). Do not tighten attaching screws at this time. Connect cable to resistor terminal at bottom of coil, and connect conduits to top of coil if necessary (par. 103).

(6) *Install engine signal sending unit conduit (left-hand cylinder head).* Position engine signal sending unit conduit on cylinder head and install two attaching screws and washers. Tighten all cylinder head screws to 70–75 foot-pounds torque.

Caution: Cylinder head screws must be tightened evenly and in the order shown in figure 38.

(7) *Install spark plugs and suppressors.* Install spark plug gasket on spark plugs. Install spark plugs and gaskets and

tighten plugs using a deep-socket wrench. Install suppressors and connect spark plug ignition cable (fig. 55).

(8) *Refill cooling system.* Install drain plug in bottom of transmission. Coat edges of drain cover under transmission with cement and install gasket, covers, and screws. Fill cooling system to proper level. Install radiator cap. Run engine until it reaches normal temperature and again check torque tightness of cylinder head screws (70–75 foot-pounds torque).

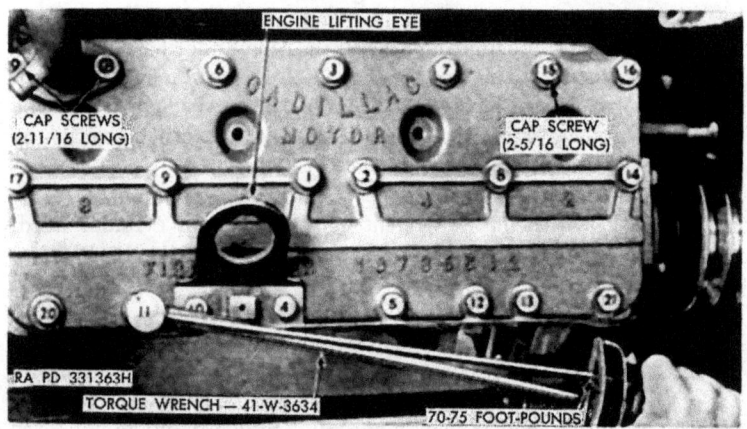

Figure 38. Tightening cylinder head screws.

(9) *Close engine compartment door.* Unhook hinged engine compartment door from turret and close door. Turn handles to locked position.

90. Crankcase Air Cleaner

a. GENERAL. Each engine is equipped with a crankcase oil bath air cleaner, mounted on the intake manifold. When engine is operating, air is drawn through the air cleaner, circulated through crankcase, and then drawn out of crankcase by manifold vacuum. It then passes through the intake manifold and out through the exhaust system. A ventilating line for the distributor is connected to the adapter.

b. SERVICING. To service air cleaner element, unscrew shell from cover (fig. 39) and wash shell and element in dry-cleaning solvent or volatile mineral spirits paint thinner. Dry parts thoroughly with air hose, fill shell to "FULL" mark with clean oil (par. 62), place element over stud, and attach shell to cover.

Caution: Be sure gasket is in good condition and in proper position.

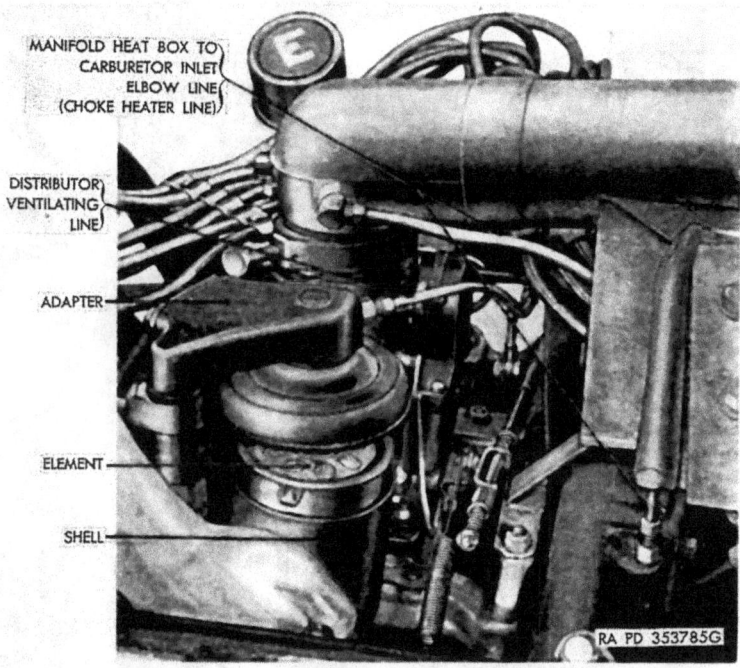

Figure 39. Removing engine crankcase cleaner element.

91. Oil Filter

a. GENERAL. Each engine is equipped with an oil filter mounted in a bracket attached to the front end of the engine. Oil enters the filter through the inlet line connected to the main oil header at the front of the engine block. After the oil passes through the filter, it is returned to the engine through the outlet line connected to the engine front cover (fig. 41).

b. SERVICING.

(1) *Open engine compartment.* Open engine compartment hinged door and remove radiator air outlet grille.

(2) *Remove filter element.* Remove retaining screw from top of oil filter cover (fig. 41) and remove cover and gasket. Discard gasket. Remove old filter element by lifting it out of filter body. If filter element is tight, be careful not to nick the gasket surface on the filter body in prying the element out.

Caution: Be sure no oil is allowed to splash on the belts during this operation.

(3) *Clean out filter.* Remove drain plug from bottom of filter

Figure 40. Removing oil filter element.

body and allow old oil to drain. Wash inside of filter body with dry-cleaning solvent, or volatile mineral spirits paint thinner, wipe dry, and install drain plug. Clean filter body and hull floor of any oil that has been spilled.

(4) *Install new element.* After removing the wrapper, thoroughly remove any wax coating with a clean cloth and dry-cleaning solvent or volatile mineral spirits paint thinner. Unless this wax coating is completely removed it will be dissolved by the passage of oil through the element and carried into the oil lines where it may cause clogging. Install the new element in the filter body. Place *new* gasket in cover, install cover, and tighten screw to 18–22 foot-pounds torque.

Note. Be sure cover is seated squarely on body before tightening screw.

(5) *Check installation.* Run engines for several minutes to allow filter to fill with oil. Inspect for oil leaks. Check engine oil level and add oil if necessary.

(6) *Close engine compartment.* Replace radiator air outlet grille and close engine compartment hinged door.

c. REMOVAL.

(1) *Open engine compartment.* Open engine compartment hinged door and remove radiator air outlet grille.

(2) *Disconnect oil pipes.* Place rags or container under oil filter to prevent oil from running over hull floor. Disconnect inlet line at top of filter by unscrewing flared tube nut from elbow (fig. 41). Disconnect outlet line at bottom of filter by unscrewing flared tube from elbow.

Caution: Be sure no oil is allowed to splash on belts during this operation.

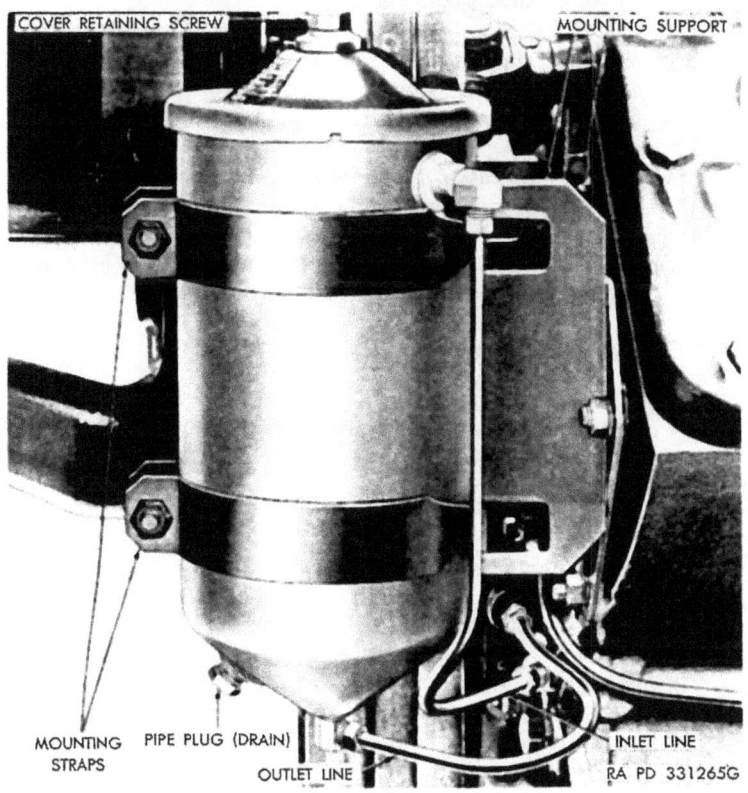

Figure 41. Oil filter installed.

(3) *Remove oil filter assembly.* Remove two nuts and bolts from mounting straps and lift out filter assembly (fig. 41).
(4) *Remove oil pipes from engine.* Disconnect inlet pipe from fitting connected to main oil header at front of cylinder block. Disconnect outlet pipe from fitting on engine front cover.
(5) *Remove oil filter mounting support.* Remove three nuts and washers holding mounting support to front of cylinder block. Remove support.

d. INSTALLATION.
(1) *Install mounting support.* Position mounting support over studs on front of cylinder block and install three nuts and lock washers.
(2) *Connect oil lines to cylinder block.* Connect outlet line to fitting on engine front cover and inlet line to fitting at main oil header on front of cylinder block.

Note. Tighten pipes fingertight.

(3) *Install oil filter assembly.* Position oil filter assembly in mounting support and install two mounting strap bolts, nuts, and lock washers. Connect outlet line to elbow on bottom of filter, and inlet line to elbow on side of filter. Tighten all connections.
(4) *Check installation.* Run engines for several minutes to allow filter to fill with oil. Inspect for oil leaks. Check engine oil level and add oil if necessary.
(5) *Close engine compartment.* Install radiator air outlet grille and close engine compartment hinged door.

92. Oil Pan and Gaskets

a. REMOVAL.
(1) *Remove rear floor cover.* Remove 18 screws in cover and remove from vehicle.

Caution: Support plate while removing last screws to prevent plate from dropping and causing personal injury.

(2) *Drain engine oil.* Remove engine oil pan drain plug and allow oil to drain. Install drain plug and tighten to 35–40 foot-pounds torque.
(3) *Remove oil pan.* Remove 25 oil pan screws. Lower rear end of oil pan and slide pan out through hull floor opening.
(4) *Remove gaskets.* Wash oil pan thoroughly with dry-cleaning solvent or volatile mineral spirits paint thinner. Remove all portions of oil pan gaskets from pan and bottom of cylinder block and discard.

b. INSTALLATION.

(1) *Install oil pan gaskets.* Coat gasket surface of oil pan lightly with gasket cement or heavy grease to hold gaskets in position. Place gaskets on oil pan.

(2) *Install oil pan.* Slide front end of oil pan through hull floor opening, being careful not to disturb gaskets on oil pan. Position front end of pan on cylinder block and raise rear end of pan into position. While holding oil pan up against block with one hand, install two or three screws to prevent pan dropping down. Install remaining screws and tighten to 7–10 foot-pounds, using a torque wrench.

(3) *Install rear floor cover.* Install rear floor cover being careful to hold cover securely while replacing the 18 screws.

(4) *Refill.* Fill with engine oil as prescribed on lubrication order (fig. 27).

93. Oil Pump Strainer

a. REMOVAL.

(1) *Remove oil pan.* Refer to paragraph 92 for removal of oil pan.

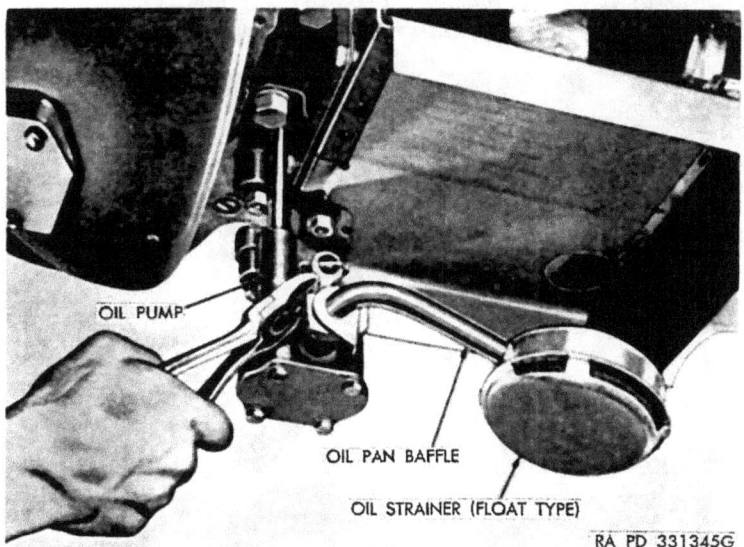

Figure 42. Removing oil pump strainer.

(2) *Remove oil pump strainer.* Remove cotter pin holding oil strainer in oil pump body (fig. 42). Slide float and strainer assembly out of oil pump.

(3) *Inspect strainer.* Inspect strainer to see that oil has not entered the float chamber. Shake assembly to detect presence of oil in float. If oil is found in float chamber, strainer must be replaced. If no oil is found in float, wash strainer thoroughly in dry-cleaning solvent or volatile mineral spirits paint thinner and blow dirt from strainer mesh.

b. INSTALLATION.

(1) *Install oil pump strainer.* Slide oil pump strainer into oil pump body. Install cotter pin.

Note. Raise and lower strainer several times to make sure it is free and that it will drop down of its own weight.

(2) *Install oil pan.* Refer to paragraph 92 for installation of oil pan.

94. Intake and Exhaust Manifolds

a. REMOVAL.
(1) *Open master battery switch.* Refer to figure 5.
(2) *Open engine compartment.* Open engine compartment hinged door and remove radiator air outlet grille.
(3) *Remove carburetor.* Refer to paragraph 112.
(4) *Remove muffler assembly.* Refer to paragraph 116.

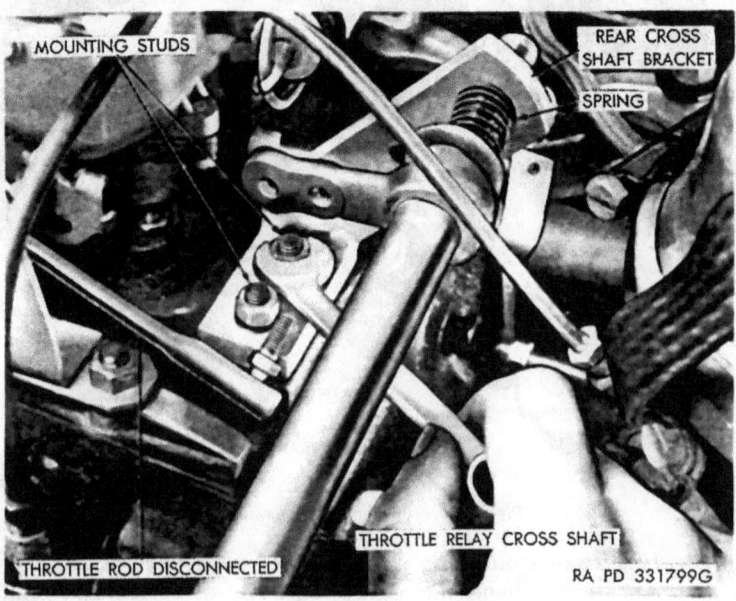

Figure 43. Disconnecting upper throttle relay cross shaft.

(5) *Remove exhaust connection.* Remove four screws holding exhaust connection to left and right exhaust manifold; lift up connection and discard gaskets.

(6) *Remove ignition cable conduits.* Disconnect spark plug ignition cables by pulling off suppressors. Remove four nuts holding ignition cable support brackets to manifolds. Lay brackets and conduits aside.

(7) *Remove crankcase air cleaner.* Disconnect distributor ventilating line at crankcase air cleaner adapter. Remove two nuts and washers holding engine crankcase air cleaner to intake manifold. Loosen lower hose clamp and remove air cleaner assembly.

(8) *Remove upper rear throttle relay cross shaft.* Remove clevis pin from lever on center of relay cross shaft. Disconnect throttle rods at cross shaft levers. Remove two nuts from left rear cross shaft bracket (fig. 43) (on engine being disassembled) and slide cross shaft out of other bracket.

Note. Do not lose springs or washers from ends of cross shaft.

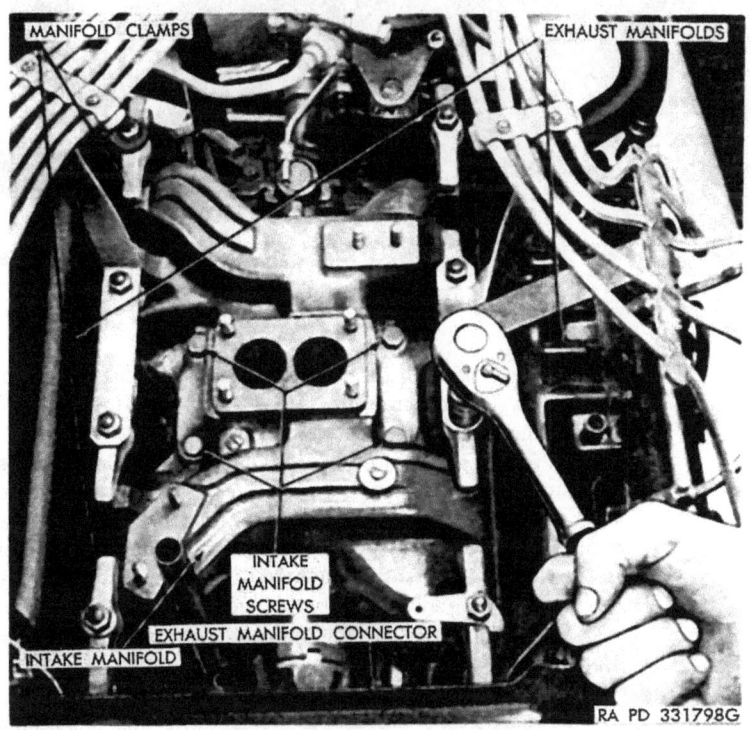

Figure 44. Removing exhaust manifold.

(9) *Remove exhaust manifolds.* Remove remaining nuts and clamps holding exhaust manifolds to cylinder block (fig. 44). Remove manifolds.

(10) *Remove intake manifold.* Bend back lock plates and remove four screws holding fan shaft universal joint to yoke at rear of intake manifold. Remove four intake manifold screws (fig. 44) and lift out manifold (fig. 45). Discard gaskets.

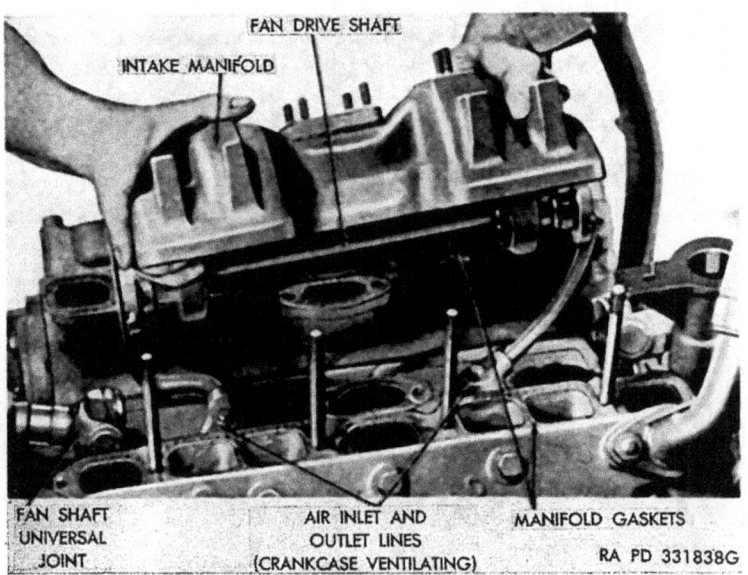

Figure 45. Removing intake manifold.

b. INSTALLATION.

(1) *Install gaskets.* Clean gasket surfaces on crankcase and manifolds. Place two *new* intake and exhaust manifold gaskets on crankcase.

(2) *Install intake manifold.* Lower intake manifold into position on crankcase. Install four screws holding intake manifold to cylinder block. Using a torque wrench, tighten screws to 25–30 foot-pounds torque. Connect fan shaft universal joint at rear of manifold and install four screws and lock plates. Bend lock plates over head of screws. Connect fan shaft universal joint at front of manifold and install four screws and lock plates. Bend lock plate over head of screws.

Note. Care must be exercised to prevent dirt getting into manifold openings.

(3) *Install exhaust manifolds.* Place both exhaust manifolds on gasket and install intake and exhaust manifold clamps and clamp nuts. Using a torque wrench, tighten nuts to 25–30 foot-pounds torque.

> *Note.* Be sure to install spark plug conduit support brackets over the center studs on both sides and to place carburetor retracting spring clip over left front stud and transmission filler pipe clip over right rear stud.

(4) *Install exhaust connection.* Install two new exhaust manifold connector gaskets on exhaust manifold and install connector and four mounting screws. Add oil to threads of these screws. Using a torque wrench, tighten screws to 25–30 foot-pounds torque. After running engine a short time, check all manifold screws for proper torque tightness.

(5) *Install crankcase air cleaner.* Slide air cleaner hose over air cleaner front pipe on valve compartment cover. Position air cleaner mounting bracket over mounting studs and install nuts and washers. Tighten lower hose clamp on air cleaner pipe. Connect distributor ventilating line to adapter.

(6) *Install carburetor.* Refer to paragraph 112.

(7) *Install upper rear throttle relay cross shaft.* Slide end of upper rear throttle relay cross shaft through mounting bracket on intake manifold.

> *Note.* Make sure spring and washer are on end of cross shaft before installing in bracket.

Place mounting bracket on opposite end of cross shaft, making sure spring and washer are in position on end of cross shaft. Install mounting bracket on studs on intake manifold and install two mounting nuts and lock washers. Connect rod from upper rear relay to lower rear relay by installing clevis pin, flat washer, and cotter pin at upper rear relay. Connect throttle rods to levers on cross shaft. Adjust throttle linkage as outlined in paragraph 112.

(8) *Install muffler and exhaust pipe.* Refer to paragraph 116.

(9) *Close engine compartment.* Install radiator air outlet grille and close engine compartment hinged door.

95. Engine Mountings

a. REMOVAL.

(1) *Remove engines.* Refer to paragraph 97.

(2) *Remove engine cushion supports.* Remove screws and washers holding engine cushion mountings to engine support brackets. Remove rubber cushions.

b. INSTALLATION.
   (1) *Install engine cushion supports.* Place engine cushion mountings over engine support brackets and install screws and washers. Using a torque wrench, tighten screws to 45–50 foot-pounds torque.
   (2) *Install engines.* Refer to paragraph 98.

Section VI. ENGINE REMOVAL AND INSTALLATION

96. Coordination with Ordnance Maintenance Unit

Replacement of the engine with a new or rebuilt engine is normally an ordnance maintenance operation, but may be performed in an emergency by the using organization, provided authority for performing this replacement is obtained from the responsible commander. Tools needed to remove an engine are issued to the using organization in accordance with ORD 7 SNL G–200.

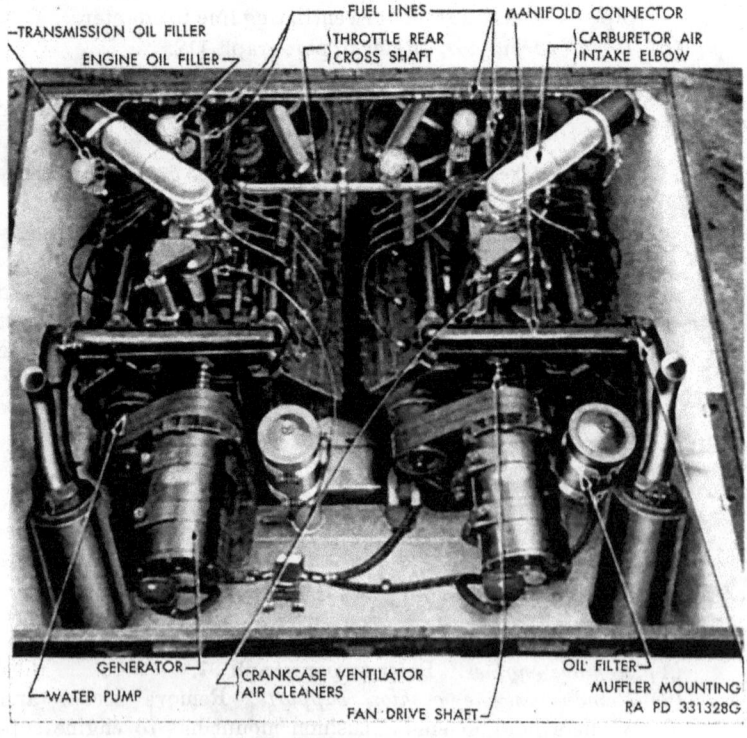

Figure 46. Engines installed.

97. Removal

a. EQUIPMENT. The only special equipment necessary for the removal of the engines consists of the engine lifting sling—41–S–3831–300 and a chain hoist.

b. PROCEDURE. The following procedure applies without change to either model engine.

   (1) *Open master battery switch.* The master battery switch is located on the front of the bulkhead near the left side wall (fig. 5). Pull out on switch knob and turn about one-quarter turn to lock in open position.

   (2) *Open engine compartment.* Open engine compartment hinged door (fig. 36). Remove six screws from outer ends of radiator air outlet grille, attach chain and hoist to lifting handles, and remove grille.

   (3) *Drain coolant.* Remove four screws from drain plate under transmission and remove plate. Drain coolant from engine by removing drain plug at bottom of transmission marked "WATER" (fig. 108). Remove radiator cap to vent system and speed draining.

   (4) *Remove exhaust pipe and muffler.* Remove two screws with nuts and washers holding muffler mounting to exhaust manifold flange.

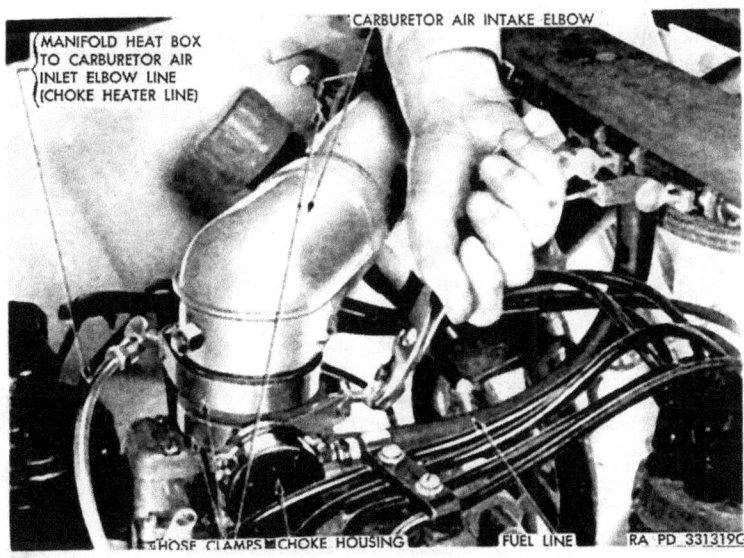

Figure 47. Removing carburetor intake elbow.

(5) *Remove carburetor intermediate air intake elbow* (fig. 47). Loosen hose clamps at carburetor. Disconnect (choke heater) line manifold to carburetor elbow and distributor ventilating line from air intake elbow. Remove ground strap. Remove two screws securing elbow to engine compartment side wall and lift out elbow and air intake pipe as a unit.

(6) *Disconnect throttle rods.* Remove nut and washer from carburetor throttle ball stud and slide out of throttle relay cross shaft (fig. 43). Remove cotter pin and clevis pin connecting lower relay rod to throttle relay cross shaft. Remove two nuts from cross shaft bracket on either intake manifold and slide cross shaft out of other brackets.

Caution: Do not lose spring and washer between brackets and cross shaft.

(7) *Disconnect fuel line at carburetor.* Loosen hose clamp on fuel line at carburetor and slide hose off nipple.

(8) *Loosen fan shaft knurled packing nut.* Loosen packing nut on fan shaft and slide nut away from fan (fig. 48). Fan shaft will slide out of spline when engine is removed.

(9) *Disconnect tachometer flexible shaft.* Disconnect tachometer flexible shaft at distributor support by unscrewing coupling and sliding core out of support. Tie shaft out of way to prevent damage.

Figure 48. Loosening fan shaft knurled packing nut.

(10) *Disconnect engine feed conduit.* Disconnect conduit at coupling at rear of left cylinder block by unscrewing coupling and pulling apart.

(11) *Remove rear hull floor cover.* Remove screws from rear hull floor cover.

Caution: Support rear end of cover while removing last screws to prevent cover from dropping down and causing injury.

Front of cover is held up by hooks on cover. Slide cover toward rear of vehicle and lower to ground.

(12) *Disconnect hose connections.* Loosen both clamps on each water outlet hose on cylinder head and push hose down on elbow until clear of radiator thermostat housing. Working under vehicle, disconnect lower hose at front end of water pump inlet tube.

(13) *Disconnect manual control shifter shaft* (figs. 49 and 106). Working through hull floor opening (fig. 49), remove clip and clevis pin that hold operating lever to connecting link (relay lever to operating lever link) on rear relay (fig. 49). The right floor opening permits access to the right lever and the left opening to the left lever.

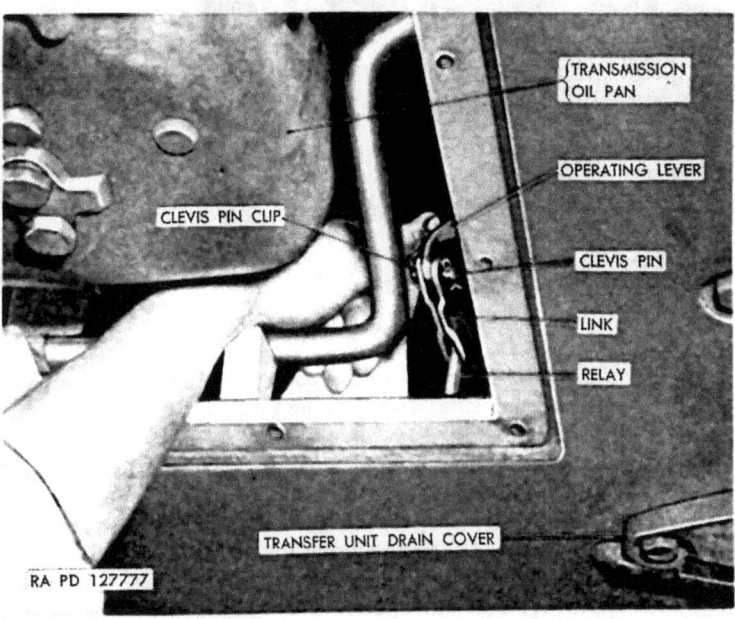

Figure 49. Disconnecting left transmission operating lever (through left rear hull floor cover).

(14) *Disconnect engine front support.* Remove two cap screws (one on each side) for the engine front support bracket on transmission.

(15) *Remove generator and starter conduits.* Unsnap cable clip and lift off generator terminal cover (fig. 50). Remove nuts and washers holding cables to terminals. Lift cables off terminals. Unscrew conduit connection at bottom of terminal shield box and pull wires out of box. Unscrew nut from starter terminal and remove starter cable.

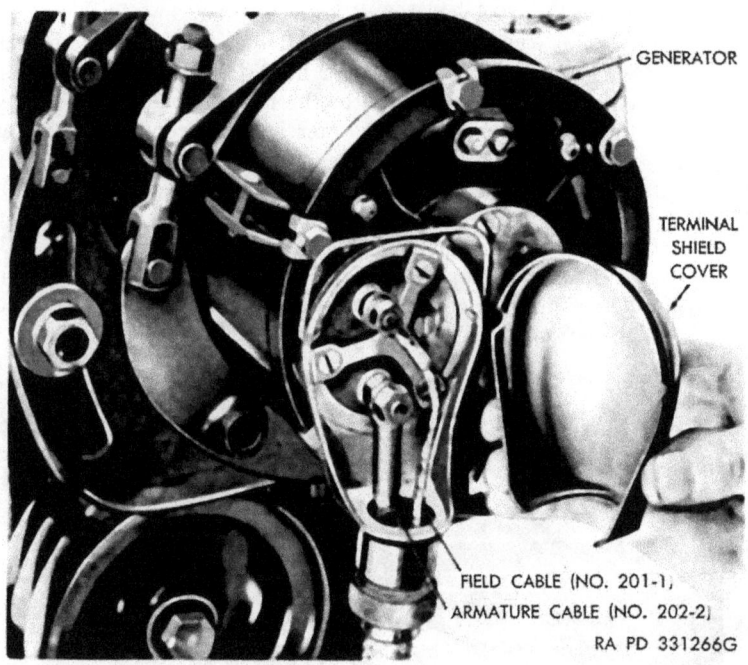

Figure 50. Generator cables.

(16) *Disconnect engine rear supports.* Remove two cap screws (one on each side) from engine rear support brackets.

(17) *Remove engine.*
   (a) Attach engine lifting sling—41–S–3831–300 to chain hoist and connect to lifting eyes on each cylinder head. Raise engine with hoist until it is clear of engine support brackets (fig. 51). Slide engine toward rear of vehicle until splined shaft on universal joint transmission slides out of transfer unit and splined shaft on fan drive is separated. Tilt engine upward at rear just enough so generator will

Figure 51. Removing engine.

clear rear hull wall; lift engine out of vehicle and place in suitable stand.

(*b*) Before returning engine for overhaul or rebuilding, remove the transmission oil cooler drain plug and wire it to the transmission water pipe. Residual water in cylinder block or transmission oil cooler may then drain, avoiding freezing, when engine is crated and in transit.

(18) *Interchange connecting parts.* If a new engine is being installed, remove fan shaft universal joint assembly from replaced engine and install on new engine. Also remove transfer unit input shaft from transmission of replaced engine assembly and install on new transmission.

98. Installation

a. EQUIPMENT. The only special equipment necessary for the installation of the engines is the engine lifting sling—41–S–3831–300 and a chain hoist.

b. PROCEDURE. The following procedure applies without change to either one of the engines.

(1) *Install engine.* Place engine sling—41–S–3831–300 on lifting eyes on cylinder heads and connect hoist to sling. Raise engine above and slightly to rear of engine compartment. Lower engine carefully with transmission end tilted down (fig. 51). Line up blind spline on fan shaft with shaft yoke. Guide splined transfer unit input shaft into transfer unit (fig. 52). Lower engine carefully until it rests on engine supports. Remove sling. Do not remove sling until all engine support bolts are installed. Sometimes it is difficult to aline the support bolts and with engine raised slightly the bolts can be started easier. Tighten packing nut on fan shaft.

(2) *Install engine supports.* Install two bolts, one on each side, in engine front support brackets. Working in fighting compartment, install two bolts in rear support brackets.

(3) *Connect generator and starter conduits.* Place starter cable over terminal and secure with nut and washer. Insert generator cables through opening in shield box on front of generator and fasten cables to proper connections (fig. 50). Install nuts and washers holding cable to terminal posts. Slide knurled nut up on conduit and fasten to bottom of box. Place terminal shield cover on box and snap locking wire in position.

(4) *Connect and adjust manual control shifter shaft.* Place transmission selector lever in "DRIVE" position. Position manual control shifter shaft in the middle detent position. If the clevis pin hole in the bottom of the operating lever does

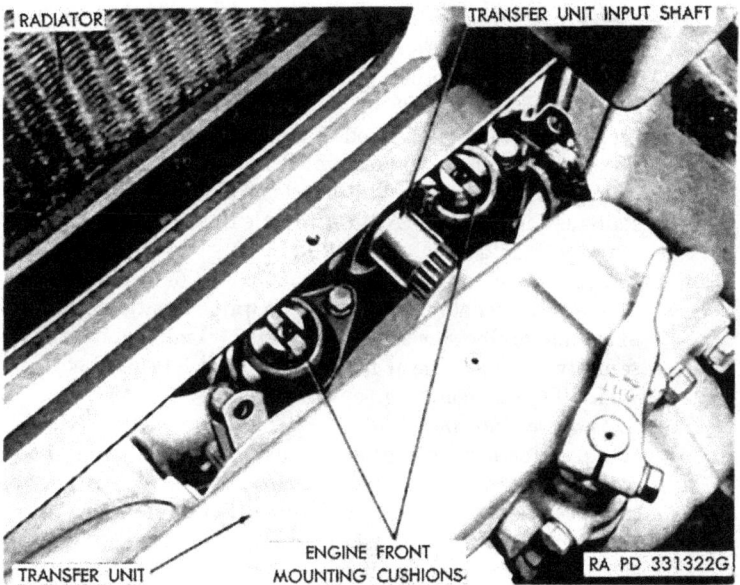

Figure 52. Inserting transfer unit input shaft.

not line up with the hole in the connecting link, loosen clamp screw on adjustable shaft and slide shaft in or out until clevis pin holes line up. Tighten clamp screw and install clevis pin and clip (figs. 49 and 106).

(5) *Install hose connections.* Working under vehicle through rear hull floor opening, install lower radiator hose between radiator and water pump inlet tube. Tighten hose clamps. Working above engine, install both outlet hoses between radiator thermostat housing and outlet connections on cylinder head. Tighten hose clamps securely.

(6) *Connect engine feed conduit.* Line up prongs on engine feed conduit with holes on other conduit. Push both ends of conduit together and connect by screwing ends of coupling together.

(7) *Connect tachometer flexible shaft.* Slide end of tachometer flexible shaft core into drive gear in distributor support and fasten flexible shaft casing to support with knurled coupling on end of casing.

(8) *Connect fuel line at carburetor.* Slide rubber fuel hose over end of nipple at rear of carburetor and tighten hose clamp securely.

(9) *Connect throttle rods.* Place springs and washers on both ends of throttle cross shaft and position cross shaft in bracket that was left on other engine. Install other cross shaft bracket and fasten to intake manifold mounting studs. Connect lower throttle relay rod to throttle cross shaft lever and install clevis pin and cotter pin. Slide throttle rod ball stud through mounting hole in throttle cross shaft lever and install nut and washer. Adjust throttle linkage as explained in paragraph 112.

(10) *Install carburetor air intake elbow.* Slide carburetor air intake hose over elbow at engine side wall. Position other end of air intake elbow on top of carburetor. Tighten hose clamps securely. Install line at intake elbow and left exhaust manifold. Tighten couplings at both ends.

(11) *Install muffler and exhaust pipe.* Lower muffler assembly into position next to engine compartment side wall (fig. 46). Position a new exhaust pipe gasket between muffler and exhaust connection. Install two screws, nuts, and washers.

(12) *Fill cooling system.* Install drain plug in bottom of transmission and install drain cover plate. Fill cooling system with coolant (each system 40 qt, aprox). Install radiator cap, making sure gasket is in good condition.

Caution: When system is filled with cold water, it may be necessary to run the engine until thermostat valve opens and permits system to be completely filled. Be sure to do this if system takes less than 40 quarts.

(13) *Install radiator air outlet grille.* Attach hoist to handles on radiator air outlet grille and lower into position on rear of vehicle. Install six mounting screws. Center muffler outlet pipes in grille opening.

(14) *Adjust carburetor.* Refer to paragraph 112.

(15) *Adjust ignition timing.* Refer to paragraph 104.

(16) *Check adjustment of both transmissions.* Set brakes. Run both engines at same time with transfer unit in "HIGH" and transmission lever in "NEUTRAL." Set hand throttle to run engines at 1,000 rpm, and move selector lever slowly toward "DRIVE." By watching tachometers, note position of lever when speed of one engine drops. Rear edge of selector lever should be approximately one-eighth of an inch in front of the front edge of slot in quadrant when engine speed drops (fig. 53). Continue moving lever until speed of other engine drops. This should be within $\frac{1}{8}$-inch additional lever travel. If difference is greater, adjust fulcrum lever of newly installed transmission (par. 149).

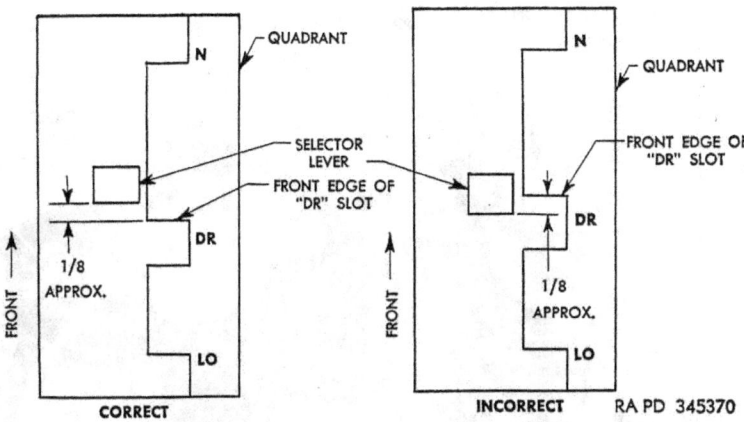

Figure 53. Correct and incorrect adjustment of selector lever.

(17) *Install lower rear hull floor plate.* Coat edges of floor plate with gasket paste and place gasket on plate. Raise front edge of plate until hooks on plate rest on hull floor. Raise rear end of plate and line up screw holes with two drift punches. Install floor plate retaining screws.

(18) *Close engine compartment.* Close engine compartment hinged door.

Section VII. IGNITION SYSTEM

99. Description

Two identical, but completely independent ignition systems are used, one for each engine. The complete ignition system is shielded both for protection against water and for eliminating radio interference. Each system consists of: an ignition coil mounted on a bracket on the center of the right cylinder head (late model engines (fig. 54)) or on a bracket on the left rear corner of the engine (early model engines (fig. 55)); a timer inside the distributor housing; a capacitor (condenser) also located inside the distributor housing; a distributor which directs the high voltage to each of the spark plugs; shielded, waterproof, two-piece spark plugs; and the necessary shielded, metal-clad cables to connect the ignition units. The wiring circuit diagram of the ignition system is shown in figure 56.

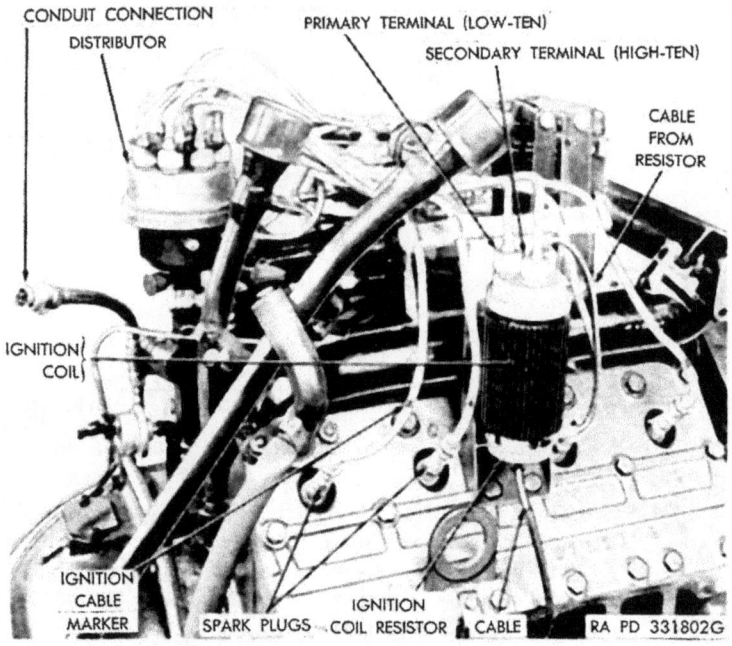

Figure 54. Ignition units (late model engine).

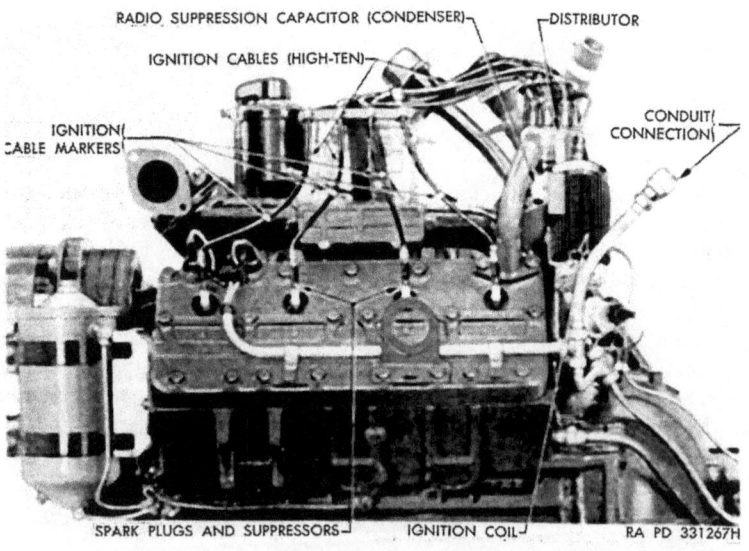

Figure 55. Ignition units (early model engine).

134

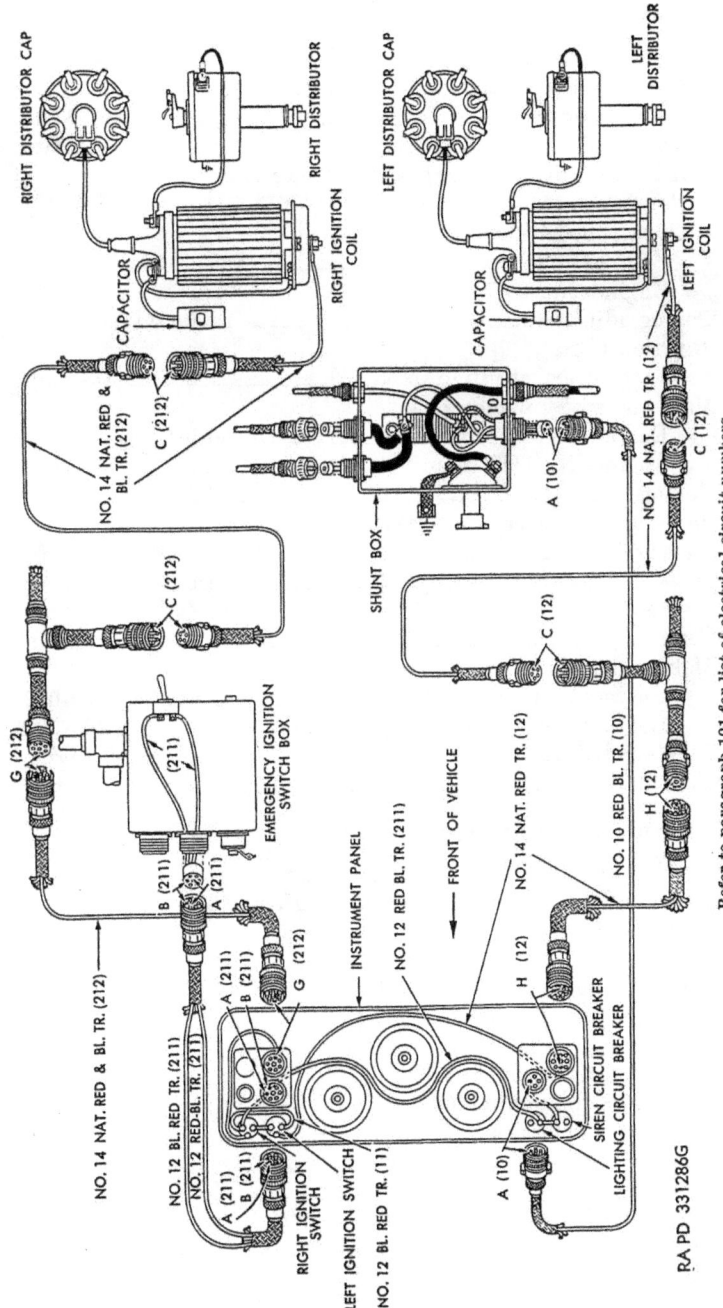

Refer to paragraph 101 for list of electrical circuit numbers.

Figure 56. Ignition circuit wiring diagram.

100. Data

a. DISTRIBUTOR ASSEMBLY.
Contact point gap_____ 0.013–0.018 in
Contact spring tension_____ 19–23 oz
Direction or rotation (from top)_____ clockwise
Dwell angle (disregard gap when setting to dwell
angle)_____ 31° for DR–1110611 and 25° for 7053275
Make and model number:
    Engines up to No. 3G7534_____ Delco-Remy 1110611
    Engines No. 3G7534 and up_____ 7053275
Timing adjustment_____ rotate on mounting

b. IGNITION COIL.
Amperage draw, at idling speed_____ 1½ amp
Make and model number:
    Engines up to No. 3G7534_____ Delco-Remy 1115281
    Engines No. 3G7534 and up_____ 7053277
Voltage_____ 24

c. IGNITION CAPACITOR (CONDENSER).
Capacity_____ 0.18–0.23 mfd
Make and model number_____ Delco-Remy 1900272

d. SPARK PLUGS.
Gap_____ 0.028–0.033 in
Make and model number:
    Engines up to No. 3G7534_____ 501012
    Engines No. 3G7534 and up_____ 7052903
Number used_____ 8 per engine
Thread_____ 10 mm

101. Electrical Circuit Number List

The following list shows electrical circuit numbers which appear on the various electrical circuit wiring diagrams (figs. 56, 63, 67, 86, 92, 93, 94, 95, 96, 103, 104, and 165):

Circuit No.   *Electrical circuit*

  1. Generator field circuit (left).
  2. Generator armature circuit (left).
  4. Generator regulator to shunt (left).
  6. Battery to master switch.
  8. Shunt to ammeter (+).
  9. Shunt to ammeter (−).
10. Shunt to instrument panel circuit breakers.
11. Ignition switch cable (left).
12. Ignition switch to ignition coil (left).
14. Starter switch circuit (including cable) to relay.
15. Main light switch cable (to BAT terminal).

| Circuit No. | Electrical circuit |

16. Light switch (HT) to service head light.
19. Light switch (BOD) to black-out driving light (including resistor).
20. Light switch (BHT) to black-out marker lights.
21. Light switch (HT) to service tail light.
22. Light switch (S) to service stop light.
23. Light switch (BS) to black-out stop light.
24. Light switch (BHT) to black-out tail lights.
25. Siren (or horn) switch (including cable) to siren (or horn).
27. Instruments feed (instruments with polarity—left).
33. Water temperature gage to sending unit (left).
34. Low engine oil pressure signal light circuit (left).
35. High water temperature signal light circuit (left).
38. Dome light circuit (including breaker and switch).
40. Instrument panel light circuit.
47. Slip ring cable.
40. Slave battery outlet (+) cable.
50. Slave battery outlet (−) cable (ground).
59. Hull ventilator circuit.
68. Battery interconnecting cables.
70. Generator regulator ground (left).
71. Windshield wiper circuit.
72. Low transmission oil pressure signal light circuit (left).
73. LH temperature gage to LH signal light (instrument panel).
74. Starter solenoid circuit (left).
75. Stop switch circuit (SW to SS on light switch).
76. Fuel pump control cable.
77. Fuel pump switch to fuel pump (left).
78. Fuel pump switch to fuel pump (right).
81. Battery (RH) to terminal blocks to shunt box.
82. Starter to terminal block (left).
93. Starter relay grounding circuit (left).
94. Starter relay auxiliary grounding circuit (left).
104. Turret motor cable.
138. Turret dome lights.
201. Generator field circuit (right).
202. Generator armature circuit (right).
204. Generator regulator to shunt (right).
211. Ignition switch cable (right).
212. Ignition switch to ignition coil (right).
227. Instruments cable (instruments with polarity—right).
233. Water temperature gage to sending unit (right).
234. Low engine oil pressure signal light circuit (right).
235. High water temperature signal light circuit (right).
270. Generator regulator ground (right).

137

941719—51——10

272. Low transmission oil pressure signal light circuit (right).
273. RH temperature gage to RH signal light (instrument panel).
274. Starter solenoid circuit (right).
282. Starter to terminal block (right).
294. Starter relay auxiliary grounding circuit (right).

102. Distributor

a. REMOVAL. Open engine compartment hinged door. Remove distributor cap. Leave cap in engine compartment suspended by ignition cables. It is not necessary to disconnect cables unless cap is to be replaced. Disconnect the two distributor housing ventilating lines. One of these lines is connected to the crankcase ventilating air cleaner and the other to the carburetor air intake elbow. Disconnect the two primary (low-tension) cables by unscrewing the coupling at side of distributor; also disconnect the end of cable inside the distributor at the terminal (fig. 57). The cable from the capacitor (condenser) connects to the same terminal. Remove the screw which holds the ground strap. Remove the cap screw which holds the clamp arm to the support base and lift out the distributor.

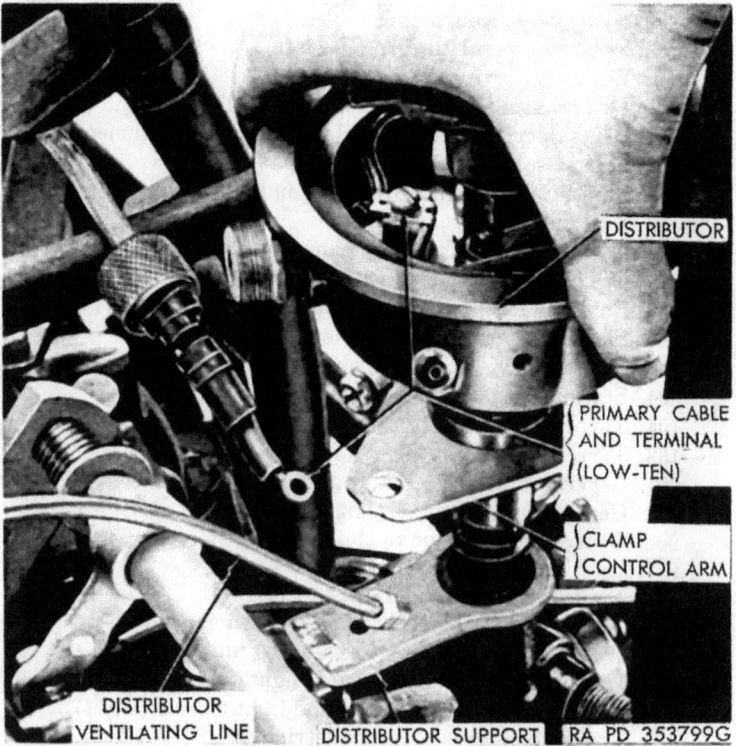

Figure 57. Removing distributor.

b. INSTALLATION. Position complete distributor assembly in support, turning rotor until lug in lower end of shaft meshes with slot in distributor drive shaft. Install cap screw holding clamp control arm to base of distributor support. Install ground strap. Connect primary circuit low-tension cable by inserting end through coupling fitting and connecting cable to terminal in distributor. Connect capacitor (condenser) to primary terminal at same time primary connection is made. Tighten knurled coupling nut at distributor housing only fingertight. If cables were removed from top of coil or distributor cap, install and tighten fingertight.

Warning: Pliers must not be used. The metal shell on distributor cap may be distorted or the high-tension towers on the bakelite cap may be cracked.

Connect the two distributor housing ventilating lines and tighten them securely. Install distributor cap, place sealing band in position, and tighten clamp screw. If ignition cables (high-tension) were disconnected from distributor cap, connect cables from coil and spark plugs. Check and adjust ignition timing (par. 104). Close engine compartment door.

c. CONTACT POINTS.

(1) *Accessibility.* Contact points are accessible for service when the distributor is either on or off the engine, in or out of the vehicle. Figures 58 and 59 show operations being performed on the bench; however, the work may be performed in the same manner in the vehicle. Remove distributor cap and rotor to service contact points.

(2) *Inspecting contact points.* The appearance of contact points reveals: first, whether the points require any service at all, and second, whether they should be cleaned or replaced. Contact point appearance can be judged by the conditions described in (*a*) through (*d*) below.

(*a*) A rough gray surface on the contact points is an ideal condition. Do not clean or replace points with this appearance.

(*b*) Oil-soaked points usually do not require dressing. Clean points thoroughly and see that the source of oil leakage is determined and corrected.

(*c*) Pitted contact points will not cause ignition failure unless the pitting is severe. Do not replace points because of minor pits or projections. Remove, clean, and smooth up with a carborundum stone.

(*d*) Oxidized points which have a blue or black scale usually do not require replacement, but the scale must be removed. Oxidized points are usually caused by a faulty coil or capacitor (condenser). Replace these units if the oxidized condition recurs frequently.

(3) *Cleaning contact points.* If points are badly pitted, burned, or worn, remove and dress carefully on a smooth carborundum stone until all trace of corrosion or pitting is removed, keeping the surfaces as square as possible. Make sure that the points line up squarely with each other and adjust gap. Retime ignition.

Caution: Never attempt to clean points with emery cloth or sandpaper.

(4) *Adjusting contact point gap.* Turn distributor drive shaft until cam is holding points at widest opening. Loosen contact support lock screw with screw driver, and adjust point opening to a clearance of 0.013–0.018 inch (0.015 is ideal) by turning eccentric screw (fig. 58) and checking clearance with feeler gage. Tighten lock screw and check clearance to make sure that points remained in adjustment.

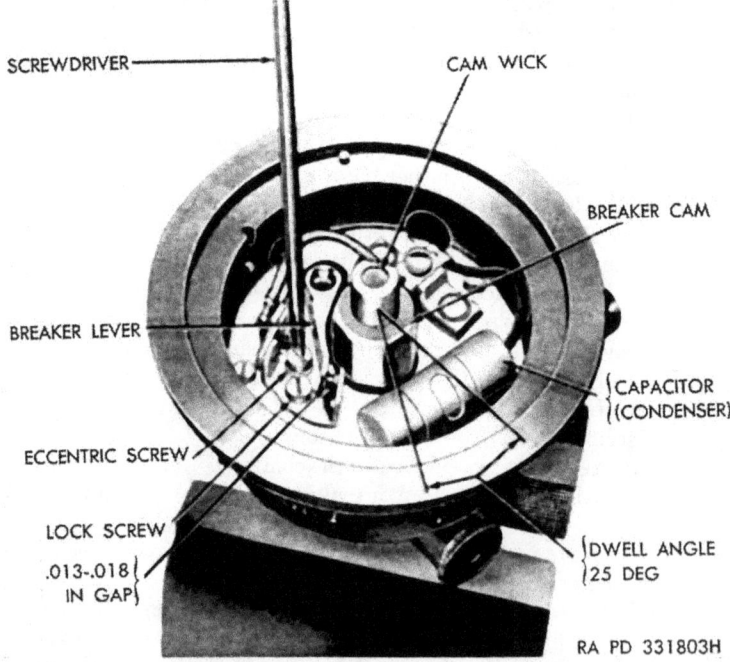

Figure 58. Adjusting contact point gap.

(5) *Checking breaker lever spring tension.* Check tension of breaker lever spring, with spring scale hooked over lever as close to points as possible and held at right angles to lever (fig. 59). Correct tension is 19–23 ounces. If spring scale is not available, check tension by feel. Tension can be adjusted

by loosening screw on breaker support fiber base and sliding the spring either forward or backward or by bending the spring. Check gap after spring tension has been verified.

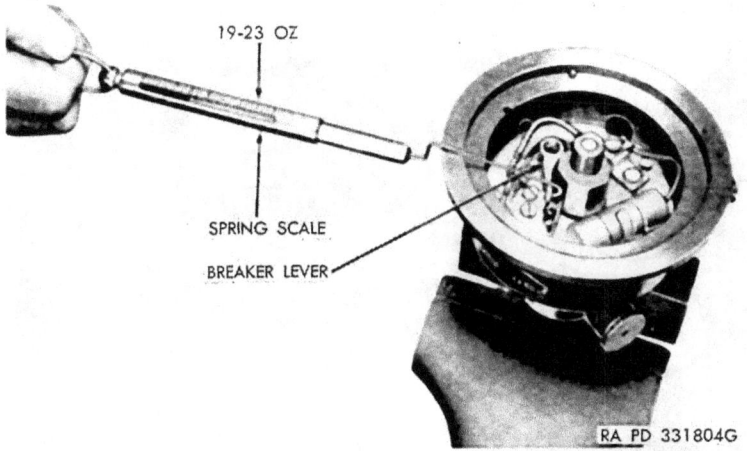

Figure 59. Checking breaker lever spring tension.

(6) *Replacing contact point set.* Remove lock screw from contact point support (fig. 58). Loosen screw which holds contact spring to connector. Remove retainer from contact point support stud. Lift out breaker lever and contact point support. Place new parts in position, install mounting screws and retainer, and adjust gap.

Note. See that fiber rubbing block on contact arm is parallel to distributor cam.

Install rotor and distributor cap. Retime ignition.

Caution: Wipe new contact point set completely free of any oil with which they may have been coated for protection.

d. CAPACITOR (CONDENSER).

(1) *Removal.* Remove distributor cap and rotor. Disconnect capacitor (condenser) cable from primary (low-tension) terminal (figs. 57 and 58). Remove screw and lock washer which holds capacitor (condenser) bracket to breaker plate, and lift out capacitor (condenser).

Note. Ignition capacitors (condensers) are serviced by replacement only.

(2) *Installation.* Position capacitor (condenser) and bracket on breaker plate and attach them securely with screw and lock washer. Connect capacitor (condenser) cable to primary (low-tension) terminal. Install rotor and distributor cap.

103. Coil

a. REMOVAL. Open engine compartment hinged door. Disconnect the primary and secondary conduits from top of coil by unscrewing the coupling nuts on the shielded cable (fig. 60). Disconnect the cable

Figure 60. Ignition coil.

from the resistor at the bottom of the coil. Remove the two cap screws and lock washers which attach coil to mounting bracket and remove coil.

Note. Ignition coils are serviced only by replacement of the complete unit.

b. INSTALLATION. Position coil on mounting bracket and attach securely with two cap screws and lock washers. Connect primary conduit (low-tension) to minus (−) tower and secondary conduit (high-tension) to center tower on top of coil. Tighten knurled coupling nuts only fingertight, plus one-half turn. Connect cable to resistor terminal at bottom of coil (fig. 60).

Caution: Pliers must not be used as too much pressure may be be applied causing coupling nuts and seals at end of conduits to be damaged.

104. Ignition Timing

a. METHOD. The ignition is timed with the aid of a timing light (fig. 61).

Note. Any necessary adjustments to distributor points must be made before timing ignition.

b. PROCEDURE.
(1) Warm up engine until a smooth idle speed of 450–475 rpm can be maintained.
(2) Stop engine. Open engine compartment hinged door and air outlet grille.
(3) Disconnect ignition conduit at No. 6 spark plug by loosening coupling nut at plug. Connect spark plug cable from timing light to the spring terminal at end of the insulator on the spark plug conduit. Connect the timing light cable to the resistor terminal at the bottom of the coil and to any good ground point such as the exhaust manifold stud (fig. 61).
Caution: Do not connect a 12-volt timing light to resistor terminal at bottom of coil. If a 12-volt light is to be used, it must be connected to one battery only, to prevent burning out the light due to high voltage.
(4) Start engine and run at a smooth idle speed at 450–475 rpm. Holding handle of timing light, aim light at pointer on engine front cover. Light should be aimed and reading should be taken as directly over timing indicator as accessibility will permit. The timing should be set so that indicator pointer on front cover lines up with "IGA" mark on the crankshaft pulley when the light flashes.
(5) Adjust timing by loosening clamp screw on distributor control arm and turning distributor until "IGA" mark on crank-

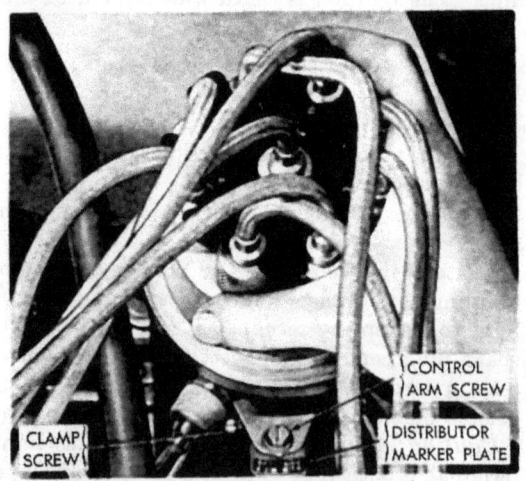

Figure 61. Adjusting ignition timing.

shaft pulley lines up with pointer on front cover. After setting has been obtained, tighten clamp screw securely (fig. 61).

(6) Disconnect timing light and connect spark plug conduit. Tighten spark plug conduit coupling nut fingertight, plus approximately one-half turn, using a small wrench.

Note. Do not force coupling nut.

(7) If detonation is heard with above setting, it indicates that the fuel used is of less than 80 octane value and the spark should be retarded to the barely audible detonation point. It is preferable, however, to obtain correct grade of gasoline.

(8) The timing procedure described in (1) through (7) above applies to either right hand or left hand engines. Both engines should be checked while timing facilities are available. If a synchroscope or timing light is not available, the timing can be set in the following manner: Connect a 24-volt lamp across the distributor breaker points. Crank the engine by hand using a socket wrench on the crankshaft pulley screw until the "IGA" mark on the pulley lines up with the pointer on the front cover. Rotate the distributor until the light just goes on. Tighten clamp screw on distributor arm. This method should not be used except when other facilities are not available.

(9) Close air outlet grille and engine compartment hinged door.

105. Wiring

a. REMOVAL OF COMPLETE IGNITION HARNESS. Open engine compartment hinged door. Disconnect spark plug ignition conduits by unscrewing coupling nut at upper portion of spark plug. Disconnect conduits at distributor cap by unscrewing knurled coupling nut at each conduit. Remove screws at straps on the conduit supports, loosen straps, and remove harness.

b. INSTALLATION OF COMPLETE IGNITION HARNESS. Connect each of the spark plug conduits, in turn, to the distributor cap, working from the wire markers and following the order indicated (fig. 62).

Note. There is a numbered disk, placed on top of the distributor cap, through which cables pass.

No. 1 terminal faces grease cup. Install conduits on supports straps and connect to spark plugs. Tighten spark plug conduit coupling nuts fingertight, plus one-half turn. Do not force coupling nuts. Tighten screws at conduit support straps and connect primary and secondary cables between distributor and coil.

c. INDIVIDUAL CABLES. Replacement of individual cables can be performed by using the procedure in *a* and *b* above as guide.

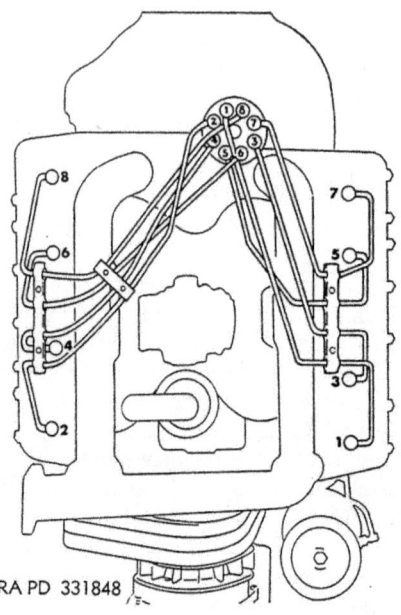

Figure 62. Engine cylinder numbering.

106. Spark Plugs

a. REMOVAL. Open engine compartment hinged door. Disconnect the ignition conduits at the spark plugs by unscrewing the coupling nut at upper portion of spark plug. Using a deep socket wrench, loosen spark plugs by turning counterclockwise and then remove plugs and gaskets.

b. CHECKING GAP. Check spark plug gap with a gage which has a round feeler for aircraft type spark plugs; or a flat, ribbon-type feeler gage for automotive type spark plugs. The correct gap is 0.028–0.033 inch (0.030 is ideal). Gap adjustment must be made by bending side electrode only. For complete information on checking and cleaning spark plugs, refer to TB ORD 313.

c. INSTALLATION. Place new gasket on spark plug and insert plug in engine. Tighten plug to a torque tightness of 7–10 foot-pounds torque. Refer to paragraph 100*d* for spark plug specifications. Check the condition of the insulator at end of each spark plug conduit. Insulator must not be cracked and spring terminal at end of insulator must not be damaged or missing. Connect the correct conduit to each spark plug (fig. 62) and tighten fingertight, plus one-half turn. Do not force these connections.

Section VIII. STARTING SYSTEM

107. Description and Data

a. DESCRIPTION.

(1) There are two independent, but identical starting systems; one for each engine. Each system includes a starter mounted on the engine; a solenoid for pinion engagement mounted on the starter housing; a starter relay mounted on the front of the bulkhead; a feed circuit through the ignition switch; a starter switch on the instrument panel; a safety circuit comprising a ground through the transmission warning signal circuit; and the necessary connecting wiring (fig. 63).

(2) To turn an engine over, the ignition switch, master battery switch, and emergency ignition switch must be turned on and the starter switch button depressed. Current then flows through the two coils of the starter relay and grounds through the transmission warning signal unit. Energizing the relay coils closes two circuits, one of which leads directly to the solenoid on the starter, while the other provides an independent ground circuit to function in case cranking action opens the transmission warning signal switch. With current flowing through the solenoid, the starter pinion is drawn into engagement with the flywheel ring gear and the circuit for the heavy starting current is closed.

(3) If the starter switch button should be depressed while the engine is running, no current will flow through the starter relay because the oil pressure built up in the rotating transmission will keep the warning signal switch open and thereby leave the relay with no ground connection. Whenever the engine stops and transmission oil pressure drops, the switch closes making the ground connection. The circuit can then be completed by pressing the starter switch button.

b. DATA.

Starter:
```
Make and model number_____ Delco-Remy 1108568
Type_____ four-pole, compound
Voltage _____ 24
```

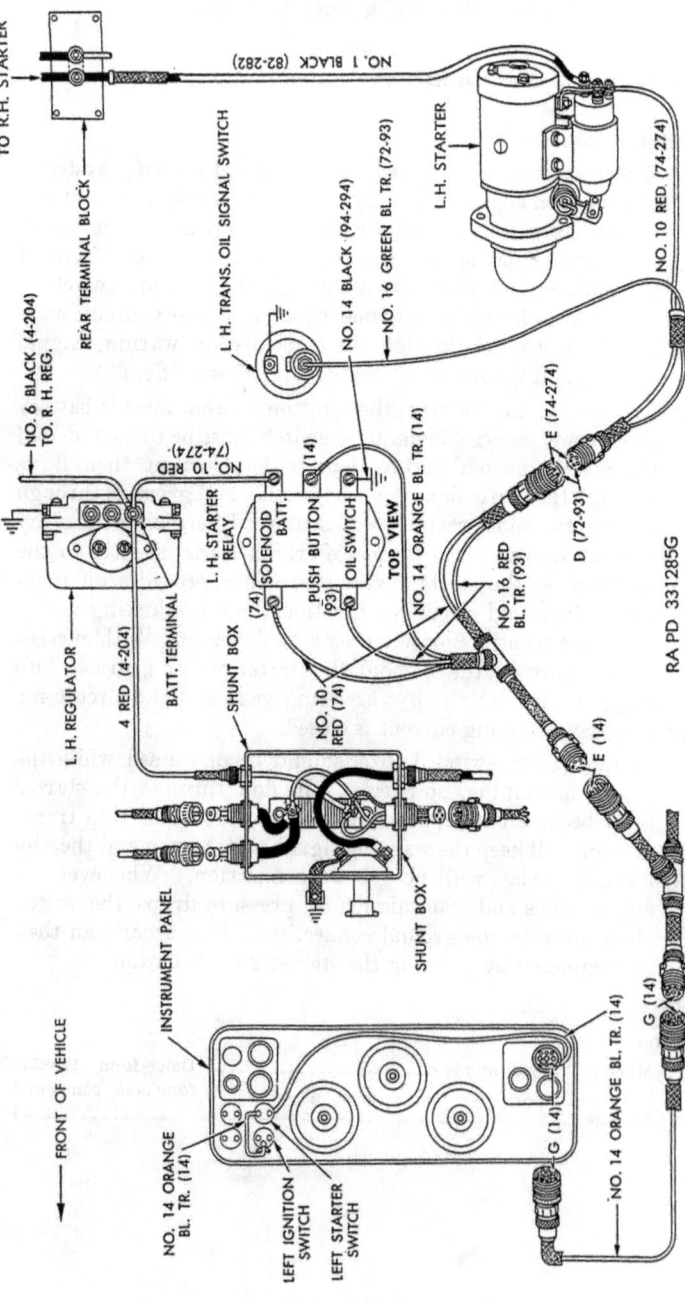

Figure 63. Starting circuit wiring diagram.

108. Starter

a. REMOVAL.
  (1) *Open master battery switch.* Open master battery switch by pulling out on knob and turn about one-quarter turn to lock in open position.
  (2) *Remove left hull floor plate.* Remove screws from left hull floor plate. Support rear end of plate while removing last screws to prevent plate from dropping down and causing injury. Front of plate is held up by hooks. Slide plate toward rear of vehicle and lower to ground.

  Note. The starters of both engines are accessible through the left floor opening.

  (3) *Disconnect battery cable.* Remove nut from battery cable and lift cable off terminal. Disconnect cable from starter solenoid by removing nut and lifting off terminal.
  (4) *Remove starter.* First remove upper mounting screw (fig. 64) which passes through flywheel bell housing; then remove bottom screw while supporting weight of starter with other hand. Remove starter through hull floor opening.

b. INSTALLATION.
  (1) *Install starter.* Reaching up through left hull opening, place starter in position in opening on flywheel bell housing so that dowel on housing fits into cut-out on starter flange. While holding starter in position with one hand, install lower

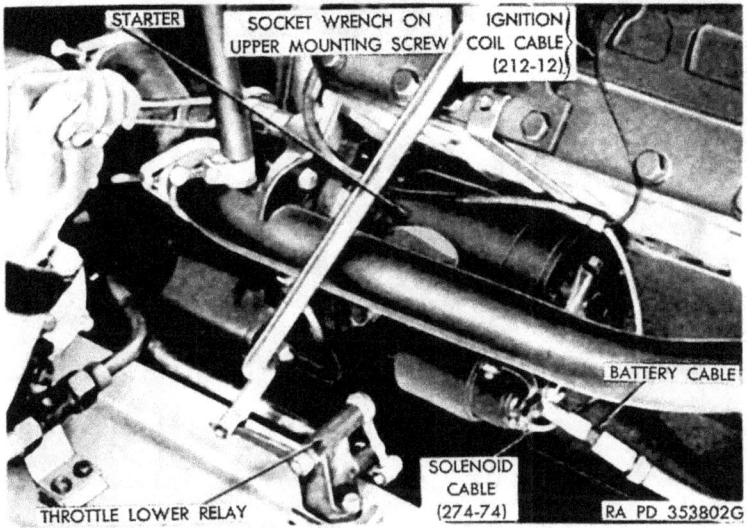

Figure 64. Removing starter.

mounting screw through flywheel bell housing and tighten screw. Install and tighten other screw.

(2) *Connect solenoid switch and battery to starter.* Connect solenoid switch cable and battery cable to starter and install retaining nuts and washers.

(3) *Install left hull floor plate.* Place gasket in position on rear hull plate. Raise front end of plate and place hooks on plate over edge of hull floor. Raise rear end of plate, aline screw holes with drift punch, and install attaching screws and washers.

109. Cranking Motor Solenoid

a. REMOVAL. Remove cranking motor solenoid only after tests (par. 73) indicate that this unit is at fault. Remove starter (par. 108). Remove cotter pin and lever pin holding solenoid plunger link to starter clutch actuating lever. Remove four mounting bolts holding solenoid to starter. Remove bolt holding ground strap to starter. Remove cables. Lift solenoid off starter assembly.

b. INSTALLATION. Place cranking motor solenoid in position on starter and install four bolts. Place clutch operating lever pin in position and lock with new cotter pin. Install cables on solenoid.

110. Relay

a. REMOVAL. Remove relay only after tests (par. 73) indicate that it is at fault. Open master battery switch. Remove relay cover. Dis-

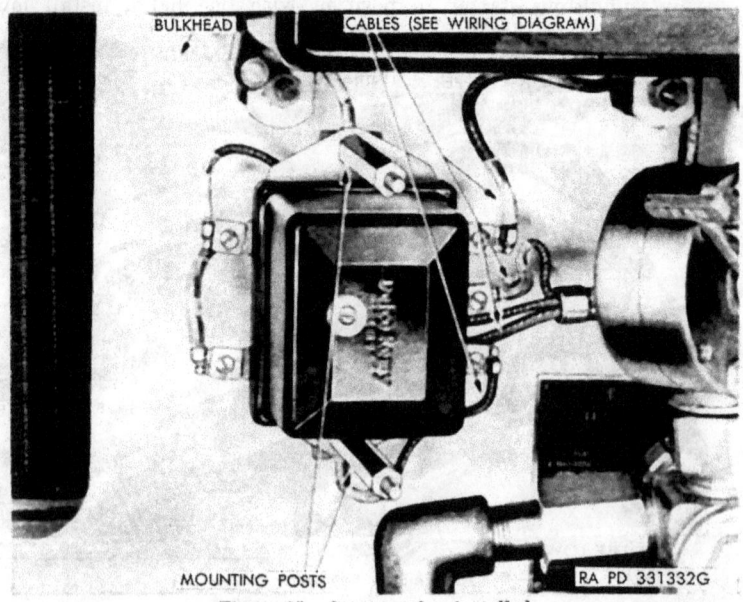

Figure 65. Starter relay installed.

connect five cables from relay terminals. Remove two hex mounting posts and remove relay assembly from mounting bracket (fig. 65).

b. INSTALLATION. Position relay on bulkhead and install hex mounting posts, with ground inserted under upper post. Connect five cables to relay according to diagram (fig. 63). Install relay cover.

Section IX. FUEL, AIR INTAKE, AND EXHAUST SYSTEMS

111. Description and Data

a. DESCRIPTION.
  (1) *Fuel system.*
   (*a*) Two fuel tanks of approximately 55-gallon capacity each are carried in narrow, deep compartments on each side of the engine compartment.
   (*b*) A screen and a filter are incorporated in the fuel pump support assembly in the fuel tanks. The screen extends from the top of the fuel tank filler opening to the top of the fuel pump cage (fig. 75). This strains the fuel as it is poured into the tank and also acts as a flame arrester. The filter surrounds the fuel pump mounting cage at the bottom of the tank. This strains the fuel in the tank as it passes to the pump and is self-cleaning due to sloshing action of the fuel. The fuel pump is mounted within the fuel pump mounting cage.
   (*c*) A fuel line extends from each fuel pump through the top of the fuel pump support assembly, to a check valve and then to the shut-off valve. From these shut-off valves, a connecting line is carried across the support at the front of the engine compartment. This line contains two tee connections, from which a rubber tube extends to the auxiliary fuel filters at the carburetor float chambers (fig. 66). This makes it possible for either tank to supply fuel to both engines.

   Note. Early production model vehicles did not have the auxiliary fuel filter, but have since been modified to include this filter.

   (*d*) The flow of fuel and selection of the tank is controlled by a pair of levers and switches at the front center of the driving compartment roof (fig. 7). From each of these levers, a control cable extends back to the shut-off valve. The cable for the pump extends from the switch to the terminal on top of the fuel pump support assembly, then to the pump in insulated conduit (fig. 67).

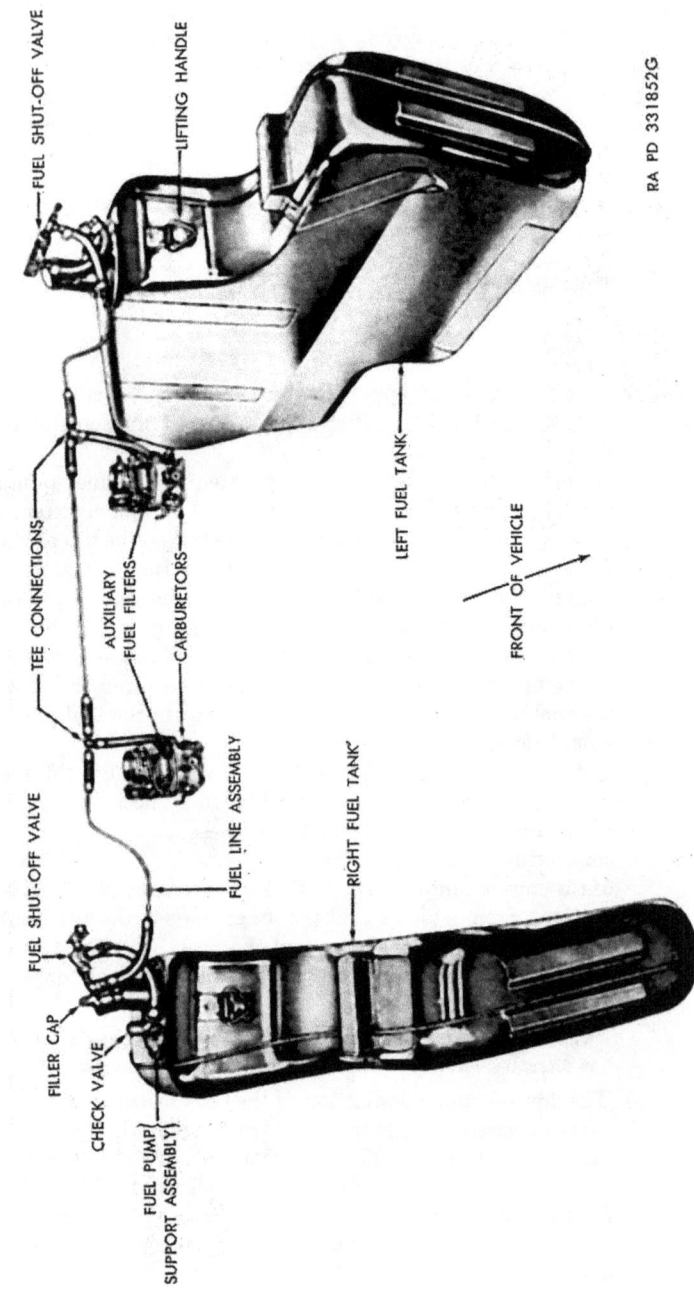

Figure 66. Fuel system.

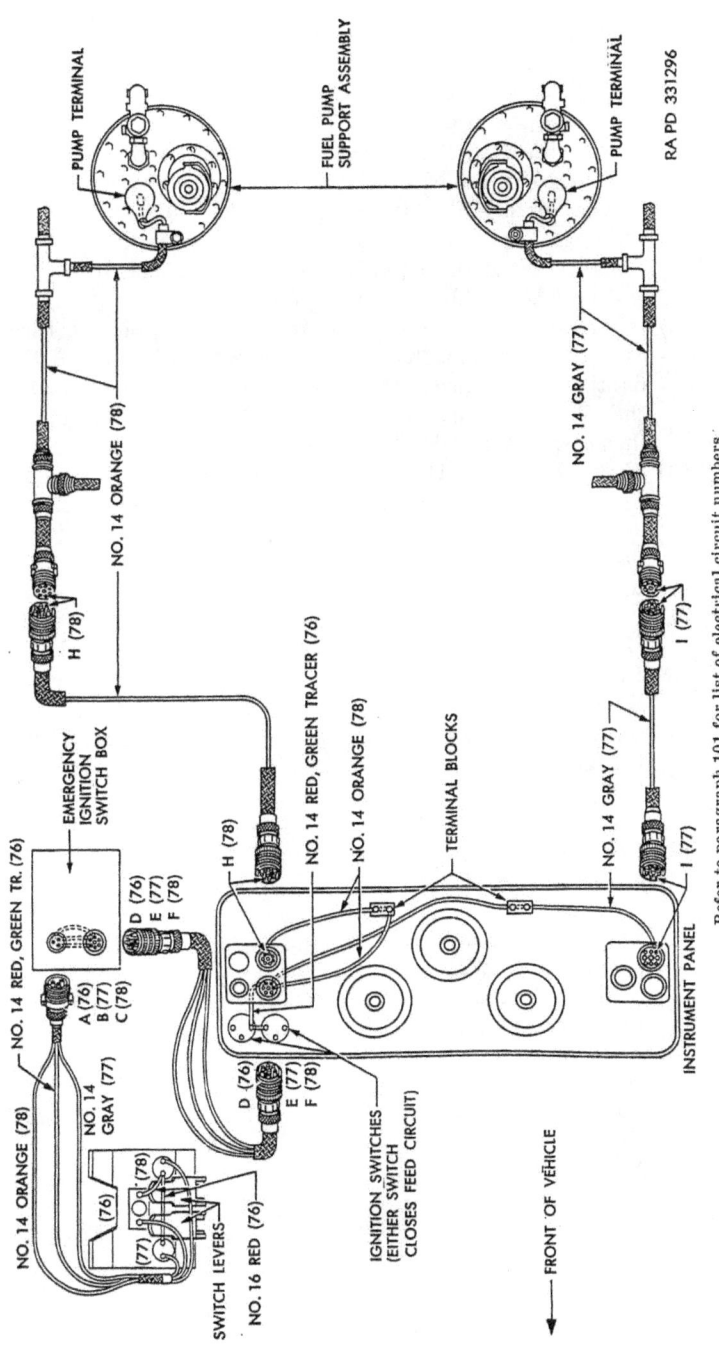

Figure 67. Fuel pump circuit wiring diagram.

(e) The carburetors are of dual down-draft design, internally vented, and fitted with dustproofing seals. Each of the dual barrels supplies the fuel mixture to four cylinders through the intake manifolds. Each carburetor is equipped with a small disk-type auxiliary fuel filter, which is mounted in a horizontal position on the carburetor float chamber fuel inlet fitting (fig. 66). This filter will retain the small particles of dirt which may pass through the main filter at the bottom of the fuel tank. The filter has a bowl removable for cleaning (fig. 73).

(2) *Air cleaners.* Two oil bath type air cleaners, one for each carburetor, are mounted on the bulkhead at the rear of the fighting compartment (fig. 77). Incoming air passes over the oil and through the filtering element, or screen, and is then carried through tubes to the carburetors.

(3) *Exhaust system.* The exhaust manifolds at the engine, manifold connection, muffler, and exhaust pipes carry off the exhaust gases and are known collectively as the exhaust system.

b. DATA.

(1) *Carburetor.*
Make _____ Carter
Type _____ WCD dual down-draft
Number _____ 2 (1 per engine)
Gas line connections _____ $\frac{3}{8}$-in. hose nipple
Idle adjustment _____ screw-type
Flange size _____ $1\frac{1}{4}$ in.
Choke type _____ automatic climatic control

(2) *Fuel pumps.*
Make _____ Carter
Location _____ in cage at bottom of fuel tank
Type _____ centrifugal, electrically driven
Number _____ 2 (1 in each tank)
Control _____ switch and manual shut-off valve
Pressure _____ 6 lb

(3) *Fuel strainers.*
Make _____ Ripley
Location _____ filler support, pump cage
Type _____ self-cleaning screens
Number _____ 4

(4) *Auxiliary fuel filter.*
Make _____ Zenith
Location _____ carburetor inlet
Type _____ disk-type
Number _____ 2

(5) *Air cleaners.*

Make	Donaldson or AC
Location	front of bulkhead
Type	oil bath
Number	2
Filter elements	steel mesh
Reservoir capacity	3 qt

112. Carburetors

a. CARBURETOR ADJUSTMENTS. Test choke shaft with fingers to see that choke operates freely (fig. 68). Run engines until thoroughly warm (160° to 210° F.) and adjust throttle stop screws (fig. 69) so engine speed is from 450–475 rpm, with the transmission selector lever in "DRIVE" position and parking brakes set. Tighten two idle adjusting screws (fig. 69) as far as possible without forcing, then loosen screws one and one-quarter turns. Engine should run smoothly without loping or stalling. Adjust screws if necessary. Check to see that choke setting mark on automatic choke housing is opposite mark on carburetor flange. If necessary, loosen three choke housing mounting screws and rotate housing until marks line up (fig. 69). Tighten screws.

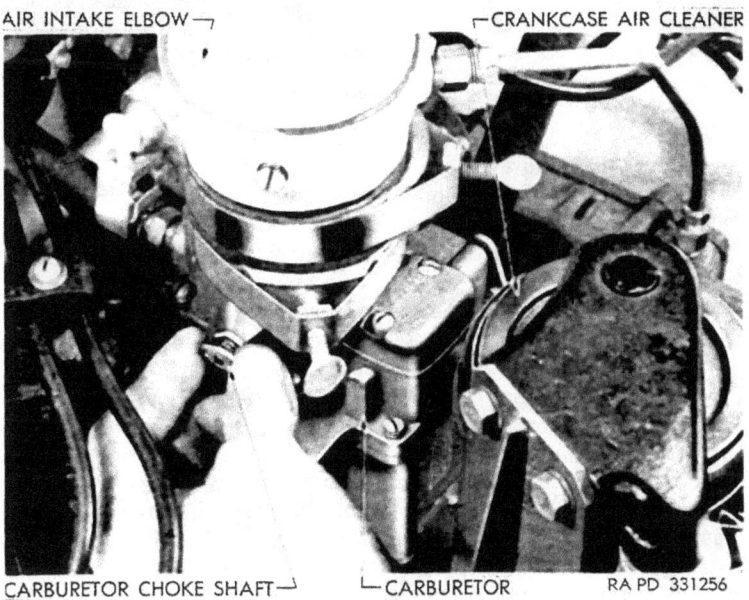

Figure 68. Testing carburetor choke shaft.

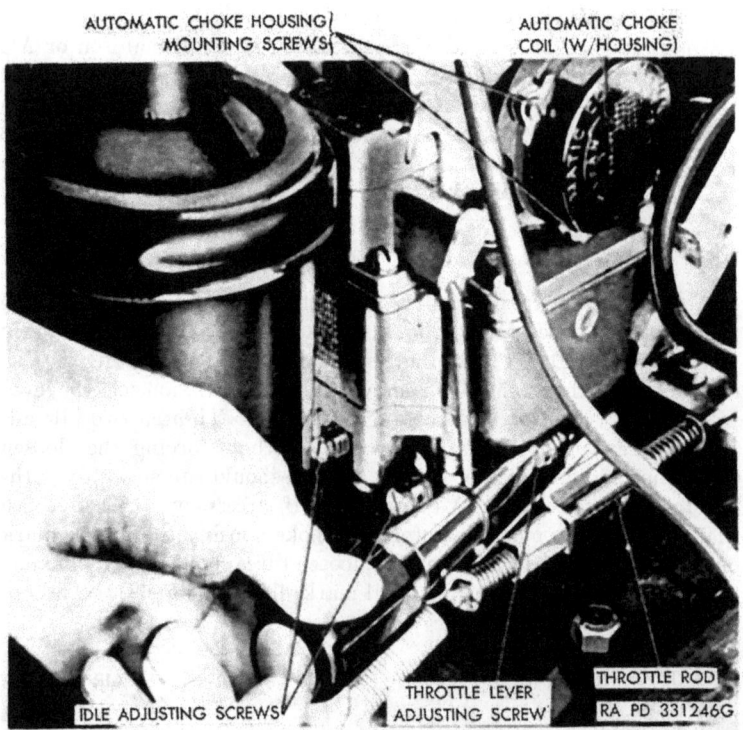

Figure 69. Adjusting carburetor.

  b. REMOVAL. Disconnect choke heater line at carburetor and at manifold and remove line. Disconnect distributor ventilating line at carburetor air intake elbow (late model engines). Loosen hose clamp at carburetor air intake elbow (fig. 46) and lift elbow off carburetor. Make sure both fuel tank control levers (in driver's compartment) are in "OFF" positions. Loosen hose clamp at carburetor auxiliary fuel filter and slide hose off connection. Loosen hose clamp on crankcase-to-air-cleaner hose at rear of carburetor and slide hose off nipple. Unhook throttle retracting spring from throttle rod, and slide rod out of carburetor throttle shaft lever. Remove four nuts which hold carburetor to intake manifold and remove carburetor and upper gasket (fig. 70). Leave insulator and lower gasket on manifold.

  c. INSTALLATION. Install a new gasket over insulator on intake manifold (fig. 70). Place carburetor over mounting studs on intake manifold, and install four retaining nuts. Slide rubber fuel hose over connection on carburetor auxiliary filter, and tighten hose clamp. Connect crankcase-to-air-cleaner hose to fitting at rear of carburetor, and tighten hose clamp. Connect choke heater line to carburetor and

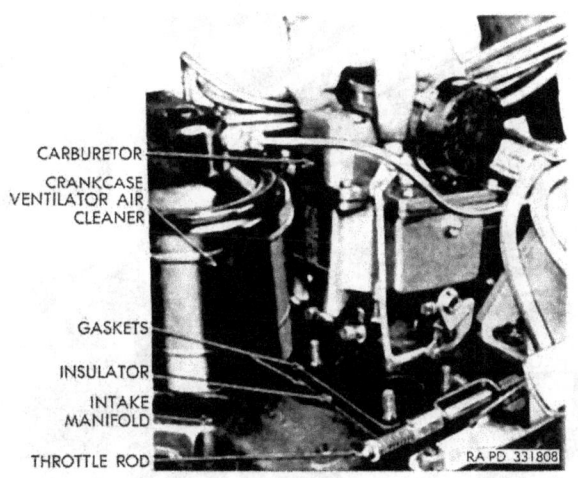

Figure 70. Removing carburetor.

to exhaust manifold by screwing nut on line into fitting on carburetor and manifold. Tighten securely. Slide air intake elbow over top of carburetor, and tighten hose clamps securely. Connect distributor ventilating line (fig. 57) to carburetor intake elbow, and tighten securely (late model engines). Slide trunnion on throttle rod through throttle shaft lever, hook retracting spring to hole in trunnion, and connect spring to bracket on manifold front clamp bolt.

    d. THROTTLE LINKAGE ADJUSTMENT (fig. 72).

        (1) Disconnect front throttle rod and lock throttle in zero position by inserting a ¼-inch drill rod pin through holes in front relay lever and front relay shaft support (fig. 71). Loosen clevis lock nut for front throttle rod and turn clevis until holes in clevis and front throttle relay shaft line up. Install clevis pin and tighten lock nut. Adjust throttle rods between front and rear intermediate throttle relay levers in the same manner. Adjust rod between rear intermediate relay lever and rear relay lever in the same manner. Connect vertical throttle rod from rear lower throttle relay to rear throttle cross shaft and lock rear throttle cross shaft in zero throttle position by inserting a pin, through relay lever and bracket on right-hand engine. Adjust clevis at upper end of rod until clevis holes line up with hole in lever and then insert clevis pins. Set idle adjusting screws (fig. 69) so that engine speed is from 450–475 rpm as noted on the tachometers when the transmission selector lever is in "DRIVE" position, transfer unit shift control lever is in "HIGH" or "LOW" position, and parking brakes are set. Disconnect each

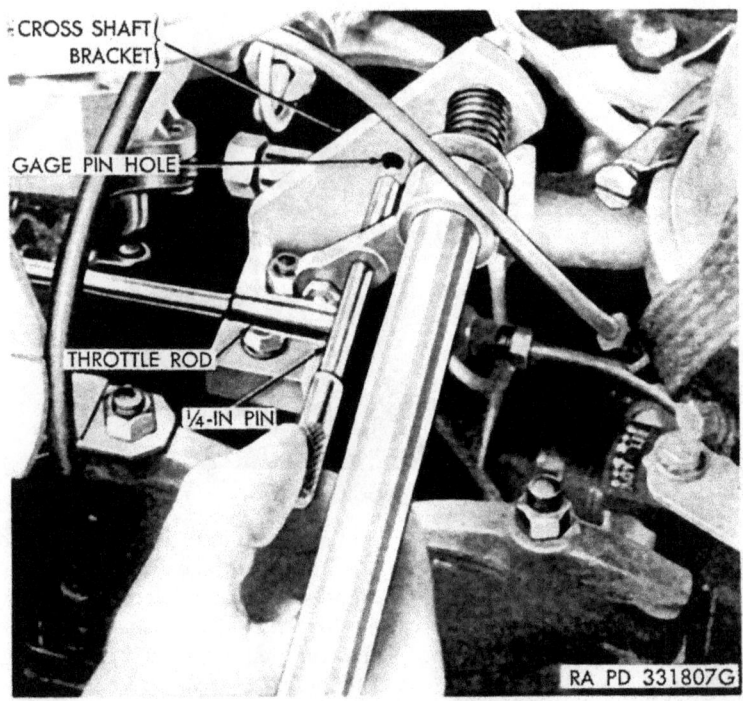

Figure 71. Checking throttle synchronization.

throttle rod at carburetor throttle levers and turn adjusting nut at rear end of each rod until trunnion pin lines up with hole in carburetor throttle lever while lever is held against the idle adjusting screw. Connect throttle rods and remove all ¼-inch drill rods from alining holes. Pull out hand throttle until speed of both engines is approximately 1,000 rpm. Speed of the two engines should not differ more than 50 rpm at idling speed or 100 rpm at 1,000 rpm.

Caution: Do not pull engine against fluid coupling for more than 1 minute at a time and allow at least 3 minutes for the oil to cool before repeating.

Pull up on rear throttle cross shaft to rear lower throttle relay rod as far as possible to see that carburetor throttle valves are wide open. Adjust if necessary.

(2) Working in driving compartment, push down on accelerator pedal (with engines not running) until throttles are in wide open position. Measure distance between accelerator pedal and front hull plate. This distance should be one-eighth of an inch with throttle linkage in full open position. If clearance is incorrect, disconnect front throttle rod (fig. 72), and

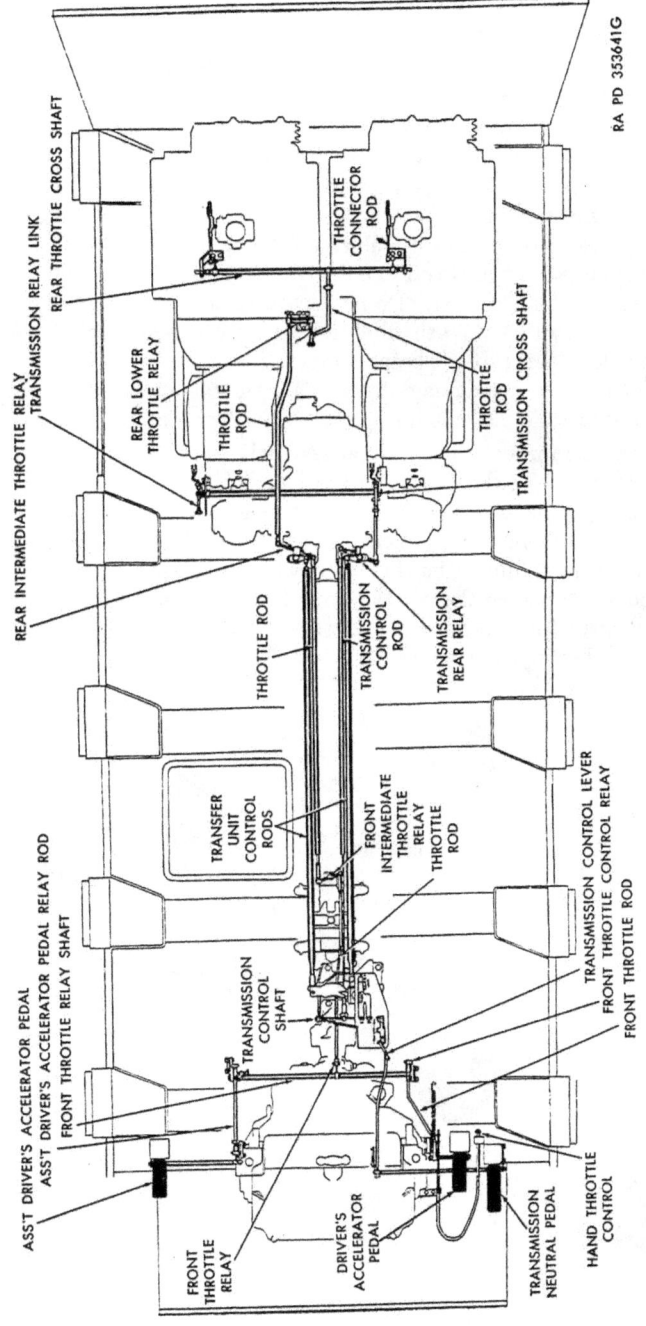

Figure 72. Throttle, transmission, and transfer controls.

turn clevis until one-eighth of an inch clearance is secured. Connect rod and tighten lock nut. Engage and adjust assistant driver's accelerator pedal in same manner. Adjust hand throttle control by turning adjusting nuts on end of hand throttle control cable until nuts are one-eighth of an inch from trunnion, with throttle control linkage in closed position. Be sure of freedom between trunnion and hand throttle control cable extension. Bend mounting bracket, if necessary, to line up cable with trunnion.

e. SERVICING AUXILIARY FUEL FILTER. Loosen bail nut and remove bowl from filter (fig. 73). Do not try to remove gasket as it may be cemented in place. Wash filter element with dry-cleaning solvent or volatile mineral spirits paint thinner.

Caution: Do not damage the disks. Do not scrape, scrub, or use compressed air. Do not disassemble the filter element.

If element cannot be cleaned satisfactorily, replace entire element. Wash sediment bowl before installing. Install sediment bowl on filter, making sure that gasket is in place and tighten bail nut. Nut should be tightened fingertight. If bail nut is overtightened, the gasket may be cut by the sediment bowl or the bail stretched to the point where a good seal between the bowl and the filter body cannot be obtained. In the event the bail has been stretched, it may be kinked slightly to allow the bail nut to contact the bowl. Check for leaks when engines are running.

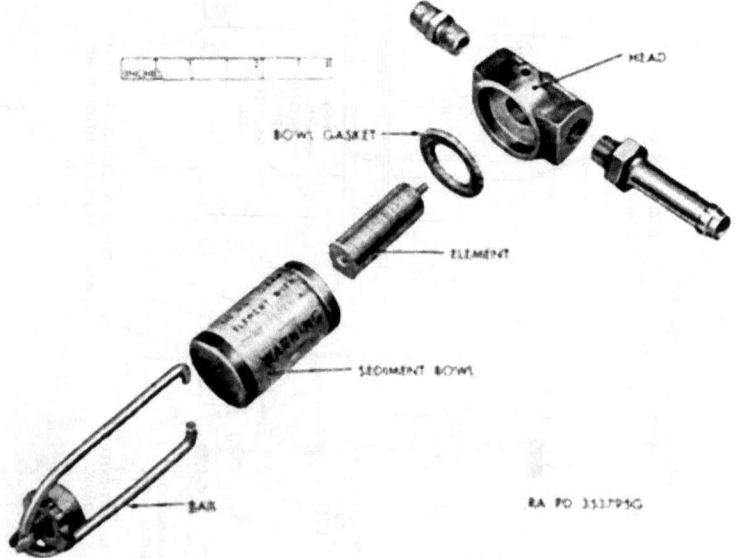

Figure 73. Auxiliary fuel filter—exploded view.

Figure 74. Fuel tank and batteries installed.

113. Fuel Pumps

a. REMOVAL. The fuel pumps are not removed as a separate item, but are removed along with the whole fuel pump support assembly (fig. 75). Fuel pumps normally require no attention in service. To replace an inoperative pump, turn master battery switch to the "OFF" position. Remove battery cover plate and fuel tank cover plate by removing the attaching screws and lift off plates, the fuel pumps are now accessible (fig. 74). Remove screw which holds pump cable to terminal and tape end of cable. Loosen hose clamp at fuel pump outlet elbow and slide hose off elbow. Remove 12 attaching screws and remove complete support assembly (fig. 75).

b. INSTALLATION. Position a new gasket on fuel tank, and slide support assembly into fuel tank and into retainer in bottom of tank. Install 12 attaching screws, mount ground strap under one screw, and tighten all screws securely. Connect fuel outlet hose to elbow and tighten hose clamp (fig. 74). Remove tape from end of cable and install it on terminal. Place rubber guard over terminal. Position fuel tank cover over opening and install attaching screws.

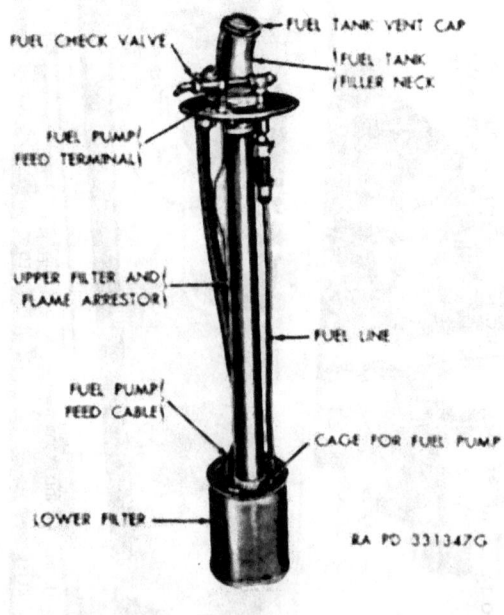

Figure 75. Fuel pump support assembly.

114. Fuel Tanks and Lines

a. DRAINING. A drain hole cover is secured to the hull floor directly under each fuel tank (fig. 34). To drain the tanks, remove the four cap screws holding the plate in position and remove the drain plug at the bottom of the tank.

b. REMOVAL (fig. 74). Remove covers over fuel tanks and batteries by removing the attaching screws and lifting off covers. Turn master battery switch to the "OFF" position. Disconnect fuel pump cable at terminal and tape cable. Disconnect terminals at batteries. Remove four battery hold-down nuts. Remove hold-downs and lift out batteries. Remove battery box retaining screws and lift out battery box. Remove screw holding conduit clamps to fuel tank cover angle brackets and lay conduit to one side. Loosen hose clamp at fuel pump assembly outlet and slide hose off connection. Remove two screws holding fuel shut-off valve to side of hull; disconnect fuel line at shut-off valve from side of hull; disconnect fuel line at shut-off valve by loosening hose clamp and slide hose off connection. Lay shut-off valve to one side. Remove screws holding compartment cover angle brackets to hull and remove brackets. Remove screws holding flat spring spacers to insulating pad on inner wall of compartment. Attach a rope sling through lifting handles on fuel tank and remove fuel tank from compartment.

c. INSTALLATION OF TANKS AND BATTERIES (fig. 74). Lower fuel tank into compartment. Install three flat spring spacers between fuel tank and inner compartment, sliding into groove provided. Install retaining screws. Position fuel tank compartment angle brackets around opening, making sure that the seals are in proper position, and install attaching screws. Lower battery box into position and install battery box mounting screws. Place batteries in battery box and install hold-downs. Connect batteries in series so that the positive terminal (right batteries) is toward rear of vehicle (fig. 90). Negative terminal on left batteries should be toward rear of vehicle. Connect terminals to batteries. Place shut-off valve against inner compartment wall and install two attaching screws. Connect fuel lines to shut-off valve and nipple on fuel pump support assembly and tighten hose clamp. Install three screws and clamps holding conduit to inner fuel compartment angle bracket. Connect ground strap to top of fuel pump support assembly and connect cable to terminal. Install rubber guard over terminal. Position fuel tank cover and battery cover over openings and install attaching screws.

d. INSTALLATION OF LINES.

   (1) The main fuel line extends from one fuel tank to the other and is carried across the support at the front of the engine compartment (fig. 46). This line contains two tee connec-

tions from which a rubber line extends to the carburetor auxiliary fuel filters. The fuel lines are connected to the tees by short hose connections. These connections are of a synthetic rubber composition and should not be replaced with ordinary rubber hose.

(2) When removing the fuel lines, be sure fuel control levers are in the "OFF" position. Remove fuel tank cover. Disconnect the fuel line at the shut-off valve at the side of the engine compartment by loosening hose clamps and sliding hose off line. Disconnect fuel hose at each carburetor auxiliary fuel filter by loosening hose clamp and sliding hose off nipple. Remove four bolts and clamps holding main line to the support bar and lift the fuel line as an assembly out of vehicle.

(3) To install the fuel line, position the assembly against the support bar and install the four bolts and clamps holding main line to bar. Slide fuel hose over nipple at filter and tighten hose clamp. Connect fuel line at shut-off valve and tighten hose clamp. Start both engines and run for a sufficient length of time to check all connections for leaks. Install fuel tank cover. Close engine compartment.

115. Air Cleaners

a. SERVICING. Loosen the two wing nuts which hold the air cleaner oil reservoir to the air cleaner. While supporting the reservoir from bottom, swing out the retaining bolts and lower the reservoir. Remove screen from Donaldson type air cleaner (fig. 76) by turning screen counterclockwise, then pulling screen down out of air cleaner. Remove element from AC type air cleaners (fig. 77) by pulling it downward. Pour out old oil and clean reservoir and screen with dry-cleaning solvent or volatile mineral spirits paint thinner. Dry all units thoroughly with compressed air if available, fill reservoir to level indicated, using approximately 3 quarts of engine oil (seasonal grade) (par. 62), and install screen or element and reservoir.

b. REMOVAL. The air cleaners are mounted to a bracket on the fighting compartment side of the bulkhead by means of two metal clamping straps that encircle the unit (fig. 77). Remove five screws holding battery compartment cover to hull roof and remove cover. Working in battery compartment, loosen hose clamp holding carburetor air intake hose to outlet connection on air cleaner. Working in fighting compartment, loosen hose clamp holding air cleaner inlet pipe to air cleaner. Remove screw holding each clamping strap around air cleaner and lift complete assembly from its mounting.

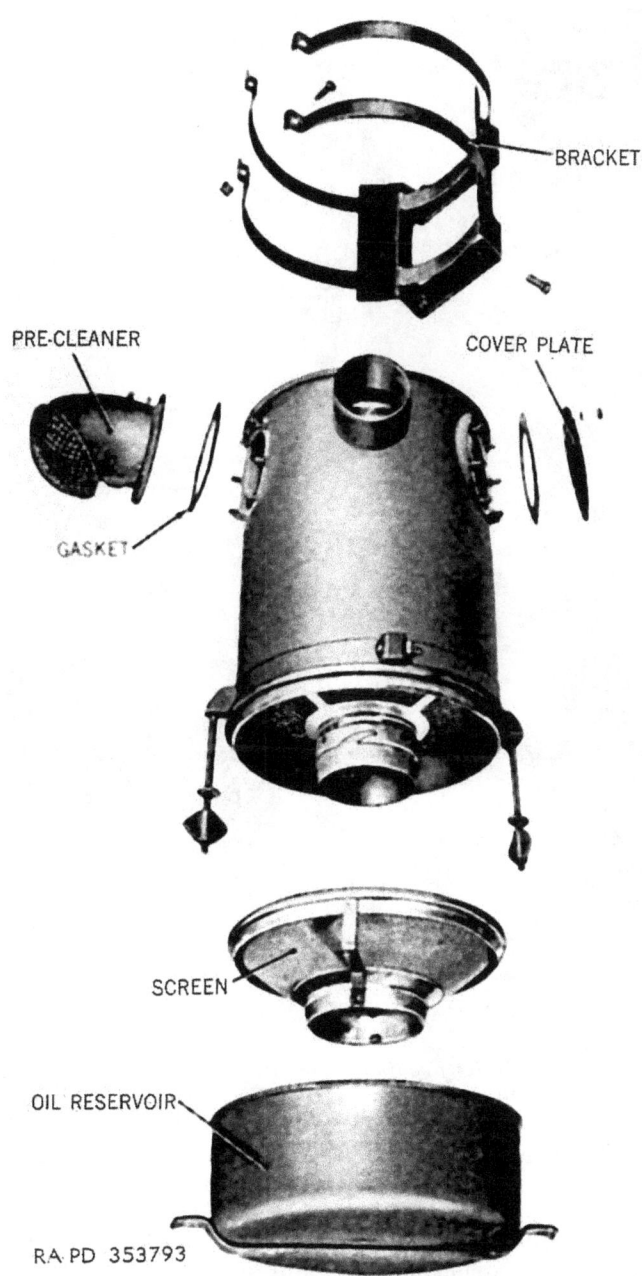

Figure 76. Air cleaner (Donaldson)—exploded view.

Figure 77. Air cleaner (AC)—servicing cleaner.

c. INSTALLATION.
  (1) *Check inlet opening.* The air cleaners are interchangeable as rights and lefts by interchanging the plate, retained by eight screws, with the precleaner inlet pipe on the cleaner. The outlet opening on the air cleaner is not removable, but is part of the cleaner itself. The outlet opening must always be connected to the carburetor air intake pipe.
    Caution: If the air cleaner is installed incorrectly or filled above level markings, oil will be drawn out of the air cleaner and into engine, or partially restrict air flow to carburetor.

(2) *Install cleaner.* Position two metal straps around air cleaner and tighten screws. Coat the inside of the carburetor air intake hose connection with gasket cement and slide outlet on air cleaner into hose. Tighten hose clamp securely. Connect air cleaner inlet pipe to cleaner and tighten hose clamp. Position cover over battery compartment and install the five retaining screws.

116. Exhaust Pipes and Mufflers

a. REMOVAL.
  (1) *Remove air outlet grille.* Open engine compartment and remove air outlet grille (par. 94).
  (2) *Remove muffler assembly.* Remove two bolts and nuts holding muffler and exhaust pipe assembly to exhaust manifold connector pipe (fig. 46). Remove muffler and discard gasket.

b. INSTALLATION.
  (1) *Install muffler.* Lower muffler and exhaust pipe assembly into position in vehicle. Place a new exhaust pipe gasket between the exhaust connection flange and muffler assembly. Install bolts and nuts holding muffler assembly to exhaust connection flange.
  (2) *Install radiator air outlet grille.*

>   *Note.* Be sure exhaust outlet pipes are in center of air outlet grille openings. Reposition muffler if necessary.

Close engine compartment door.

Section X. COOLING SYSTEM

117. Description and Data

a. DESCRIPTION.
  (1) This vehicle has two identical, but completely independent cooling systems; one for each engine and transmission (fig. 78). Each system contains a radiator, water pump, thermostat and necessary connections, engine fan and fan drive, and an oil cooler for the hydramatic transmission.
  (2) The coolant is drawn from the radiator by the water pump and forced into the engine water jackets. After circulating through both cylinder blocks and cylinder heads, the heated coolant is forced up through hose connections to the top tank of the radiator. A bimetal thermostat, with double poppet valves located in the radiator inlet housing permits free flow when the coolant is hot, but causes cold fluid to recirculate

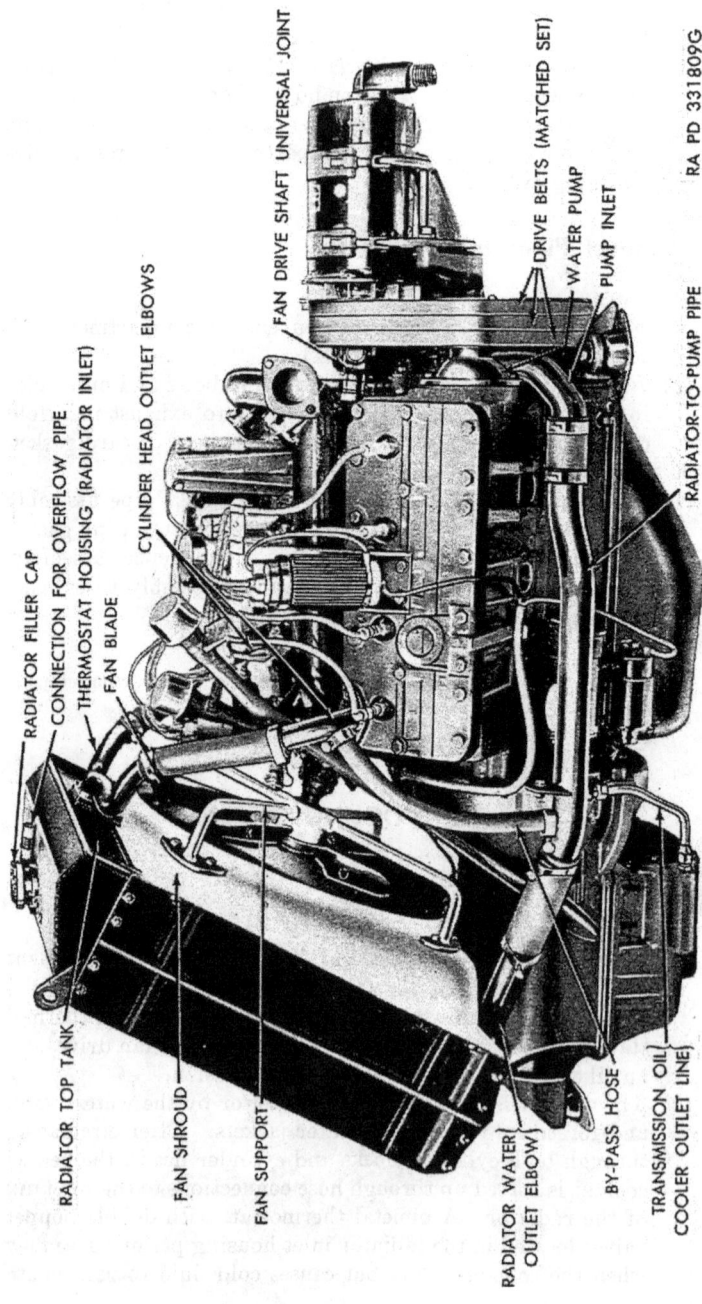

Figure 78. Cooling system.

through a bypass hose back to the water pump and through the engine until the engine reaches an efficient operating temperature.

(3) The coolant is also pumped through an external pipe to an oil cooler located in the transmission oil pan, where it cools the transmission oil, and is then returned to the water pump.

(4) The capacity of each cooling system is 40 quarts.

(5) The engine fan is carried in a mounting spider attached to the fan shroud at rear of radiator. The fan is driven by a drive shaft mounted on ball bearings under the intake manifold and through a universal joint and yoke connected to the generator pulley.

b. DATA.

(1) *Fan.*
   Blades, angle _____ 32 deg
   Blades, diameter _____ 21 in
   Number of blades _____ 4
   Drive _____ belts, drive shaft, and universal joints
   Drive ratio _____ 1.1 to 1
   Make _____ Hayes

(2) *Radiator.*
   Core area (each) _____ 540.5 sq in
   Type _____ tube and fin
   Make _____ Harrison

(3) *Thermostat.*
   Location _____ radiator inlet housing
   Opening temperature _____ 141° to 146° F.
   Type _____ bimetal
   Make _____ Dole

(4) *Water pump.*
   Drive _____ triple belt
   Lubrication _____ fitting
   Packings _____ spring-loaded, chevron type
   Type _____ centrifugal
   Location _____ right front of engine block

118. Maintenance

a. ADDING COOLANT.

(1) *Vent filler cap before removing.*

   Caution: Before removing the radiator filler cap from a hot system, always vent the radiator long enough to allow all pressure in the system to escape; otherwise, there is a possibility of serious personal injury from steam and hot water.

To vent the radiator, turn the cap to the left (counterclockwise) until the first stop is reached (fig. 79). Leave the cap in this position at least ½ minute or long enough to vent the system thoroughly. Then press down the cap to clear the stop and turn further to the left to remove.

(2) *Install cap securely.* After filling with fluid to 2 inches below filler neck, install the radiator filler cap and be sure to turn it as far as possible by hand so that the entire cooling system will be sealed while operating. If this is not done, there may be excessive loss of coolant while operating under heavy service.

Note. Make sure radiator cap gasket is in good condition and in place before installing cap.

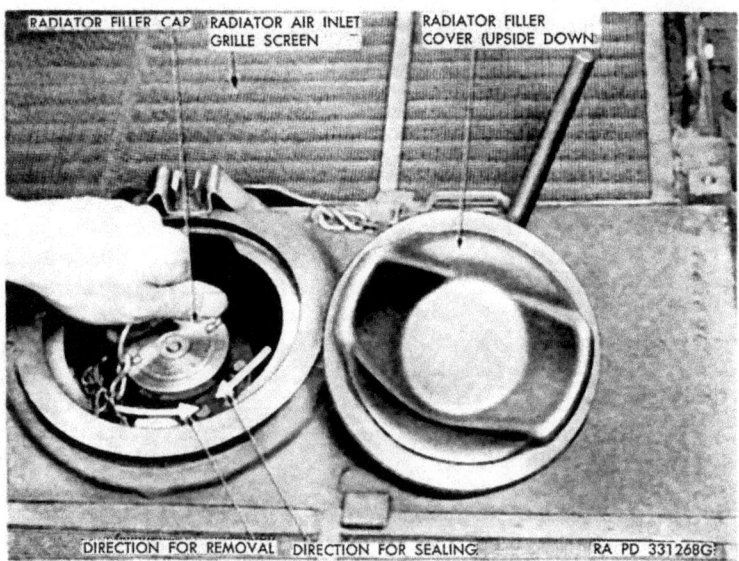

Figure 79. Removing radiator filler cap.

b. DRAINING. Each cooling system is drained at one point only, a plug marked "WATER," in the bottom of the transmission oil pan (fig. 108). Remove radiator cap for rapid and complete draining of the system.

Caution: Open valves or petcocks which control flow of coolant to heaters or other accessories, to facilitate complete draining.

c. CLEANING.

(1) Open the petcocks which shut off the coolant from the heaters or other accessories, if used, to allow for complete circulation

during the cleaning, flushing, and draining. Run the engine, with the air inlet covered, if necessary, until the temperature is within operating range. Stop the engine, remove the radiator filler cap (fig. 79) and drain the system.

(2) Allow the engine to cool. Close the drain plug, pour water slowly into the radiator until the system is approximately half full (20 qt.), then run the engine at idling speed. Add the cleaning compound—51–C–1568–500 in the proportion of one container of cleaner to every 4 gallons of cooling system capacity. Then complete filling the system with water.

Caution: Never mix the water and the cleaning compound before putting them into the system. Do not spill the solution on skin, clothing, or painted portions of the vehicle.

(3) Replace the radiator filler cap and run the engines at fast idling speed, covering the air inlet if necessary, until the coolant reaches a temperature above 180° F. Do not drive the vehicle.

(4) Stop the engine after it has run for 30 minutes at 180° F. Then remove the radiator cap and drain the system completely.

d. NEUTRALIZING.

(1) Allow the engine to cool. Install drain plug, pour water slowly into the radiator until the system is approximately half full (20 qt.), then run the engine at idling speed. Add the neutralizer compound in the proportion of one container of neutralizer to every 4 gallons of cooling system capacity. Each cleaning compound container contains a separate portion of neutralizer. Then fill the system with water.

(2) With the air inlet covered, let the engine idle for at least 5 minutes at the normal operating temperature. Then stop the engine.

(3) Drain the system completely.

e. FLUSHING.

(1) Allow the engine to cool. Install drain plug. Pour water slowly into radiator until system is approximately half full (20 qt.), then run engine at idling speed, and fill system completely.

(2) Run the engine, keeping air inlet covered, if necessary, until the coolant is heated to normal operating temperature.

(3) Drain the system by removing drain plug and radiator cap. Repeat flushing operation until drain water is clear.

(4) Again allow the engine to cool and clean all sediment from the radiator cap valves and overflow pipe. Blow insects and dirt from radiator core air passages with compressed air,

blowing from the rear. Use water, if necessary, to soften obstructions.

f. LEAKS. After completing the flushing operation, make certain that the engine has been allowed to cool again. Install the drain plug. Pour water slowly into the radiator until the system is approximately half full (20 qt.), then run the engine at idling speed and fill the system to 2 inches below bottom of filler neck. Stop the engine. Examine the entire cooling system for leaks. This is important because the cleaning solution uncovers leaks which may have been plugged with rust or scale. Leaks that cannot be corrected by the using organization should be reported immediately to higher authority. After leaks have been remedied, proceed as in *g* below.

g. COOLANT SERVICE.

(1) When servicing the vehicle for summer, fill the system nearly full with clean water. Add corrosion-inhibitor compound in the proportion of 1 container of inhibitor to each 4 gallons of cooling system capacity. Then complete filling the system with water.

(2) When servicing for winter (for operation in arctic regions, refer to par. 55), fill the system about one-quarter full of clean water. Add sufficient antifreeze compound (ethylene glycol type) for protection against the lowest anticipated temperature (par. 55). Add water until the system is nearly full, then run the engine until the normal operating temperature is reached. Then add sufficient water to fill the system to 2 inches below bottom of filler neck.

119. Hoses and Connections

a. REMOVAL. Drain cooling system (par. 118*b*). Loosen screws in hose clamps and slide hose off ends of radiator, cylinder head and water pump elbows, and bypass fittings.

b. INSTALLATION. The hoses are installed without gasket cement. Install clamps on hoses, slide in place on connection elbows and fittings, and tighten clamp screws securely.

120. Radiator Thermostat

a. REMOVAL. Drain approximately 2 gallons of fluid from cooling system and install drain plug. Open engine compartment door. Remove five screws holding metal seal to radiator cross angle and remove seal. Loosen hose clamps on overflow pipe at filler neck and engine compartment side. Remove overflow pipe clip screw and remove pipes. Loosen hose clamps and disconnect the two hoses at radiator thermostat housing on radiator top tank. Remove four cap screws and washers holding thermostat housing to radiator and remove housing. Remove thermostat and discard gasket (fig. 80).

b. INSPECTION. Check thermostat by placing it, with bimetal coil down on a brick or stone, in a pan of water which contains a thermometer.

Caution: Do not place either the thermostat or the thermometer on bottom of pan because of concentration of heat at bottom when pan is heated over a burner.

Heat water until thermostat valve begins to open. The temperature at which this occurs depends on the heat range for which the thermostat is designed. The operating temperature is stamped on the housing flange of most thermostats. If the thermostat does not open at all or does not open at a temperature close to the markings found on the thermostat flange, it should be replaced. If the thermostat does not open and close completely, does not function freely, or is badly rusted, it should be replaced.

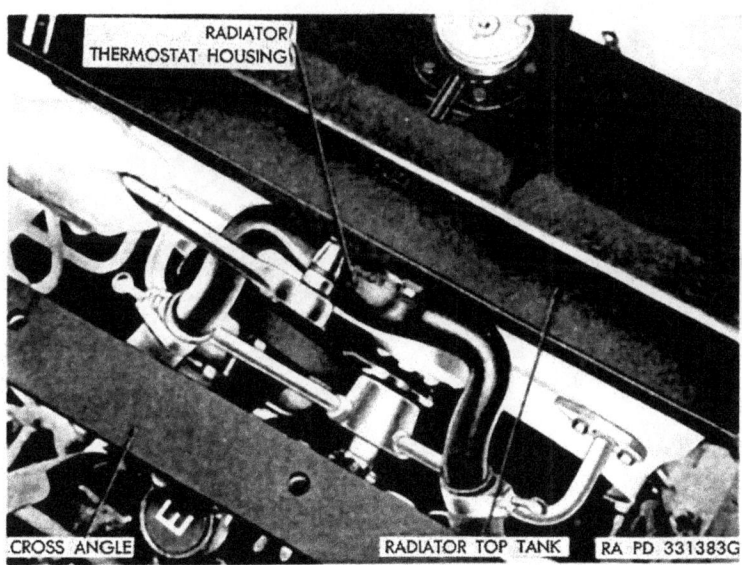

Figure 80. Removing thermostat and radiator inlet housing.

c. INSTALLATION. Place a new housing gasket on radiator top tank, and position thermostat on gasket with valve extending into tank (fig. 81). Position housing and install four mounting screws and washers. Using a torque wrench, tighten cap screws to 18 foot-pounds torque. Connect hose to housing and tighten hose clamps. Position metal seal on radiator and cross angle and install five mounting screws. Place overflow pipe in position, connect hose connections at filler neck and engine compartment side, and tighten hose clamps. Install overflow pipe clip screws and tighten. Fill cooling system (par. 118) and close engine compartment door.

Figure 81. Installing radiator thermostat.

121. Radiator and Fan

a. REMOVAL.

(1) *Drain cooling system.* Refer to paragraph 118.

(2) *Remove radiator cover and engine compartment door.* Traverse turret 90° from the straightforward position. Remove the six screws that hold radiator cover assembly to hull and the six screws that hold cover and engine compartment door to engine compartment cross angle (fig. 82). Using a chain, rope, or cable as a sling, attach to handles on cover inlet, then attach sling to hoist. Unlatch engine compartment door. Lift the cover and door together as a unit off vehicle.

Figure 82. Radiator cover.

(3) *Remove radiator.* Disconnect radiator overflow pipes at hose connection at filler neck and under metal seal at engine compartment sides and remove pipes. Remove screws which hold metal seal to radiator cross angle and remove seal. Disconnect oil cooler hoses at each end of cooler where connected to tubes and plug pipes to minimize oil spillage (fig. 78). Remove oil cooler mounting bracket screws and lift oil cooler, hanger, and hoses from vehicle as an assembly. Loosen hose clamps on radiator thermostat housing at top of radiator and outlet pipe at bottom of radiator and pull hose off connections. Remove four screws which hold fan shaft universal joint bearing trunnions on fan drive shaft yoke. Slide slip joint back to clear yoke. Remove two long bolts which hold upper part of radiator to center and side supports. Back out one long bolt which holds lower part of radiator to side support.

Note. This bolt cannot be removed completely until radiator has been lifted out.

Working on rear side of radiator, remove short bolt which holds radiator to support at center. Hook sling to lifting rings on top of radiator (fig. 83), connect hoist to sling, and lift radiator, shroud, and fan assembly out of vehicle (fig. 84).

175

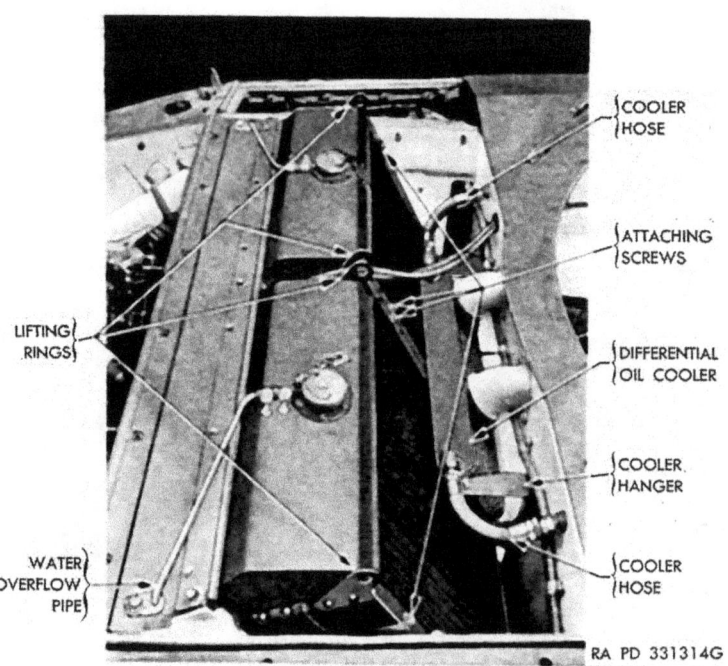

Figure 83. Radiator and transmission oil cooler mountings.

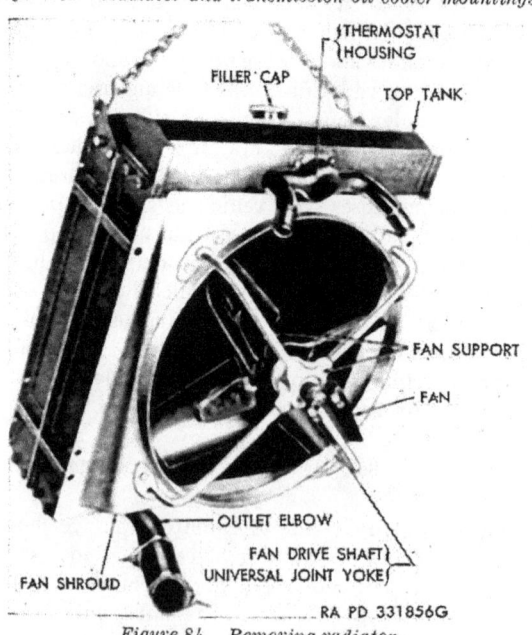

Figure 84. Removing radiator.

Tilt radiator forward, to clear cross angle with fan drive shaft yoke. Remove other radiator in same manner. Remove eight bolts and washers which attach fan support to fan shroud and remove fan. Remove four bolts, nuts, and washers which attach fan shroud to radiator and remove shroud.

b. INSTALLATION.

(1) *Install radiator.* Place fan shroud on radiator and install four attaching bolts, nuts, and washers. Place fan assembly on fan shroud and install eight bolts and washers which attach fan spider to shroud. Connect sling to radiator lifting rings, attach hoist to sling, slide long, lower, outer bolt in position, and lower radiator into position in engine compartment. Tilt top of radiator toward rear of vehicle to clear cross angle with fan drive shaft yoke (fig. 84). Install one long bolt through upper part of radiator and into tapping nut on center support. Screw long, lower, outer bolt into side support. Working at rear of radiator, install short bolt through center support and into tapping nut on radiator. Connect hose connections at radiator thermostat housing at top of radiator and outlet housing at bottom of radiator and tighten hose clamps. Line up fan drive shaft yoke with universal joint bearing trunnions and install four screws and lock plates. Tighten screws to a torque tightness of 16–25 foot-pounds and lock in place by bending lock plate ears over flat of screw. Place oil cooler on bulkhead and install hanger screws. Connect oil cooler hoses at each end of oil cooler. Position metal seal on radiator cross angle and radiator, and install five mounting screws with washers. Place radiator overflow pipe on top of radiator, connect short hose at filler neck and hose under metal seal at engine compartment side to pipe, and tighten hose clamps. Install overflow pipe clip screw.

(2) *Install radiator cover and engine compartment door.* Attach sling to handles on radiator cover assembly, attach hoist to sling, and lift cover and engine compartment door together as a unit. Place over engine compartment. Fasten the six screws that hold these two items to the engine compartment cross angle. Fasten the six screws that hold the radiator cover assembly to the hull. Turret can now be traversed back to its customary position.

(3) *Fill cooling system.* Refer to paragraph 118.

122. Water Pump

a. REMOVAL. Drain cooling system (par. 118). Install drain plug. Open engine compartment door and fasten to turret. Remove six cap screws that hold engine compartment rear air outlet to mounting angles and remove outlet. Loosen locking nut holding generator cradle in position and lower generator to lowest position. Slip belts off water pump pulley *by hand*. Loosen hose clamp that holds inlet connection hose on pump body inlet. Remove five screws and washers that hold water pump body to cylinder block (fig. 85). Remove pump and discard gasket.

b. INSTALLATION. Place new gasket on water pump and position pump on cylinder block with pulley under generator and water pump belts. Install five pump mounting screws and lock washers and draw up evenly (fig. 85). Connect inlet connection hose and tighten hose clamps. Install belts on water pump pulley, prying in place *by hand only*. Using large screwdriver or small utility bar between generator and generator cradle, pry up on generator and adjust belts until there is from five-eighths to three-quarters of an inch deflection in belts, measured midway between generator and water pump pulleys (fig. 37). Hold generator in this position and tighten generator cradle locking nut. Lift or hoist engine compartment rear air outlet into position on mounting angles and install the six screws. Close engine compartment door. Fill cooling system (par. 118).

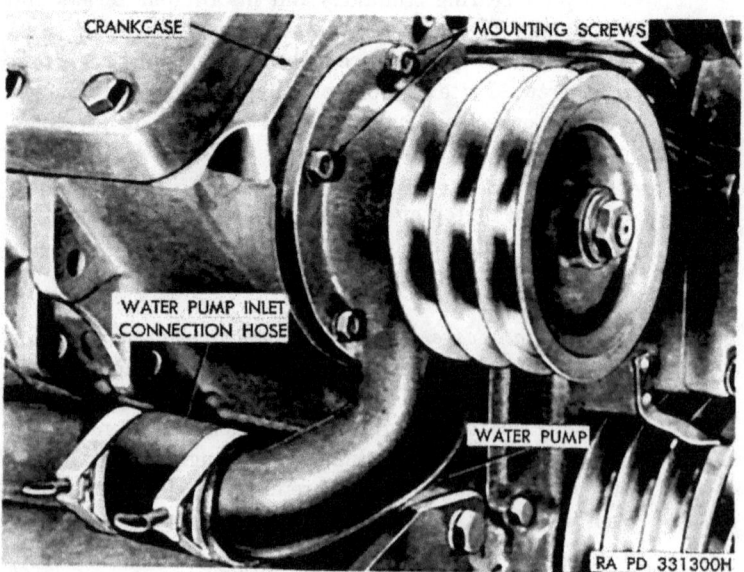

Figure 85. Water pump installed.

Section XI. GENERATING AND CHARGING SYSTEM

123. Description and Data

a. DESCRIPTION.

   (1) *Charging circuit.* Since there are two generators, one on each engine, there are two charging circuits, identical in circuit and each connected so as to charge all of the batteries. Each charging circuit consists of a generator, a four-unit generator regulator, and the cables and conduits required to connect these units to the batteries (fig. 86).

   (2) *Generators.* Two generators are used, one mounted on each engine. Each generator is a 24-volt, 50-ampere, 4-brush, 4-pole, shunt-wound type, with sealed, self-lubricated ball bearings supporting the armature. The generator is clamped in a mounting cradle and bracket and driven from the crankshaft by three matched belts. The cradle is adjustable to provide correct belt tension.

   (3) *Generator regulators.* The two generator regulators are mounted on the front of the bulkhead, one on each side (fig. 87). Each regulator consists of four units: a voltage regulator, a current regulator, a circuit breaker relay, and an actuating relay. These last two relays function together to perform the same duty as a cut-out relay, namely to open the circuit when generator voltage falls below battery voltage and to close the circuit when generator voltage is sufficient to charge the battery.

b. DATA.

   (1) *Generator:*
   Make and model number _____ Delco-Remy 1117309
   Output (cold) _____ 48–50 amp
   Rotation (drive end view) _____ counterclockwise
   Winding _____ 4-pole, shunt

   (2) *Generator regulator:*
   Make and model number _____ Delco-Remy 1118501
   Type _____ 4-unit, 24-v

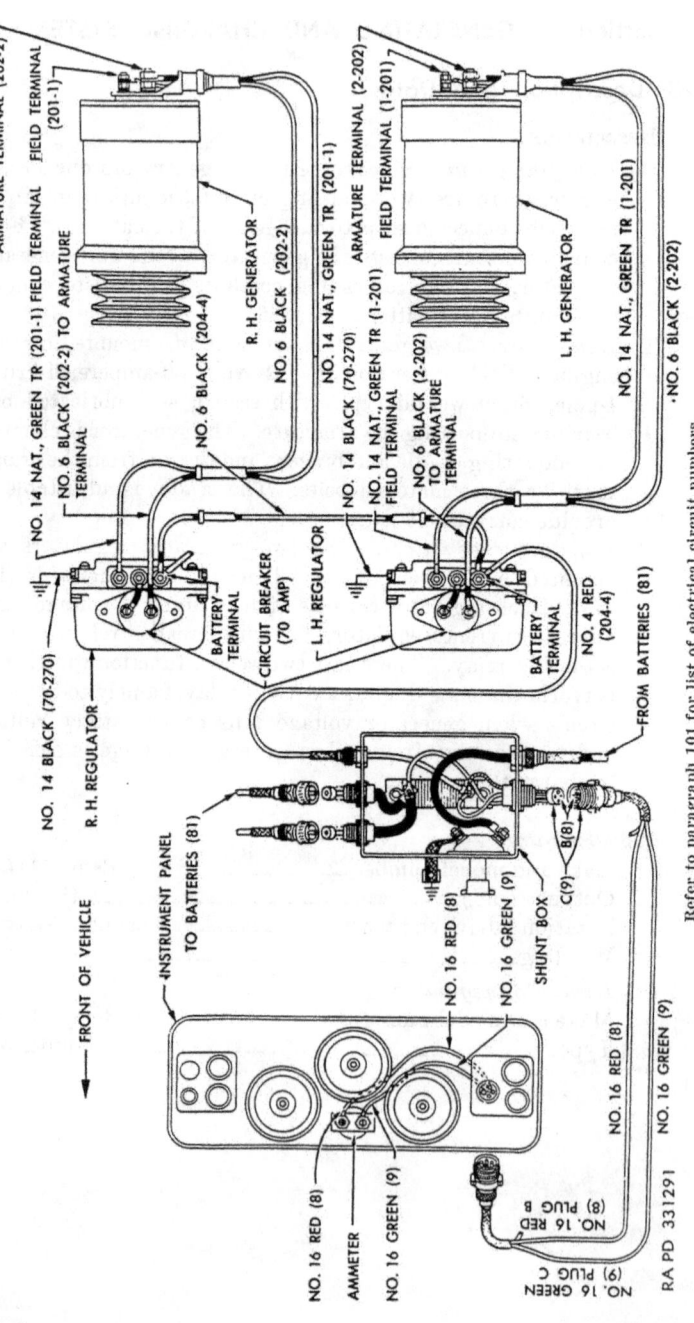

Refer to paragraph 101 for list of electrical circuit numbers.

Figure 86. Charging circuit wiring diagram.

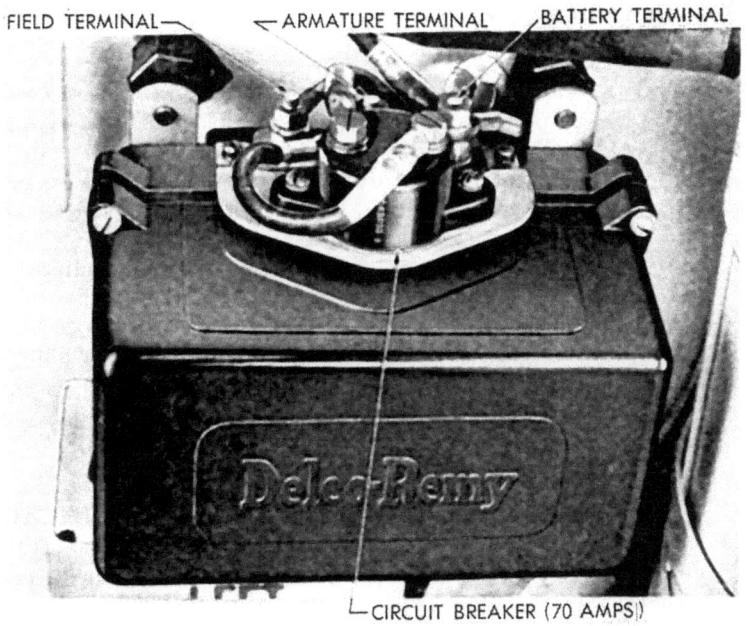

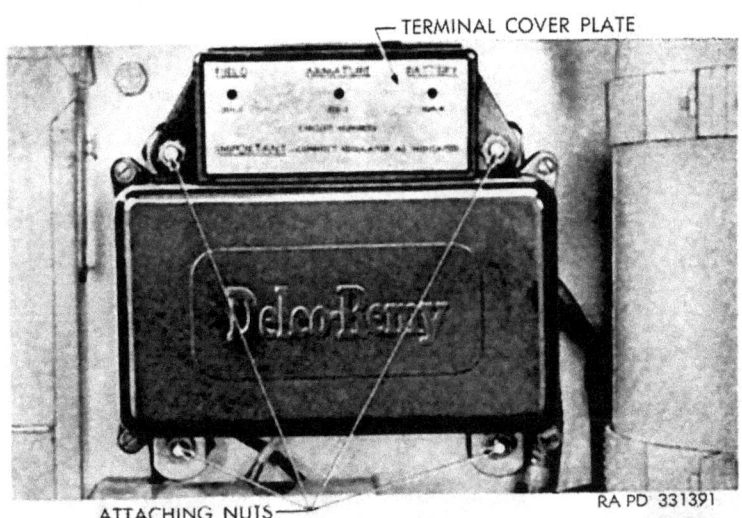

Figure 87. Generator regulator installed.

124. Generators

a. REMOVAL.

(1) *Test generator.* Make certain, by following test procedures given in paragraph 76, that generator requires replacement.

(2) *Open engine compartment.* Open master battery switch (fig. 5). Turn handles to unlocked position and open hinged engine compartment door. Remove six mounting screws and lift off rear outlet grille.

(3) *Remove generator belts.* Loosen generator cradle adjusting nut and allow generator and cradle to drop to lowest position. Loosen knurled packing nut on front fan shaft yoke at generator pulley end. Slide three drive belts off generator pulley.

Caution: Never pry belts off pulley with a screwdriver or other sharp tool, as pulleys may be accidentally nicked and burred.

(4) *Remove generator.* Remove two nuts and spacers holding generator mounting strap clamps to cradle clevis (fig. 88) and swing clamps back. Remove terminal shield and disconnect generator cables. Lift generator out of cradle.

b. INSTALLATION.

(1) *Mount generator in cradle.* Lift generator into mounting cradle on engine, at same time sliding fan shaft behind generator pulley into universal joint yoke splines, and seating

Figure 88. Generator installed.

locating lug on cradle in locating hole in generator housing (fig. 89). Swing generator mounting strap clamps into position on generator and insert clevis in toggle. Install spacer and nuts or clevis and tighten. Connect generator cables (fig. 50) and install terminal shield.

(2) *Install and adjust belts.* Lift each belt in turn up into its position on pulley. Tighten knurled packing nut at front fan shaft universal joint yoke. Raise generator and cradle until there is five-eighths to three-quarters of an inch slack in the belts, measured midway between generator and water pump pulleys (fig. 37).

(3) *Close engine compartment.* Lift rear outlet grille into position and install six mounting screws securely. Close hinged engine compartment door and turn handles to locked position.

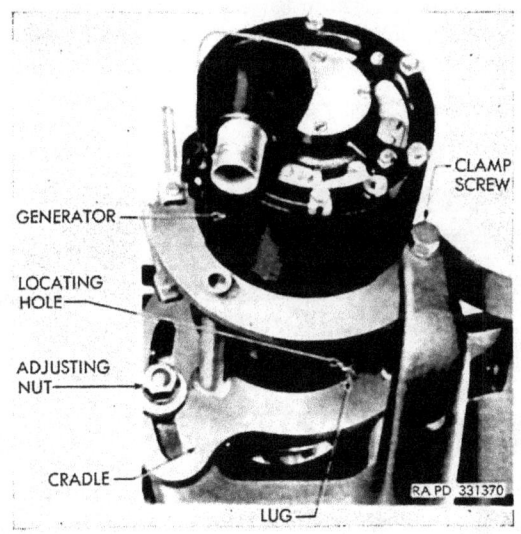

Figure 89. Generator (underside).

125. Generator Regulators

a. REMOVAL. Make certain, by following test procedures given in paragraph 77, that generator regulator requires replacement. Open master battery switch. Working in fighting compartment, remove four nuts (fig. 87) that hold regulator assembly to bulkhead. Pull generator regulator and conduit assembly away from bulkhead. Disconnect three wires at regulator terminals and remove regulator.

b. INSTALLATION. Connect three wires to regulator terminals, as shown in figure 87. Close master battery switch and connect a jumper *momentarily* between field terminal and battery terminal of regulator. This allows a momentary surge of battery current to flow to the generator and polarize it correctly. Position regulator assembly on bulkhead and install four nuts holding regulator to bulkhead (fig. 87).

Note. Never attempt to adjust generator regulators; service by replacement only.

Section XII. BATTERIES AND LIGHTING SYSTEM

126. Description and Data

a. DESCRIPTION.

(1) *Battery.* A 24-volt electrical system is employed in this vehicle. The battery consist of four separate 6-volt, 118-ampere-hour, 3-cell units connected in series. Two 6-volt batteries are carried in battery boxes on each side of the vehicle at the bulkhead above the fuel tanks (fig. 90) and are connected in series by a cable clipped to the bulkhead.

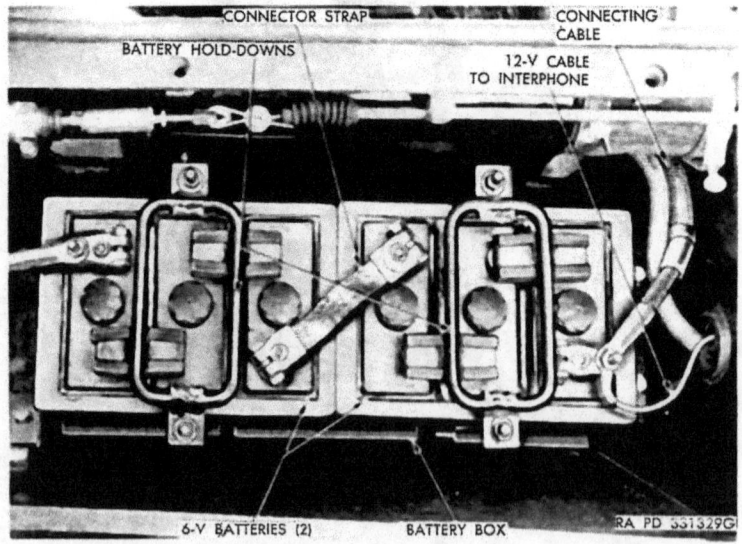

Figure 90. Battery installed.

(2) *Master battery switch.* The master battery switch, located on the left side of the engine bulkhead between the air cleaner and hull side wall (fig. 5), is provided to open the battery ground circuit and thus cut off all electrical circuits.

Caution: This switch must be turned "OFF" while filling the fuel tanks.

(3) *Slave battery receptacle.* The slave battery receptacle is located on the right of the bulkhead on the fighting compartment side (fig. 91). This receptacle is provided for plugging in a battery or battery charger when the batteries are too low to function properly. The circuit is shown in figure 92. If an external charger is employed to recharge the batteries it must be connected to this receptacle by cable—17–C–568 (fig. 26).

Note. In an emergency, a cable may be connected from another tank to supply current to start the vehicle by use of the slave battery receptacle.

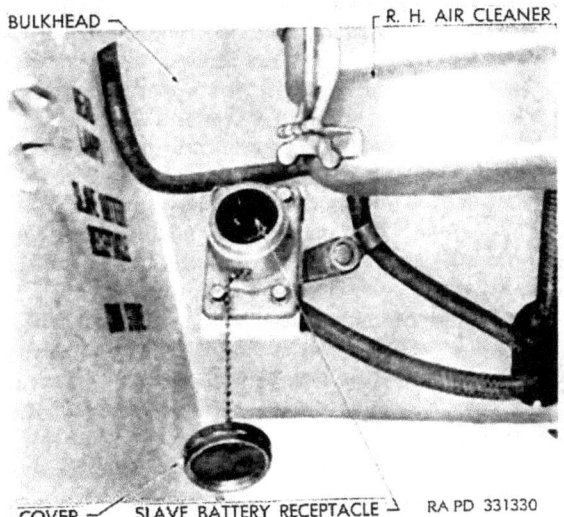

Figure 91. Slave battery receptacle.

(4) *Lighting system.*
   (a) *Driving lights.* The driving lights consist of two head lights, with blackout marker lights incorporated, a blackout driving light, and two dual-purpose tail lights. The left tail light contains a blackout tail light and a double filament combination service tail and stop light. The right tail light contains a blackout tail light and a blackout stop light. These circuits are shown in figures 93 and 94.
   (b) *Blackout driving light.* The blackout driving light when not in use is carried in a bracket mounted on the left hull wall in back of the driver's seat and is put in use by substituting for the head light in the left head light mount. A resistor is inserted in the circuit (fig. 94) to reduce the

voltage to the light from 24 to 6 volts. This reduction in voltage is necessary, due to the special design of the blackout driving light, in order to obtain the proper illumination. The resistor assembly is mounted behind the instrument panel at the junction of the upper and lower deck of the hull.

(c) *Instrument panel lights.* There are two lamps in the instrument panel that provide illumination for all the instruments. These lamps are controlled by a rheostat switch located in the upper left-hand corner of the panel (fig. 6). These lamps are available by removal of the access plugs.

(d) *Dome lights.* Four dome lights are provided. Two are located on the turret roof just forward of the cupola door and turret door. Two are located in the driving compartment of the hull, one on the upper deck just above and forward of the assistant driver and the other on the bottom of the emergency ignition switch box. Each dome light has two lamps and two lenses (clear and red). This circuit is shown in figure 96.

(e) *Spot light.* On tanks of early manufacture, a spot light is mounted on a base forward of the cupola. The operating handle and switch extend inside the turret. The sealed-beam unit can be replaced by removing the shield and door.

(5) *Circuit breakers.* Eleven circuit breakers are incorporated in the electrical system in place of the conventional fuse arrangement. The circuit breakers (except those in the charging circuit) are all of the manual reset type, which means that the protected circuits can be closed again, after the circuit breaker has operated, simply by pressing the reset button. The charging circuit breakers are fully automatic. The various circuit breakers are located as follows:

(a) *Lighting.* The circuit breaker for the driving lights is located in the lower left corner of the instrument panel (fig. 6).

(b) *Siren (or horn).* The circuit breaker for the siren (or horn) is also located in the lower left corner of the instrument panel.

(c) *Fuel pump.* The circuit breaker for the fuel pump is located in the center of the fuel pump switch bracket (fig. 7).

(d) *Ventilating blower.* The circuit breaker for the ventilating blower motor is located on the blower assembly, on the driver's side, toward the front of the vehicle.

(e) *Hull dome lights.* The circuit breaker for the hull dome lights and the windshield wiper is located on the front face of the emergency ignition switch box (fig. 12).

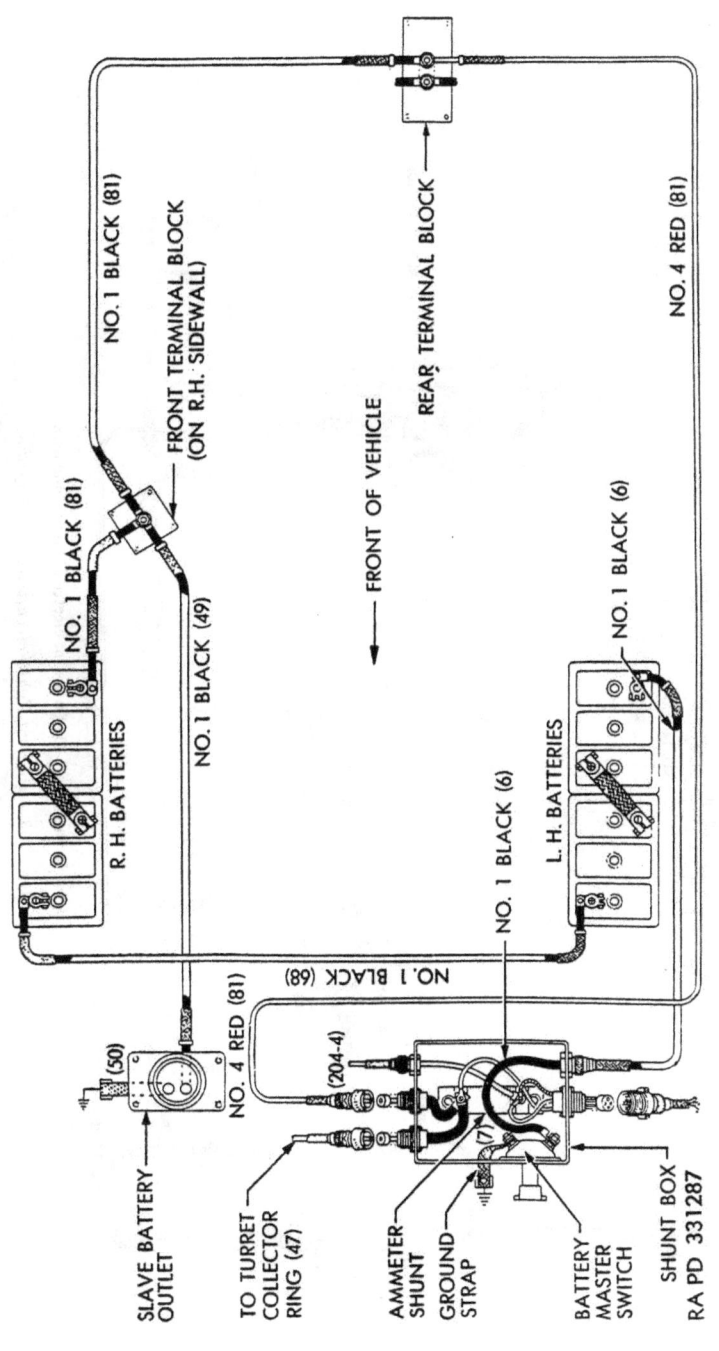

Refer to paragraph 101 for list of electrical circuit numbers.
Figure 92. Battery circuit wiring diagram.

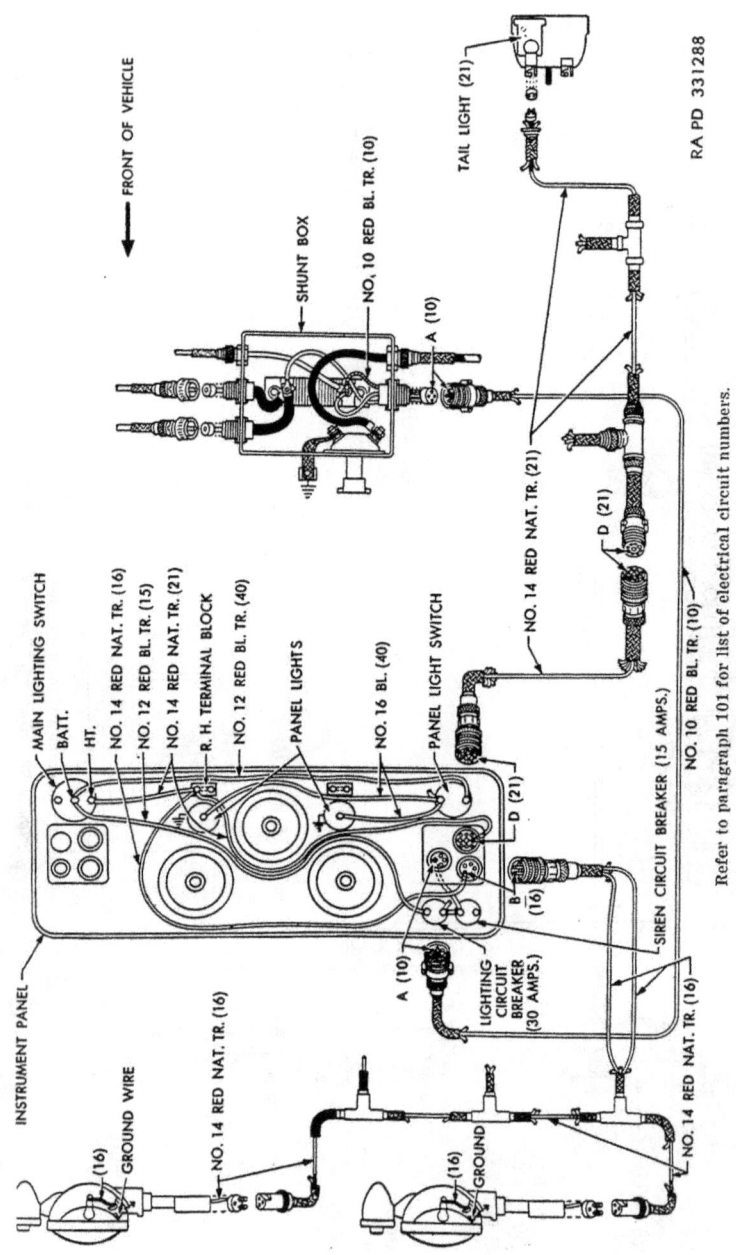

Figure 93. *Driving lights circuit wiring diagram.*

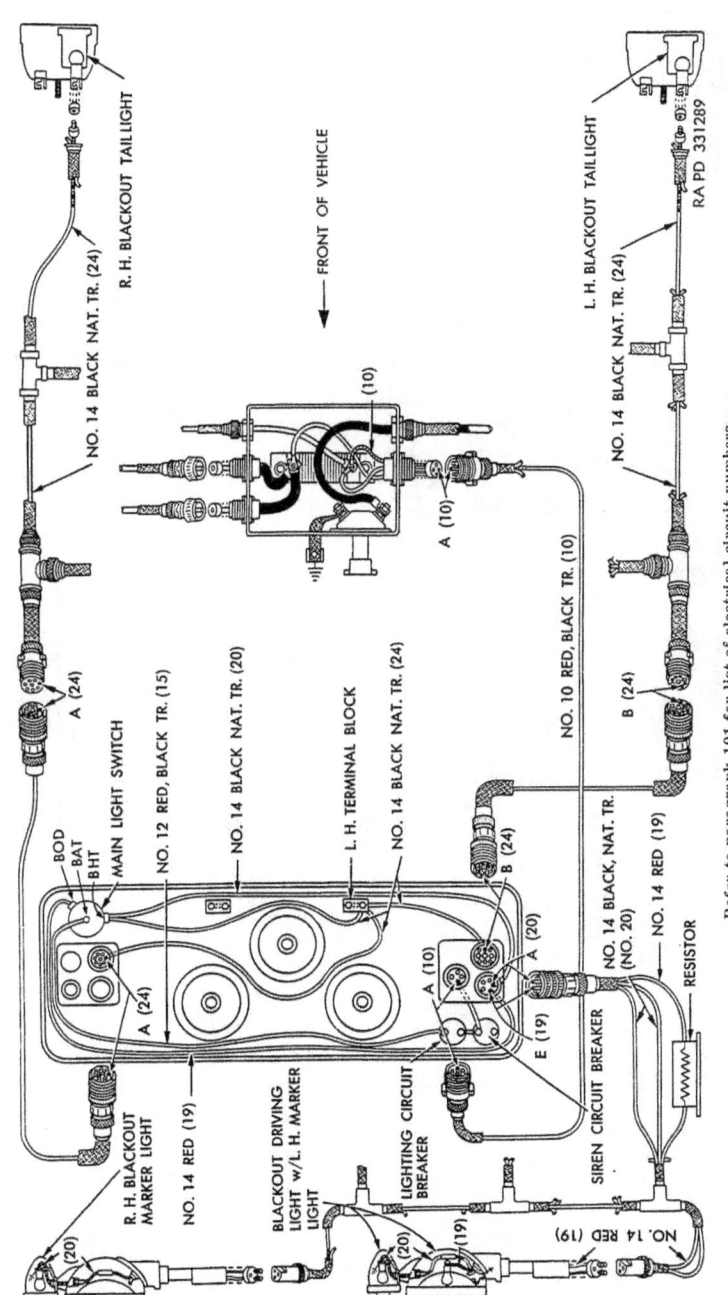

Figure 94. Blackout light circuit wiring diagram.

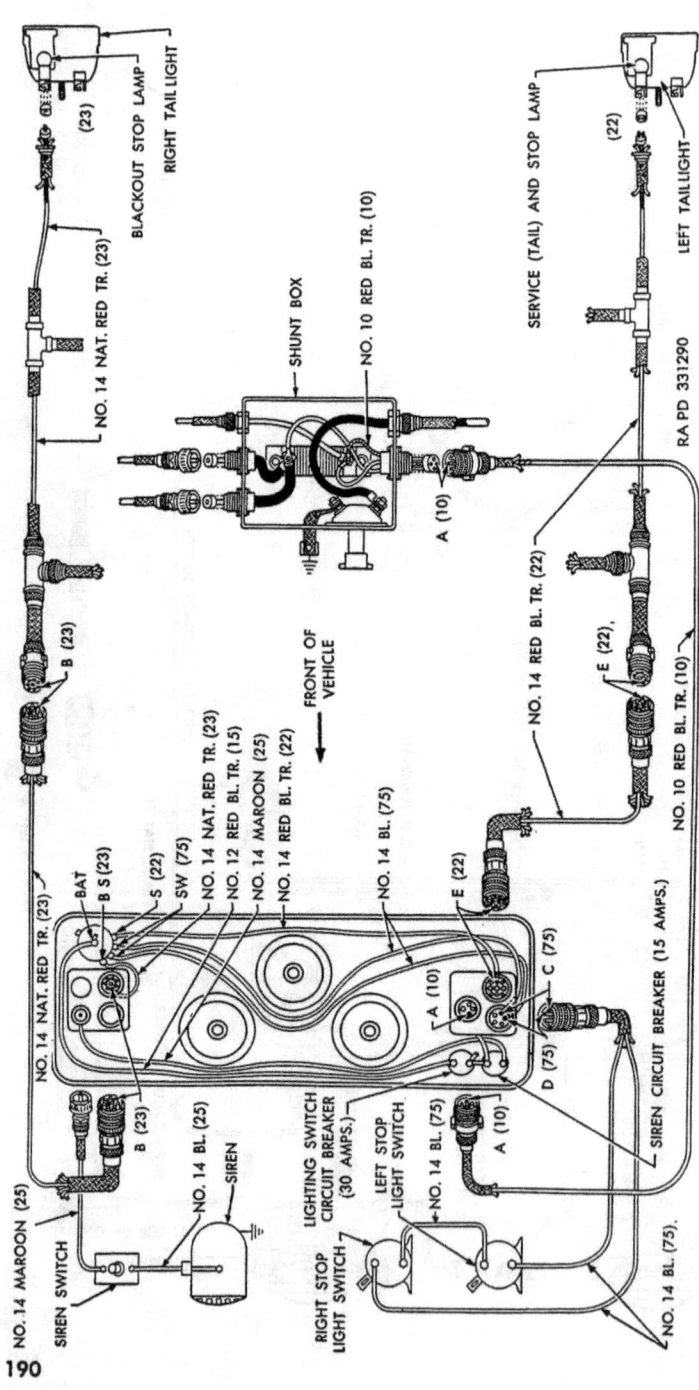

Figure 95. *Stop light and siren circuit wiring diagram.*

Refer to paragraph 101 for list of electrical circuit numbers.

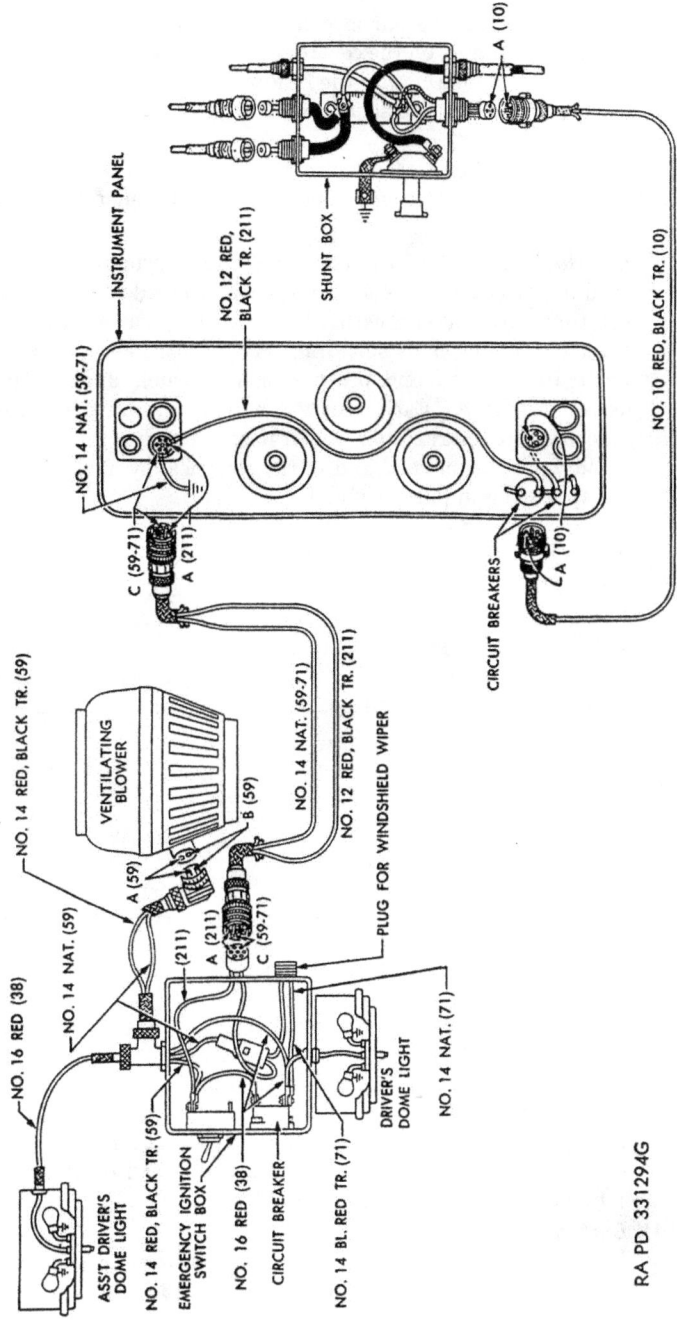

Figure 96. Dome lights, ventilating blower, and windshield wiper circuits wiring diagram.

(f) *Turret circuits.* Four circuit breakers for the turret circuits are located on the control box above the collector ring (fig. 19). These control the following circuits: turret dome lights, turret traverse motor, stabilizer, and gun firing solenoids.

(g) *Charging circuit.* Two circuit breakers in the generator armature circuit are mounted one on each generator regulator.

(6) *Conduits and cables.* Electrical wiring is carried in shielded conduits or cables. Several cables are grouped together and run through a metal sheath with index plugs at each end to form a conduit. This shielding serves to protect the cables from dirt, grease, and other elements which might cause defective wiring. Each circuit is numbered, and the circuit numbers are marked on illustrations (wiring diagrams) throughout this manual and in the electrical symbol number list appearing in paragraph 101.

b. DATA.
  (1) *Batteries.*
      Type _____ 3-cell (coml)
      Voltage (each cell) _____ 6
      Voltage (entire system) _____ 24
      Number of batteries _____ 4
      Number of plates per cell _____ 17
  (2) *Master battery switch.*
      Type _____ spst
  (3) *Circuit breakers.*
      (a) Capacity:
          Siren _____ 15 amp
          Driving lights _____ 30 amp
          Fuel pump _____ 15 amp
          Ventilator _____ 15 amp
          Dome lights and windshield wiper _____ 15 amp
          Generator _____ 70 amp
          Stabilizer control box (for dome and spot lights) _ 20 amp
          Stabilizer _____ 15 amp
          75-mm and cal. .30 guns _____ 25 amp
          Turret motor _____ 105 amp
      (b) Type (all except charging circuit) _____ bimetal disk
      (c) Type (charging circuit) _____ automatic reset
      (d) Reset (all except charging circuit) _____ manual
  (4) *Lights.*
      Type _____ standard quartermaster and ordnance
      Blackout marker light _____ 3-cp, sgle-contact

Service tail light _____ 3-cp, dble-contact
Blackout stop lights _____ 3-cp
Service head lights _____ 40 w
Type _____ sealed beam
Service stop light _____ 21-cp, dble-contact
Instrument panel lights _____ 3-cp, sgle-contact
Signal lights _____ 3-cp, dble-contact
Blackout tail lights _____ 3-cp
Blackout driving light _____ 6-v w/resistor
Dome lights _____ 6-cp, sgle-contact
Spot light _____ sealed beam

127. Batteries

a. GENERAL. The service which the using organizations can perform on the battery includes recharging or replacement, replacing cables, adding water, checking specific gravity, and cleaning.

b. CHECKING BATTERIES. Never use matches or flame as a light when checking the battery. Distilled or clean water can be added to bring the fluid level just above the battery plates. If the specific gravity of any cell is 1.200 or below, the battery should be recharged. A fully charged battery should have a specific gravity reading of from 1.275 to 1.300. Batteries can be charged in the vehicle by connecting an outside source to the slave battery outlet (fig. 91). Master battery switch must be "ON" when recharging batteries by this method.

c. BATTERY TERMINAL CORROSION. Warm water poured slowly over the terminals will loosen any copper sulphate that has been deposited, so that it can be brushed off and flushed away.

Caution: This should never be done with the battery in the vehicle. Battery posts and terminals should be wiped clean with a cloth saturated in an alkaline solution such as ammonia or a solution of bicarbonate of soda and water. Corrosion can be retarded by coating with grease. If battery fluid has overflowed or has spilled, all affected metal parts should be flushed with an alkaline solution and wiped dry. Remove drain plate on bottom of hull under fuel tanks to allow water to drain from fuel tank compartment.

Caution: Care must be exercised to prevent alkaline solution getting into battery electrolyte.

d. REMOVING BATTERY CABLES. Turn master battery switch to "OFF" position. Remove five screws holding battery box cover in position and remove cover.

Note. It is not necessary to remove batteries from vehicle to replace cables or connector strap.

Remove cable from the shunt box to battery by removing nut holding cable to battery post and pulling cable into fighting compartment.

Back off knurled nut attaching cable to the shunt box and lift out cable. Loosen screws holding connector cable to posts on batteries and remove cable from attaching clips.

e. INSTALLING BATTERY CABLES. Install connector cable in clips on rear of bulkhead and attach to foremost battery post on each side (fig. 92). Attach master switch cable to shunt box, thread cable through opening in bulkhead, and connect to rearmost negative battery terminal post. Install battery box cover.

f. REMOVING BATTERY. Turn master battery switch to "OFF" position. Remove five screws holding battery box cover in position and remove cover. Remove hold-down bolts and lift hold-downs out of position (fig. 90). Remove connector cable from battery post by loosening bolt in connection. Place cable out of position. Remove master battery switch cable from battery. Remove bolt holding connector strap to post connection. This will leave strap connected to the other battery. Lift battery out of battery box and remove from vehicle.

g. INSTALLING BATTERY. Turn master battery switch to "OFF" position. Lift two batteries into place in battery box with one positive and one negative battery post together in the center (fig. 90).

Note. On the right side of the vehicle, the negative post of the front battery should be toward the front of the vehicle. On the left side of the vehicle, the positive post of the front battery should be to the front.

Connect front and rear battery connector strap. Install battery hold-down clamps.

Caution: Do not draw up hold-down clamps too tightly or battery cell covers may crack.

Install battery connector cable and master battery switch cable and tighten securely. Lift battery box cover into position on hull side angles and install five mounting screws.

128. Master Battery Switch

a. REMOVAL. Place master switch handle in "OFF" position (fig. 5). Disconnect ground strap by removing bolt holding strap to hull side wall. Loosen knurled nut holding upper front conduit to shunt box and disconnect conduit. Loosen knurled nuts holding two bottom front conduits to box and disconnect conduits. Remove four bolts, two above and two below, that anchor box to bulkhead and pull shunt box toward front of vehicle. Remove four screws holding box side cover in position and remove cover.

Note. If additional clearance is required to remove screws, the clips holding two remaining conduits on the bulkhead and hull side wall may be removed.

Remove nut holding cable to master battery switch terminal, remove cable and tape end. Remove screw holding switch handle knob and remove knob. Remove two bolts holding master battery switch in position. Working from inside of box, turn master switch assembly about 90° and remove switch, pulling ground strap out of slot in shunt box.

b. INSTALLATION. Place ground strap through slot and position switch in shunt box. Install two bolts through box and switch and tighten securely. Remove tape from cable end and install cable on master battery switch terminal. Position shunt box side cover and install four screws. Place shunt box in position on bulkhead and install four bolts, two above and two below. Connect upper front conduit and tighten knurled nut securely. Install switch handle knob. Connect ground strap to hull side wall.

129. Lights

a. GENERAL. All tail light and head light assemblies are composite assemblies of lamps, lenses, and reflectors and are sealed against dirt and moisture. Therefore, the only service operations on the lights is replacement of the sealed assembly. The complete head light assemblies can be removed for safety in combat zones. Service operations on the lighting switches are covered in paragraph 113.

b. REMOVAL OF COMPLETE HEAD LIGHT (fig. 97). To remove the head lights, turn head light retainer inside hull to the left to loosen. Back off knurled nut on lighting conduit. Remove conduit from head light assembly. Lift assembly out of mounting.

c. INSTALLATION OF COMPLETE HEAD LIGHT (fig. 97). See that light switch on instrument panel is "OFF." Place head light in position on hull front plate. Secure in position by turning head light retainer. Place lighting conduit in position and tighten knurled nut securely.

d. REMOVAL OF HEAD LIGHT SEALED LAMP-UNIT. Remove screw holding head light door to head light body (fig. 97). Swing lower end of door outward and lift door from body. Pull sealed lamp-unit assembly outward until terminal screws holding cables to sealed lamp-unit assembly can be removed. Disconnect cables and remove sealed lamp-unit assembly.

e. INSTALLATION OF HEAD LIGHT SEALED LAMP-UNIT. Connect red cable in head light body to terminal on center of sealed lamp-unit assembly. Connect black cable in head light body to sealed lamp-unit assembly. Position head light door case by inserting lug on top of door through slot in top of body. Install screw holding head light door to body.

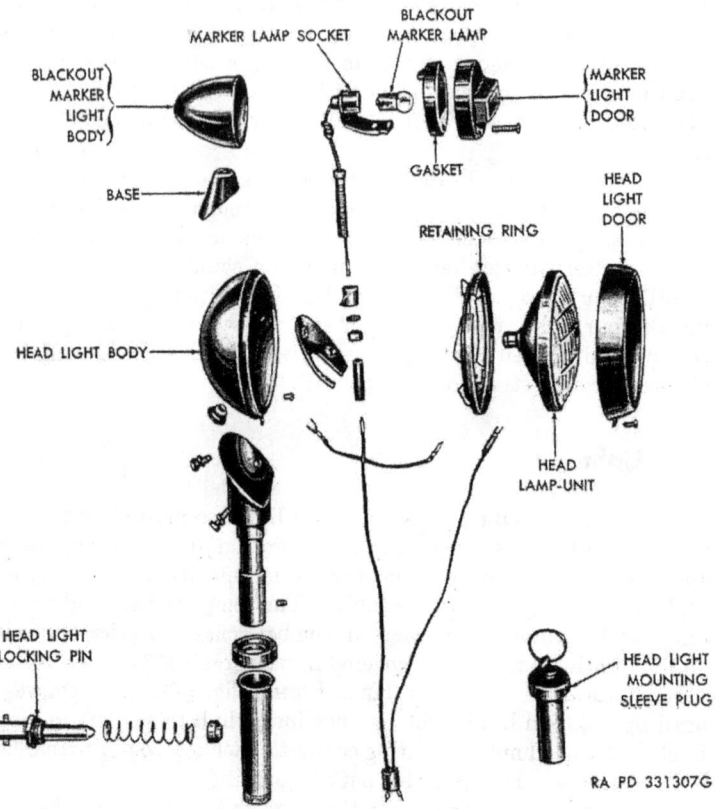

Figure 97. Head light assembly—exploded view.

  f. REMOVAL OF COMPLETE TAIL LIGHT ASSEMBLY (fig. 98). Remove three screws holding base of tail light assembly to hull. Turn tail light assembly over and remove two nuts and lock washers holding body to base. Slide base away from body and disconnect both cables from body by turning and pulling connections until they slip out of plugs.

  g. INSTALLATION OF COMPLETE TAIL LIGHT ASSEMBLY (fig. 98). Connect cables to tail light by inserting cables in sockets and turning them so that they will be locked in place.

Note. Be sure to connect cables to right sockets to assure proper operation of lighting system. Refer to wiring diagrams.

Slide tail light base on body and install two nuts and lock washers. Position tail light on hull and install three screws holding base to hull.

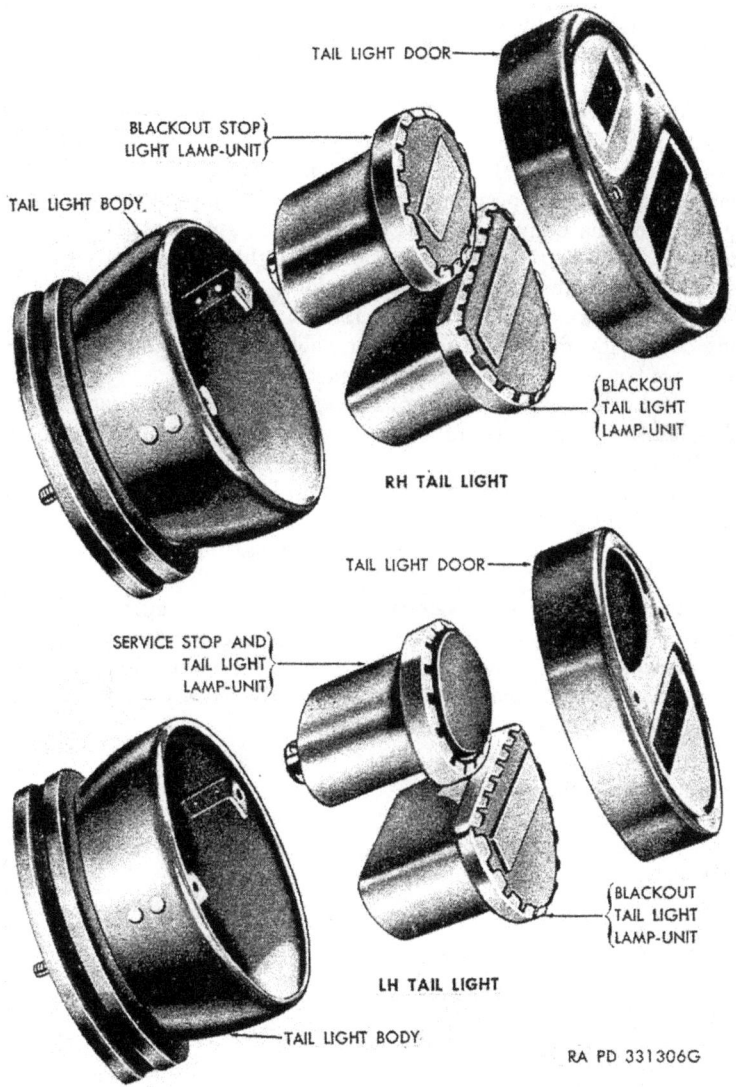

Figure 98. Tail light assemblies—exploded view.

h. REMOVAL OF TAIL LIGHT LAMP-UNIT. Remove two screws holding tail light door to light body and pull door off body. On right-hand tail light, pull the right black-out lamp-unit from the lower portion of the body and the black-out stop lamp-unit from the upper portion of the body. These are sealed assemblies. On left-hand tail light, remove the black-out lamp-unit from the lower portion of the body and the service tail and stop lamp-unit from the upper portion of the body.

i. INSTALLATION OF TAIL LIGHT LAMP-UNIT. On right-hand tail light, position the black-out tail lamp-unit in lower portion of body and the black-out stop lamp-unit in the upper portion of the body. On left-hand tail light, position the black-out tail lamp-unit in the lower portion of the body and the service tail and stop lamp-unit in the upper portion of the body. Position tail light door on body and install two mounting screws and lock washers.

j. REPLACEMENT OF DOME LIGHT LAMPS. Remove two screws holding dome light door and lift off door, rubber gaskets, and both lenses. Replace lamp-unit. Install door, gaskets, and lenses.

k. REMOVAL OF WARNING SIGNAL AND INSTRUMENT PANEL LIGHTS. The lamps for the warning signals and for panel illumination can be replaced readily without disturbing the panel itself. First use a screwdriver to pry the access plug for the panel lamps or the ruby lens for the warning signals (fig. 100) from the panel. Then depress the lamp slightly, turn counterclockwise to release bayonet type catch, and remove lamp.

l. INSTALLATION OF WARNING SIGNAL AND INSTRUMENT PANEL LIGHTS. The lamps used for panel lighting and for warning signals are 24-volt, 3 candlepower. Line up ribs on lamp with slots in socket, depress lamp, and turn clockwise until locked securely. Install access plug or ruby lens by snapping into place.

130. Circuit Breakers

a. GENERAL. Inoperative circuit breakers can be serviced by replacement only. They should not be disassembled.

Note. When replacing circuit breakers, be sure master battery switch is in "OFF" position.

b. REMOVAL.
  (1) *Turret (dome lights, stabilizer, 75-mm and cal. .30 guns, and turret motor).* These circuit breakers can be removed by removing two screws for the circuit breakers and disconnecting the cables.
  (2) *Dome lights.* Remove four screws holding dome light in position on emergency ignition switch box and lower dome light and bottom cover of box (fig. 12). Remove two screws holding circuit breaker in position in box, remove circuit breaker and disconnect cables.
  (3) *Fuel pumps.* Remove four nuts holding control assembly to mounting bracket and lower assembly to obtain enough clear-

ance to disconnect the two cables and remove two screws holding circuit breaker in the assembly. Remove circuit breaker.

(4) *Lights and siren.* Remove instrument panel (par. 133). Both circuit breakers are now accessible and can be removed by removing two mounting screws for each, then disconnecting cables from terminals.

(5) *Ventilating blower.* Remove two mounting screws. Remove both cables and circuit breaker from assembly.

(6) *Charging circuit.* Remove generator regulator terminal cover. Disconnect two cables and remove two screws and washers holding circuit breaker to top of generator regulator box (fig. 87).

c. INSTALLATION.

(1) *Turret (dome lights, stabilizer, 75-mm and cal. .30 guns, and turret motor).* Install cables on "A" and "B" terminals of circuit breaker or breakers and tighten screws securely. Position circuit breaker in place and install two mounting screws.

(2) *Dome lights.* Install cables on circuit breaker terminals and tighten securely. Install two mounting screws holding circuit breaker in position on emergency ignition switch box. Place dome light and bottom cover in position on box and secure with four screws.

(3) *Fuel pumps.* Install cables, lock washers, and screws on circuit breaker. Place circuit breaker in position on control assembly and install two mounting screws. Position control assembly on mounting bracket and install four nuts.

(4) *Lights and siren.* Install cables, lock washers, and screws and tighten securely. Place circuit breaker or breakers in position on panel and install two mounting screws. Install instrument panel (par. 133).

(5) *Ventilating blower.* Install cables, lock washers, and screws. Position circuit breaker on ventilating blower assembly and install with two screws.

(6) *Charging circuit.* Position circuit breaker on top of generator regulator box and install two screws and washers. Connect short cable from regulator to right-hand terminal and cable from generator to left-hand terminal (fig. 87). Install generator regulator terminal cover.

131. Conduits and Cables

a. REMOVAL. All electrical conduits are removed in essentially the same manner. Unscrew the connectors at each end and detach connector plugs. Disconnect the cables at the terminals. Remove the retaining clips, after taking out any stowage items or sheet metal parts that interfere, and remove conduit from vehicle.

b. INSTALLATION. Position conduit in hull or turret and install retaining clips. Assemble connectors snugly into place. Observe the following precautions:

(1) Make sure conduits are connected to proper receptacles. Refer to pertinent wiring diagrams.

(2) When installing connector plugs, make sure that the letters on the plug and receptacle coincide and that the tongue and groove line up. Do not force plugs. If plug is properly lined up, it will slip into receptacle without forcing.

(3) Do not twist or kink conduit. See that all retaining clips are tightened in proper position.

Section XIII. INSTRUMENT PANEL, INSTRUMENTS, SWITCHES, AND SENDING UNITS

132. Description

a. INSTRUMENT PANEL CONTROLS AND INSTRUMENTS. The following controls and instruments, which are mounted on the instrument panel, are described in paragraphs 13, 16, and 33 through 40: panel light switch, driving lights switches, siren circuit breaker reset button, driving lights circuit breaker reset button, ignition switches, starter switch buttons, ammeter, warning signal lights, engine temperature gages, speedometer, tachometers, two panel light access plugs, and an electrical outlet for connecting auxiliary electrical equipment.

b. SENDING UNIT AND SIGNAL SWITCHES (fig. 99).

(1) *Engine water temperature sending unit.* This unit is mounted at the upper front of the left cylinder head. This is essentially an electrical resistor which changes resistance with changes in temperature. These changes are indicated by the gage on the instrument panel which measures the resistance in the gage circuit.

(2) *Engine water high temperature warning signal switch.* This switch is mounted at the upper front of the left cylinder head, just to the rear of the engine water temperature sending unit. This is a gas expansion type switch in which the gas expands as the temperature increases until, at 238°–242° F., contact points close, establishing a completed circuit and causing the warning signal light to burn.

(3) *Engine low oil pressure warning signal switch.* This switch is mounted at the rear of the engine cylinder block. It is a

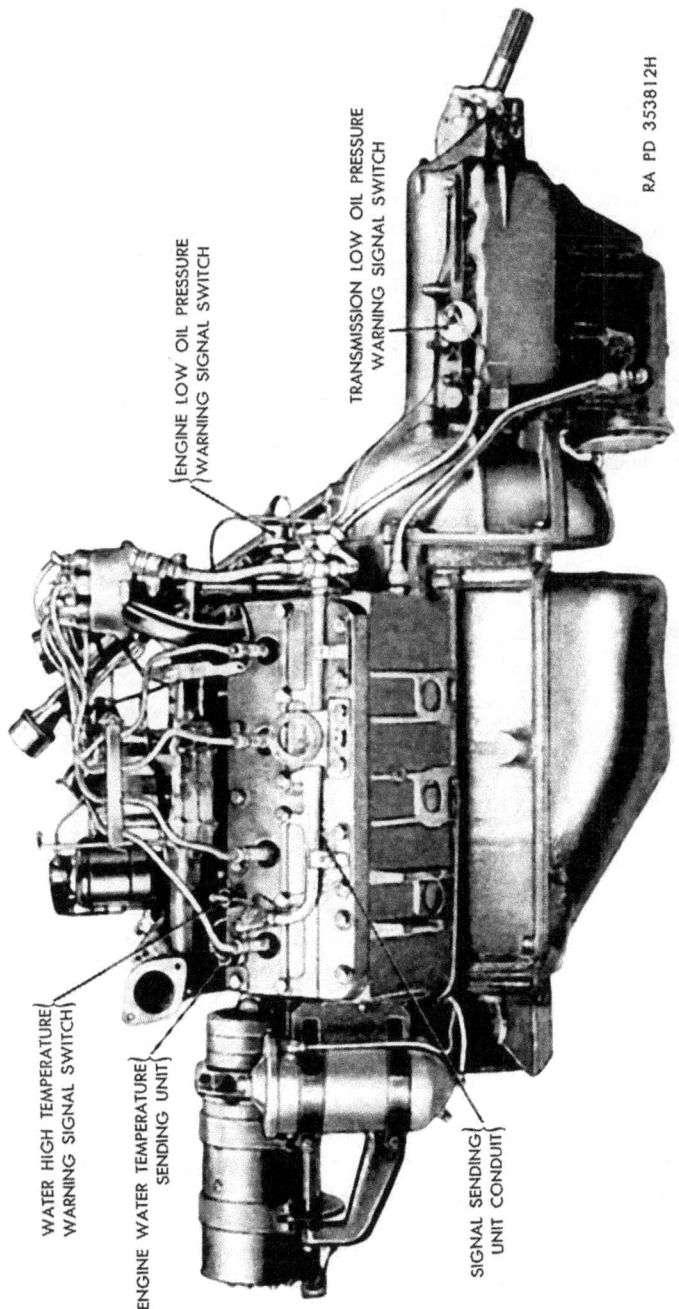

Figure 99. Signal switches and sending unit.

low pressure switch. With little or no pressure the switch contacts are closed, causing the warning signal light to burn.

Note. This warning signal light is the same one that is operated by the engine water high temperature warning signal switch.

This switch breaks the circuit at 8–10 psi.

(4) *Transmission low oil pressure signal switch.* This switch is mounted on the upper left of the transmission case. It is a low pressure switch. With little or no pressure the switch contacts are closed, causing the appropriate warning signal light to burn. This switch breaks the circuit at 62–68 psi.

c. DOME LIGHT SWITCHES. Dome lights in the hull are controlled by individual switches on the fixtures.

d. SIREN (OR HORN) AND SIREN SWITCH. The siren (or horn) is mounted on the front deck next to the left head light. It is operated by a spring-loaded momentary contact switch on the hull floor in front of the driver (fig. 7).

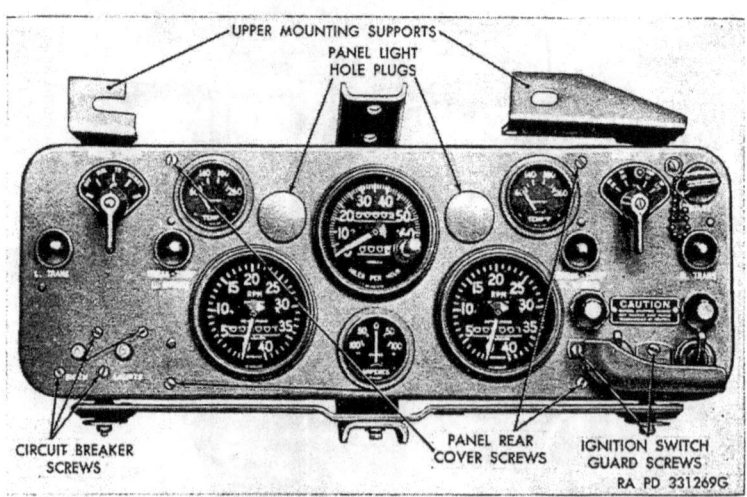

Figure 100. Instrument panel.

133. Instrument Panel

a. REMOVAL. Working from driver's seat, remove cap screw and lock washer at lower center mounting bracket. Remove two screws and lock washers at upper center mounting bracket and one at upper right support. Loosen upper left mounting screw, slide panel and assembly to left, and pull forward. Place panel face down in lap, disconnect three flexible shafts to speedometer and tachometer and six flexible electrical conduits. Remove panel from vehicle.

b. INSTALLATION. Install three flexible shafts for speedometer and tachometers. Connect six electrical conduits (figs. 101 and 102), making certain that plugs are correctly indexed. Position panel against and to the left of tapping blocks on hull front wall and slide to right, engaging slot at upper left bracket behind flat washer and lock washer of mounting cap screw. Install remaining cap screws: two at upper center bracket, one at upper right, and one at lower center. Tighten all screws securely.

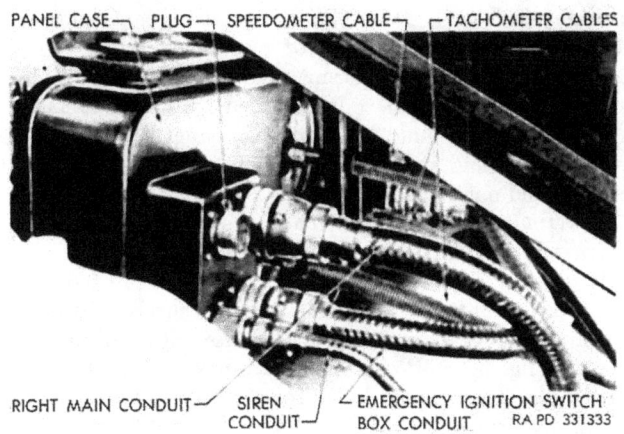

Figure 101. Cables and conduits at rear of instrument panel.

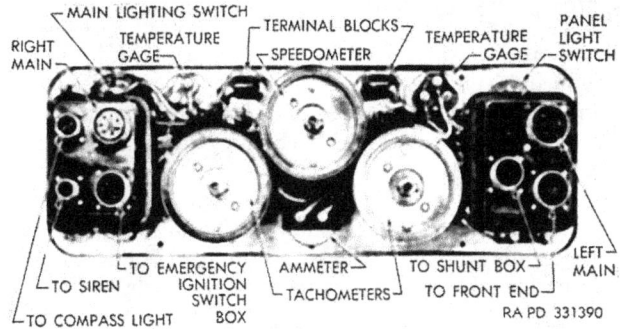

Figure 102. Instrument panel connections rear cover removed.

134. Ammeter

a. REMOVAL. Remove instrument panel from vehicle (par. 133). Remove two terminal nuts (fig. 102) and disconnect cables from ammeter. Remove two mounting nuts from terminal studs, slide mounting clamp off studs, and remove ammeter from front of panel.

b. INSTALLATION. Position ammeter in opening in panel and install mounting clamp at rear. Install two mounting nuts on terminal studs, position cables on studs (fig. 102), and install terminal nuts. Install panel in vehicle (par. 133).

Note. Cables have circuit numbers marked at ends to aid correct assembly.

135. Ignition Switch

a. REMOVAL. Turn master battery switch "OFF." Remove instrument panel (par. 133). Take out three mounting screws for right-hand conduit box at back of instrument panel (fig. 102) and pull box out far enough to permit access to rear of ignition switches. Disconnect cables that are to be removed. Turning to face of panel, remove two screws holding guard to panel (fig. 100). Unscrew knurled nut around switch lever and remove nut. Push switch and lever back and out of panel.

b. INSTALLATION. Insert switch in opening from rear of panel and position with switch lever pointing up. Install knurled nut over switch lever and turn down snugly against face of panel. Install switch guard, tightening attaching screws securely. Turn panel over and connect cables to switch in accordance with wiring diagram and markings (fig. 56). Position conduit box on mounting studs and install three attaching screws. Install instrument panel in vehicle (par. 133).

136. Emergency Ignition Switch

a. REMOVAL. Turn master battery switch "OFF." Remove four cap screws and lock washers that hold emergency ignition switch box cover in place and remove cover and dome light from box as an assembly. Disconnect cables to switch and remove switch after taking out hex nut at front.

b. INSTALLATION. Install switch through opening in switch box and attach with hex nut. Connect cables to switch in accordance with wiring diagram and markings (fig. 56). Position cover and dome light assembly against switch box and install four cap screws and lock washers.

137. Starting Switches

a. REMOVAL. Turn master battery switch "OFF." Remove instrument panel from vehicle (par. 133). Take out three screws for right-hand conduit box at rear of instrument panel (fig. 102) and pull box out far enough to permit access to starting switches. Disconnect cables at rear of switch to be removed. Turning to face of panel, remove rubber dust seal, loosen large hex nut surrounding switch button, and remove nut. Push switch back and remove from rear of panel.

b. INSTALLATION. Insert switch in opening from rear of panel and install large hex nut and rubber dust seal at front, turning down until snug against face of panel. Turning to back of panel, connect cables to switch terminals according to wiring diagram and markings (fig. 63). Position conduit box on mounting studs and install three attaching screws. Install instrument panel in vehicle (par. 133).

138. Lighting Switches

a. REMOVAL. Both the panel light switch and the main light switch are removed by the same procedure. Turn master battery switch "OFF." Remove instrument panel (par. 133). Disconnect cables from terminals at rear of switch. Remove hex nut from switch lever shaft, remove switch plate, and take switch out from rear of panel. Take screw out of center of switch lever (fig. 6) and remove lever from face of panel.

b. INSTALLATION. Insert switch in opening from rear of panel, position switch plate against face of panel, and install retaining nut on switch lever shaft. Place switch lever on shaft and secure in position with round head screw. Test lever for proper operation. Connect cables to switch terminals in accordance with the wiring diagram (fig. 93). Install instrument panel in vehicle (par. 133).

139. Fuel Switches

a. REMOVAL. Turn master battery switch "OFF." Disconnect both fuel valve rods at switch levers. Disconnect electrical conduit at emergency ignition switch box. Remove four cap screws and lock washers and remove switch assembly from hull roof (fig. 7). Remove individual switches or circuit breaker by disconnecting cables and removing attaching screws.

b. INSTALLATION. Install individual switches or circuit breaker and connect cables as indicated in diagram (fig. 67). Position switch assembly against hull roof with mounting holes alined and install four cap screws and lock washers. Connect conduit to receptacle at rear of emergency ignition switch box. Connect fuel valve rods to switch levers. Check operation of fuel valves and adjust rods if necessary.

140. Siren (or Horn)

a. REMOVAL. Remove siren switch from bracket assembly (par. 141) and remove cable from siren by removing terminal nut. Loosen screw holding siren conduit clip to hull and slip conduit out of clip. Remove nut from lower end of siren mounting tube and pull siren and conduit out of opening in front hull plate.

b. INSTALLATION. If rubber seal at siren opening is damaged or torn, replace it. Slip conduit through opening and lower siren into position. Slide mounting nut over conduit and tighten or lower end of mounting tube. Position conduit under clip and tighten clip. Connect conduit to switch and install switch.

141. Siren (or Horn) Switch

a. REMOVAL. Open master battery switch. Remove two screws and lock washers holding siren switch bracket to hull floor (fig. 7). Turn switch and bracket assembly over and remove two nuts holding switch conduits to terminals. Loosen siren conduit clip mounting stud nut and slide both conduits out of clip. Remove two screws holding switch to bracket and remove switch.

b. INSTALLATION. Position siren switch on bracket assembly and install two mounting screws. Connect both siren cables to terminals on switch and tighten terminal nuts.

Note. Cables may be installed on either terminal.

Slip both conduits under clip on bracket and tighten clip screw. Position entire assembly on hull floor and install two mounting screws and lock washers.

142. Engine Temperature Gages

a. REMOVAL. Remove instrument panel from vehicle (par. 133). Remove two terminal nuts and take off two gage cables. Remove two nuts, one of which is located on the terminal stud at the side of the gage. Pull mounting clamp back off studs and remove gage from front of panel. This procedure applies for either gage.

b. INSTALLATION. Position temperature gage in opening in panel and place mounting clamp over studs at rear. Install two mounting nuts on studs and tighten snugly. Connect gage cables according to circuit numbers and markings on gage unit. Install terminal nuts. Install panel in vehicle (par. 133). This procedure applies for either gage.

143. Speedometer and Tachometers

a. REMOVAL. Remove instrument panel from vehicle (par. 133). Remove two nuts and washers (fig. 102) and remove mounting bracket from rear of instrument. Remove instrument from front of panel. This procedure applies to speedometer and both tachometers.

b. INSTALLATION. Position instrument in proper opening in front of panel. Place mounting bracket on rear of instrument with holes for screws and mounting studs properly lined up. Install two screws and washers. Install panel in vehicle (par. 133).

144. Signal Switches and Sending Unit

a. ENGINE WATER TEMPERATURE SENDING UNIT.

(1) *Removal.* Open engine compartment door. Disconnect cable from sending unit. Remove water drain plug at bottom of transmission (fig. 108) and drain coolant below level of cylinder heads. Remove sending unit from cylinder head.

(2) *Installation.* Install sending unit in opening in upper front corner of left cylinder head.

> *Note.* Sending unit is cadmium-plated to distinguish it from warning signal switch, which is brass.

Do not use gasket cement (plastic type) or sealer on threads of sending unit, but tighten sufficiently to secure a good seal. Connect cable according to wiring diagram (fig. 104). Fill cooling system (par. 118) and close engine compartment door.

b. ENGINE WATER HIGH TEMPERATURE SIGNAL SWITCH.

(1) *Removal.* Open engine compartment door. Disconnect cable from signal switch. Remove water drain plug at bottom of transmission (fig. 108) and drain coolant below level of cylinder heads. Remove signal switch from cylinder head.

(2) *Installation.* Install signal switch in opening in upper front corner of left cylinder head.

> *Note.* Warning signal switch is brass to distinguish it from engine water temperature sending unit, which is cadmium-plated.

Do not use gasket cement (plastic type) or sealer on threads of unit, but tighten sufficiently, to secure a good seal. Connect signal switch cable according to wiring diagram (fig. 103). Fill cooling system and close engine compartment door.

c. ENGINE LOW OIL PRESSURE SIGNAL SWITCH.

(1) *Removal.* Open engine compartment cover. Reaching down from top of vehicle, disconnect cable from terminal on switch; then remove switch from "L" connection at rear of engine vee.

(2) *Installation.* Reaching down from top of vehicle, install switch on "L" connection, turning down first by hand, and then tightening with wrench only enough to secure a good seal. Connect cable to terminal on switch (fig. 103). Close engine compartment door.

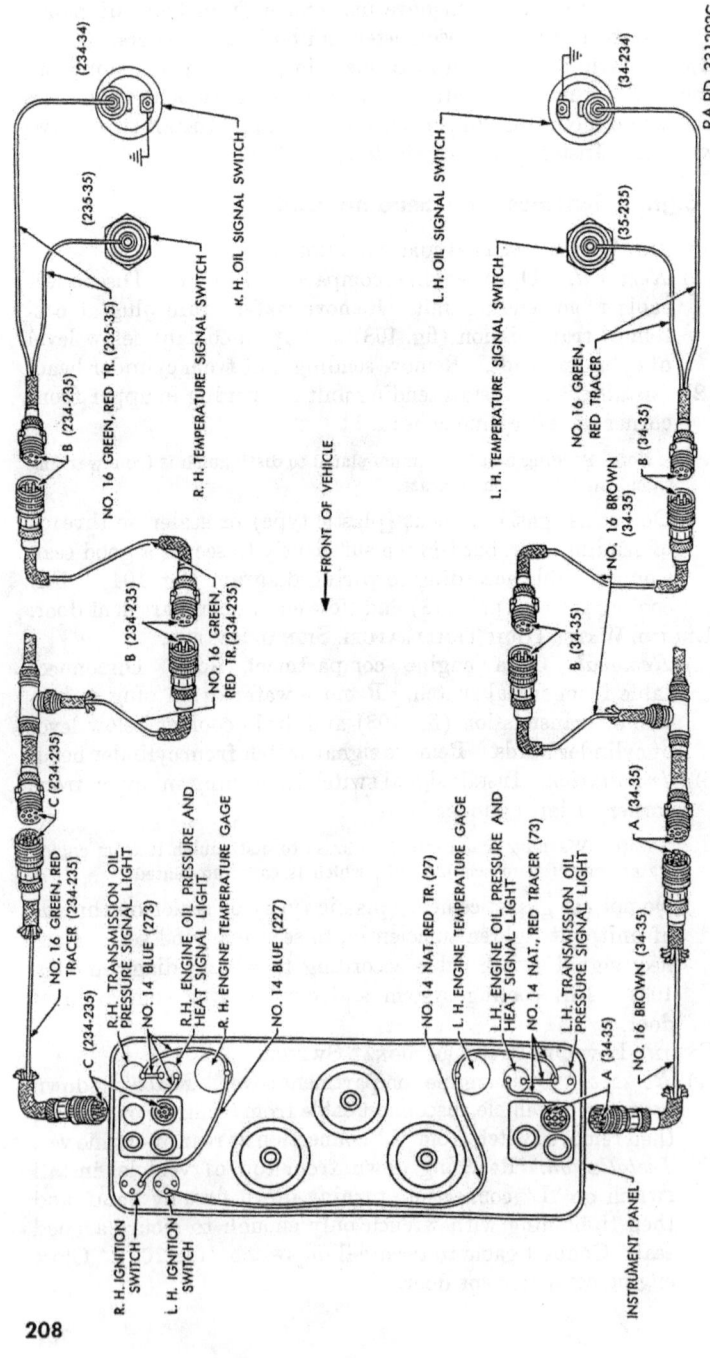

Figure 103. Engine low oil pressure and engine water high temperature circuits wiring diagram.

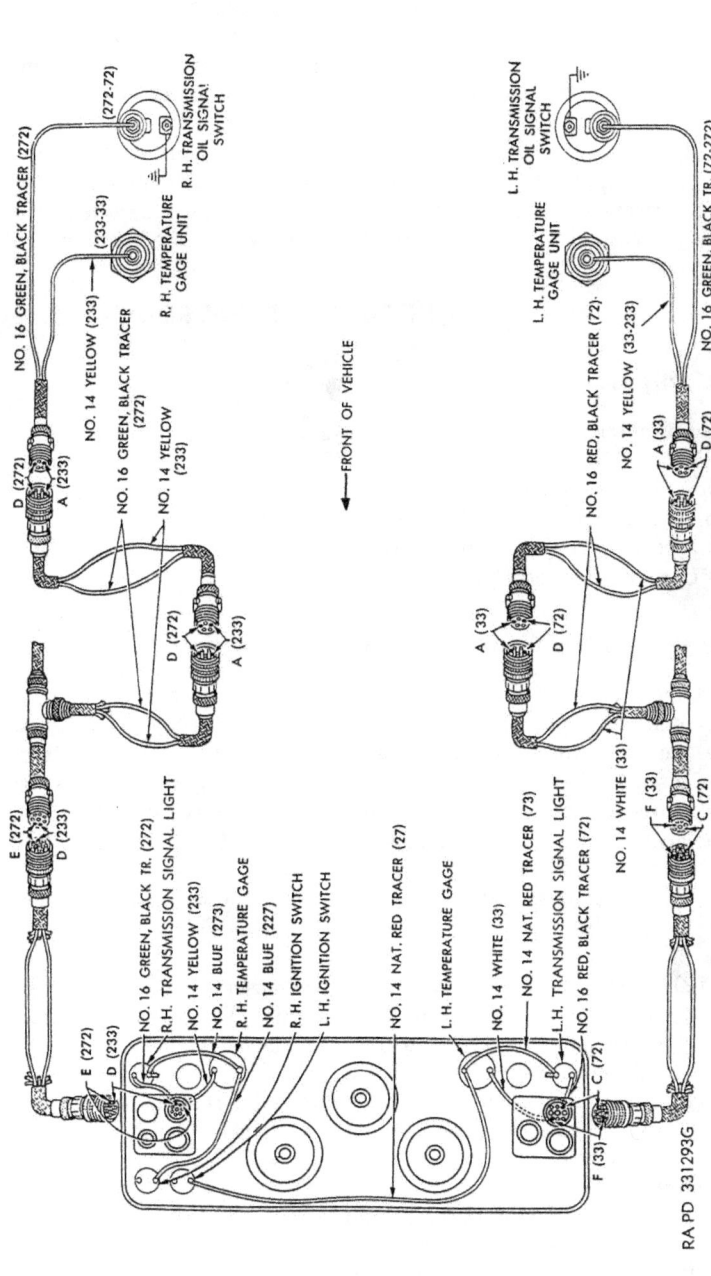

Figure 104. *Engine water temperature sending unit and transmission low oil pressure circuits wiring diagram.*

d. TRANSMISSION LOW OIL PRESSURE SIGNAL SWITCH.
   (1) *Removal.* Remove cover plate from floor of hull underneath engine. Reaching up through opening, remove dirt from around switch, disconnect cable from terminal on switch, and unscrew switch.
   (2) *Installation.* Reaching up through opening in hull, install switch, turning down first by hand and then tightening with wrench only enough to secure a good seal. Connect cable to terminal on switch (fig. 104). Install cover plate in hull floor.

Section XIV. RADIO INTERFERENCE SUPPRESSION

145. Purpose

Radio interference suppression is the elimination or minimizing of the electrical disturbances which interfere with radio reception or disclose the location of the vehicle to sensitive electrical detectors. It is important, therefore, that vehicles with, as well as vehicles without, radios be suppressed properly to prevent interference with radio reception of neighboring vehicles.

146. Description (Early Model Vehicles)

  a. GENERAL. These vehicles were equipped with the resistor capacitor (condenser) type of suppression, wherein spark plug suppressors were used to reduce the high-tension disturbances, and capacitors (condensers) were used to bypass radio frequency disturbances at their source of origin within the vehicle electrical system. Flexible metal hose was used for mechanical protection only.
  b. IGNITION SYSTEM.
   (1) A suppressor of 10,000 ohms resistance was used in each spark plug cable and in the high-tension cable of the coil and distributor.
   (2) An 0.1-microfarad capacitor (condenser) was used on the battery terminal of each ignition coil. This capacitor (condenser) prevented radio frequency surges from being transmitted into the wiring system.
   (3) These engines have since been modified to include new wiring and nonmetallic conduit for the suppression of excessive radio interference.
  c. GENERATING SYSTEM.
   (1) In each generator, a 0.25-microfarad capacitor (condenser) is connected to each armature brush.
   (2) In each regulator, a coaxial capacitor (condenser) of 0.1-

microfarad capacity is connected to the armature terminal.
(3) In each regulator, a coaxial capacitor (condenser) of 0.1-microfarad capacity is connected to the battery terminal of each regulator.

Note. A coaxial capacitor (condenser) is one in which the cable terminals function as the conductor of the circuit to which it is added.

d. BOND STRAPS AND FASTENERS.
  (1) *Function.* These bonds served a dual function: to complete the electrical circuits and to provide a low-resistance path to ground for any radio frequency circuits that might be generated with components to which they were attached.
  (2) *Sheet metal.* Sheet metal within the vehicle was held together by tooth-type lock washers. These teeth dug through the paint and bonded each section to the other, thereby preventing discharge disturbances from one member to the other and providing a uniform shield against radiation within the vehicle.

147. Description (Late Model Vehicles)

a. GENERAL. These late model vehicles are equipped with a completely shielded, waterproof ignition system, together with capacitors (condensers), resistor suppressors, and bonding jumpers. In addition, metal parts in the vicinity of the engine are formed into a shield by use of mesh screen, metal doors, and grilles confining electrical disturbances so they cannot disturb receiving equipment.

b. IGNITION SYSTEM.
  (1) *Ignition coil.* Each ignition coil is integrally shielded and suppressed as follows:
    (*a*) Ferrules are incorporated at electrical connections for terminating shielding of external wiring with exception of cable to coil where shielding terminates approximately 10 inches from the coil.
    (*b*) One 0.1-microfarad, 100-volt, dc capacitor (condenser) is built into each coil and connected to the battery terminal of each coil.
    (*c*) Each coil is mounted on a bracket bolted to the cylinder head and both the coils and brackets are grounded by means of tooth-type lock washers.
  (2) *Distributor.* Each distributor is integrally shielded and suppressed as follows:
    (*a*) Ferrules are provided at electrical connections for terminating loose braid shielding of external wiring.
    (*b*) One 5,000-ohm resistor-suppressor is molded in each tower of each distributor (18 used).

(c) Each distributor is grounded to the radiator support cross-member by a bonding jumper.

(3) *Spark plugs.* Each spark plug is integrally shielded and suppressed as follows:

(a) Ferrules are provided for terminating shielding of high-tension wiring.

(b) One 5,000-ohm resistor-suppressor is internally mounted in each spark plug.

(4) *High-tension wiring.* Each high-tension cable from ignition coil to distributor and distributor to spark plugs is inclosed in loose braid shielding and each end terminated at ferrules.

(5) *Low-tension wiring.*

(a) The low-tension cable from ignition coil to distributor is inclosed in loose braid shielding and each end terminated at ferrules.

(b) The low-tension cable to the ignition coil is inclosed in flexible metal hose shielding terminating approximately 10 inches from the coil but grounded to the cylinder block through a pressure clip.

c. GENERATOR AND CHARGING SYSTEM.

(1) *Generator.*

(a) Two 0.1-microfarad, 100-volt, dc feed-through capacitors (condensers) are mounted inside the housing of each generator and connected to the positive brush terminals of each generator (four used).

(b) Each generator is grounded to the generator mounting bracket by means of a bonding jumper.

(2) *Regulator.*

(a) One 0.1-microfarad, 100-volt, dc feed-through capacitor (condenser) is mounted in the subbase and connected to the armature terminal of each regulator.

(b) One 0.1-microfarad, 100-volt, dc feed-through capacitor (condenser) is mounted in the subbase and connected to the battery terminal of each regulator.

(c) One 0.0019-microfarad, 100-volt, dc capacitor (condenser) is mounted in the subbase and connected in series with a 4-ohm resistor to the field terminal of each regulator.

(d) Each regulator is grounded to the bulkhead by a bonding jumper from each mounting leg across rubber shock mounts to the bulkhead (eight used).

(3) *Charging system wiring.*

(a) Regulator terminals are covered on the sides and top by a metal shield secured to the regulator by bolts and tooth-type lock washers. The rear of the shield is open.

(b) Generator terminals are covered by a shield secured over the terminals by a spring clamp.

(c) Armature and field cables between regulator and generator are inclosed in flexible metal-hose shielding terminated at a ferrule at the generator, but terminated approximately 10 inches short of the regulator terminals.

d. RADIO TERMINAL BOX AND MISCELLANEOUS BONDING.

(1) One 0.1-microfarad, 100-volt, dc capacitor (condenser) is mounted in the radio terminal box connected to the 24-volt terminal.

(2) The engine rear air outlet grille is grounded to support brackets by tooth-type lock washers.

(3) The top front of engine compartment grille is shielded by mesh screen.

(4) The radiator support is grounded to the engine compartment side wall by two bonding jumpers.

(5) Each carburetor air intake pipe is grounded to the engine compartment side wall by a bonding jumper.

Section XV. HYDRAMATIC TRANSMISSION

148. Description and Data

a. DESCRIPTION. The hydramatic transmission (fig. 105) consists of a fluid coupling and an automatic transmission having four speeds forward. No reverse gearing is incorporated in the transmission, as

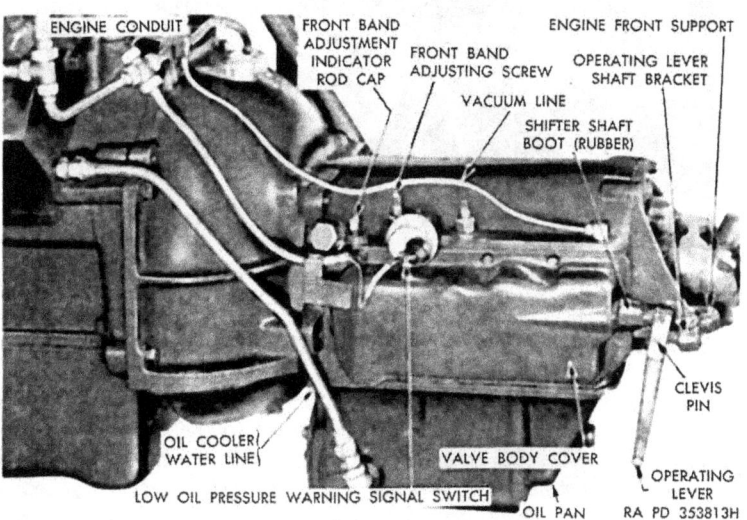

Figure 105. Hydramatic transmission.

this is provided in the transfer unit. Slippage in the fluid coupling at engine idling speeds eliminates the need for a clutch. Gear changes are made automatically by hydraulic pressure and are governed by the speed of the vehicle and the extent to which the driver depresses the accelerator. In this section, the fluid coupling end of the transmission will be referred to as the "front" end and the output shaft end as the "rear" end.

 b. DATA.

  Clutch type_____ fluid coupling
  Number of speeds_____ 4
  First speed gear ratio_____ 3.92 to 1
  Second speed gear ratio_____ 2.53 to 1
  Third speed gear ratio_____ 1.55 to 1
  Fourth speed gear ratio_____ 1 to 1
  Type gearing_____ planetary

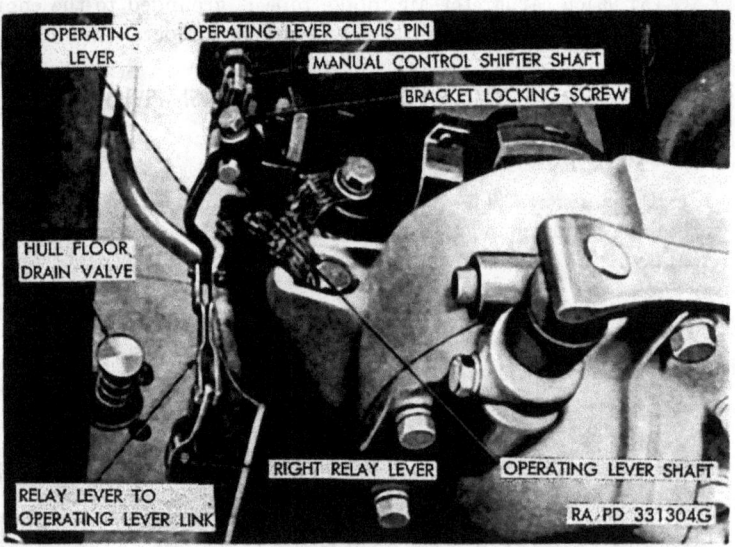

Figure 106. Transmission control linkage.

149. Manual Control Linkage Adjustment

 a. GENERAL. The transmission manual control linkage from the neutral pedal in the driving compartment to the cross shaft under the bulkhead is set at assembly of the vehicle and normally requires no further adjustment. If engines are replaced or it becomes necessary to synchronize transmission control linkage, adjustment should be made at the operating lever mounting bracket on the transmission rear bearing support (fig. 106) as outlined in *b* and *c* below.

b. ADJUSTMENT PROCEDURE. Place transmission selector lever in "DRIVE" slot of quadrant (fig. 9). Open door in bulkhead extension cover over transfer unit. Working through opening, loosen locking screw on operating lever shaft (fig. 106). Move operating lever and attached levers to the middle detent position. Holding shaft in position, tighten locking screw. Adjust other transmission linkage in same manner.

c. CHECK ADJUSTMENT OF BOTH TRANSMISSIONS. Set brakes. Run both engines at same time with transfer unit in "HIGH" and transmission selector lever in "NEUTRAL." Set hand throttle to run engines at 1,000 rpm, and move selector lever slowly toward "DRIVE." By watching tachometers, note position of lever when speed of one engine drops. Rear edge of selector lever should be approximately one-eighth of an inch in front of front edge of slot in quadrant when engine speed drops. Continue moving lever until speed of other engine drops. This should be within one-eighth of an inch additional lever travel. If difference is greater, adjust operating lever of newly installed transmission.

150. Test and Band Adjustment

a. GENERAL. The transmission should be tested for unusual noise, oil pump pressure, and band adjustment.

b. TEST FOR UNUSUAL NOISE. During the course of testing the transmission, any unusual noises should be reported to ordnance maintenance personnel.

c. CHECK OIL PUMP PRESSURE. Remove the low oil pressure warning signal switch (fig. 107) and install the transmission oil pressure gage—41-G-446. Run engine until the fluid in transmission is at normal operating temperature. With the manual control rod in the middle detent position ("DRIVE" position), the oil pressure, as indicated on the pressure gage, should not be less than 85 pounds at 1,000 rpm or over. If pressure is below 85 pounds, refer to ordnance maintenance personnel.

d. BAND ADJUSTMENT.

(1) *General.* The transmission front band is provided with an adjustment indicator rod (fig. 107) and can be adjusted in the vehicle. The rear band is provided with an automatic band adjuster and does not incorporate an indicator rod. If tests indicate that the rear band requires adjusting, notify ordnance maintenance personnel.

(2) *Band adjustment procedure.* Remove 18 mounting screws and remove hull rear floor cover under engine and transmis-

sion to be checked. Set vehicle brakes, place transmission selector lever in "NEUTRAL," and transfer unit selector lever in "HIGH" or "LOW" position, and start engine.

Note. Transmission should, if possible, be at normal operating temperature before making band adjustments.

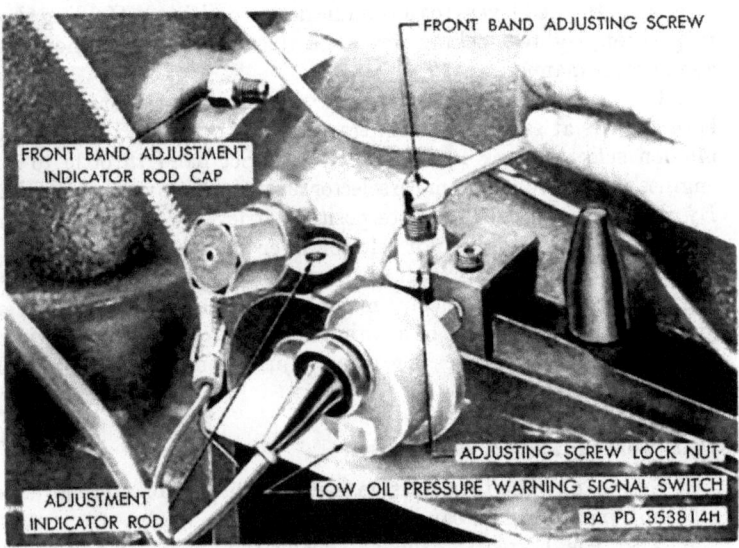

Figure 107. Adjusting transmission front band.

Working through the engine compartment floor opening, remove transmission front band indicator rod cap (fig. 105). Run engine for a few minutes at idling speed; then move transmission selector lever to "DRIVE" position and set hand throttle so that engine is running at approximately 1,000 rpm. Hold a straightedge or scale across front band indicating rod boss, and check to see if indicator rod end is flush with rod boss on transmission case. Band is correctly adjusted when indicator rod end is flush with rod boss on transmission case (fig. 107). If band adjustment is incorrect, move lever back to "NEUTRAL," loosen band adjusting screw lock nut, turn front band adjusting screw to estimated correct adjustment, move lever again to "DRIVE," and check. Repeat until indicator rod end is flush with rod boss. Idle engine. Lock band adjusting screw and check adjustment. Install indicator rod cap, being sure seal is in place.

Caution: Run one engine only; do not pull the engine against the fluid coupling for more than 1 minute at a time and allow at least 3 minutes for the oil to cool before re peating.

151. Transmission Replacement

a. REMOVAL. Replacement of this major assembly with a new or rebuilt unit is normally an ordnance maintenance operation, but may be performed in an emergency by using organizations, provided authority for performing this replacement is obtained from the responsible commander. Tools needed for this operation which are not carried in the using organization, replacement transmission, and necessary instructions, may be obtained from the supporting ordnance maintenance unit.

(1) *Remove engine.* Remove engine and transmission assembly (par. 97).

(2) *Remove lower flywheel housing.* Remove six screws that hold lower flywheel housing to upper housing. Remove housing and discard gasket.

(3) *Remove starter.* Disconnect solenoid cable at terminal on solenoid relay. Remove two mounting bolts holding starter to flywheel housing and remove motor.

(4) *Remove lines and electrical cables.* Disconnect right and left oil cooler lines at transmission oil pan fittings (fig. 105), loosen upper connections at crankcase and water pump inlet line, and move lines out of the way. Loosen hose clamp on hose connecting transmission and engine oil filler tubes at transmission filler side. Remove screw that holds transmission filler tube lower mounting to transmission case. Remove nut holding filler tube upper mounting bracket to rear manifold clamp stud and remove tube. Disconnect vacuum line at fitting on intake manifold and at elbow on transmission case (fig. 105). Remove flywheel housing mounting screw holding vacuum line clips and remove line. Disconnect low oil pressure warning signal switch cable at switch on transmission and pull conduit out of clip on transmission side cover.

(5) *Drain transmission.* Install socket wrench on engine crankshaft pulley mounting screw and turn crankshaft until torus cover drain plug is at the bottom. Remove drain plug in torus cover and in transmission oil pan (fig. 108) and drain transmission. Install drain plugs.

(6) *Remove transmission.* Install lifting eye bolt —41–B–1586–300 in top of transmission case. Connect hoist to eye bolt and take up weight of transmission. Remove 30 screws that hold flywheel cover to flywheel.

Note. Rotate crankshaft with socket wrench on engine crankshaft pulley mounting screw.

Remove remaining six screws that hold flywheel housing to crankcase and pull engine conduit out of the way. Pull transmission back to slide flywheel housing off dowels in crankcase and main shaft pilot out of bearing in end of crankshaft and remove transmission.

Figure 108. Transmission drains.

b. INSTALLATION.

(1) *Install transmission.* Position new torus cover gasket on flywheel.

Caution: Do not use gasket sealer of any kind.

Install lifting eye bolt—41–B–1586–300 in transmission case, attach hoist to eye bolt, and lift transmission into position behind engine crankcase. Push transmission toward engine, entering dowels in crankcase in holes in flywheel housing and mainshaft pilot in bearing in end of crankcase. Install six flywheel housing mounting screws, placing clips for engine conduit, and vacuum pipe clip under upper screws. Tighten screws to 45–50 foot-pounds torque, using torque wrench, and tighten lower screws first.

(2) *Install torus cover.* Pull torus cover toward flywheel and line up dowels in flywheel with holes in cover. Hold cover in position and install one mounting screw. Turn flywheel 180° (using socket wrench on crankshaft pulley mounting screw) and install another screw. Tighten these two screws to 12–15 foot-pounds torque, making sure that cover seats properly on flywheel and flywheel dowels. Install remaining 28 screws and tighten all screws to 25–30 foot-pounds torque, using a torque wrench. Then tighten all screws to 40–45 foot-pounds, proceeding in rotation around the flywheel. Disconnect hoist and remove lifting eye bolt.

(3) *Install starter.* Lift starter into position on flywheel housing and install two mounting screws through housing and into starter mounting flange. Connect solenoid cable to terminal on solenoid relay.

(4) *Connect lines and electrical cables.* Connect transmission low oil pressure warning signal switch cable to switch and push conduit into clip on transmission side cover. Connect vacuum line to intake manifold and to elbow on transmission case. Position filler pipe in transmission case and push upper side outlet into hose connection on engine oil filler tube and upper mounting bracket over engine manifold rear stud. Install screw through lower filler tube mounting bracket and into transmission case. Install nut holding upper tube bracket to manifold stud and tighten hose clamp on transmission to engine filler tube hose. Connect oil cooler lines to fittings on transmission oil pan and tighten upper and lower connections.

(5) *Install lower flywheel housing.* Install new gasket on lower flywheel housing and lift into position on upper flywheel housing and install six mounting screws.

(6) *Install engine.* Install engine and transmission assembly in vehicle (par. 98).

(7) *Fill transmission.* Proceed as directed on lubrication order (par. 62). Start engine and run for approximately 5 minutes to fill fluid coupling (fig. 109); then add additional oil to bring level up to full mark on gage.

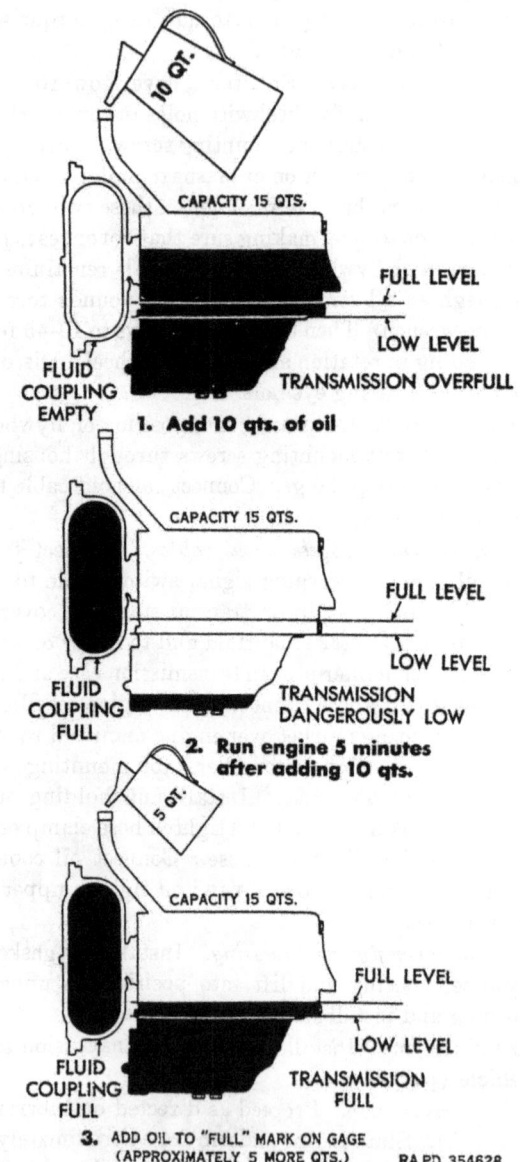

Figure 109. Filling transmission.

152. Torus Replacement

a. REMOVAL.
   (1) *Remove engine.* Remove engine and transmission assembly from vehicle (par. 97).
   (2) *Remove transmission.* Remove transmission assembly from engine (par. 151).
   (3) *Remove torus members.* Straighten main shaft nut lock and remove nut and lock holding driven torus member on shaft. Pull driven torus off main shaft. Remove snap ring holding driving torus on intermediate shaft, using snap ring pliers (fig. 110). Pull out on driving torus and remove from intermediate shaft.

b. INSTALLATION.
   (1) *Install torus members.* Push driving torus on splines of intermediate shaft and install snap ring (fig. 110), using snap ring pliers. Push driven torus on transmission main shaft splines and see that oil pressure regulator on inner hub slides over end of intermediate shaft. Install driven torus retaining nut and new lock on main shaft and bend lock back over nut.
   (2) *Install transmission.* Install transmission assembly on engine (par. 151).

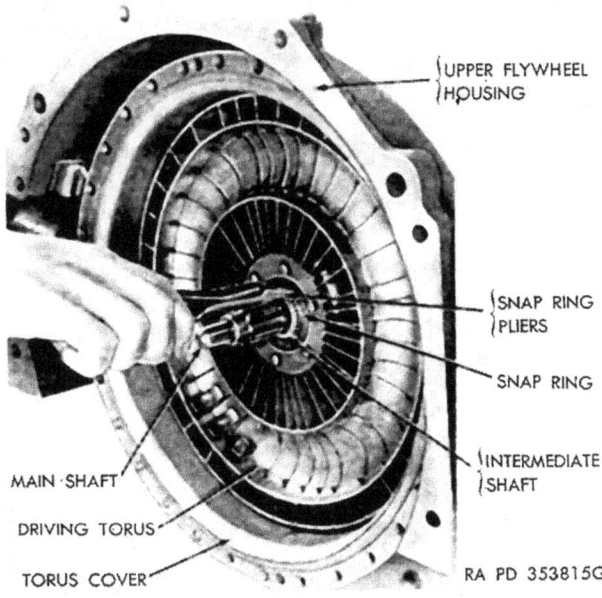

Figure 110. *Removing driving torus snap ring.*

(3) *Install engine.* Install engine and transmission assembly in vehicles (par. 98).

153. Valve Body

a. REMOVAL.

(1) *Remove hull floor cover.* Remove screws from hull floor cover under transmission. Front of cover is held up by hooks on cover. Slide cover toward rear of vehicle and lower to ground.

Caution: Support rear end of cover, while removing last screws, to prevent cover from dropping down and causing injury.

(2) *Remove side cover.* Working through the floor opening, remove hairpin lock and clevis pin which hold operating fulcrum lever to adjustable mounting bracket on rear bearing support (fig. 106) and lower lever out of the way.

Note. Do not disconnect operating lever at lower end.

Remove the two cap screws which hold the lever shaft bracket to the transmission and remove the bracket, by pulling it straight out, to clear the two locating dowel pins. Pull transmission low oil pressure signal unit conduit out of retaining clip on side cover. Disconnect cable on signal unit and remove signal unit. Pull rubber boot off end of manual control shifter shaft and back shaft bushing nut out of cover. Remove 11 screws which hold side cover to transmission case (fig. 105) move cover out of engine end to clear valve body; then pull toward front of vehicle to clear manual control shifter shaft end and remove cover.

Caution: When removing side cover, be very careful to prevent any dirt from falling into transmission through case openings.

(3) *Remove valve body.* Disconnect control valve vacuum control line at valve body and elbow on transmission case and remove line (fig. 111). Remove four control valve body mounting screws, pull valve body assembly toward engine to move oil tubes out of governor sleeve, and remove assembly (fig. 112).

Note. The only operations the using organizations are authorized to perform on the valve body are removal and installation.

An improperly operating valve body should be reported to ordnance maintenance personnel.

b. INSTALLATION.

(1) *Install valve body.* Make sure valve body mounting surface on transmission case and valve body are free of nicks and

Figure 111. Transmission valve body installed.

burs and thoroughly clean. See that three oil lines are in position in valve body (fig. 112) and place valve body in position on transmission case, entering the three oil lines into governor sleeves. Hold in position and install four mounting screws, tightening evenly to 6-8 foot-pounds torque, using a torque wrench. Position vacuum line on valve body and transmission case fittings and connect.

(2) *Install side cover.* Be sure governor plug is in place in governor sleeve. Position a new side cover gasket on transmission case.

Caution: Do not use gasket sealer of any kind.

Place side cover in position by holding valve body end of cover away from transmission and sliding opposite end of cover over manual control shifter shaft, then move cover toward engine block and over valve body. Install 11 cover mounting screws and copper washers. Slide shifter shaft bushing nut over shaft, tighten into side cover, and install rubber boot over shaft, seeing that outer end of grommet seats in groove in shaft. Install transmission low oil pressure warning signal switch. Push conduit into clip on side cover and connect wire to switch terminal. Install operating lever shaft bracket to transmission, making sure holes in bracket line up with locating dowels on transmission. Install two attaching screws and tighten securely.

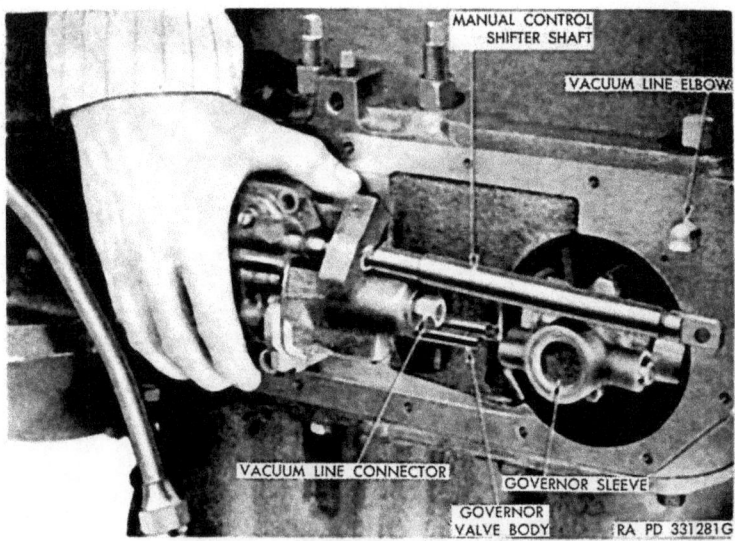

Figure 112. Removing transmission control valve body.

(3) *Adjust manual control linkage.* Connect and adjust linkage at operating lever shaft bracket (par. 149).

(4) *Install hull floor cover.* Coat edges of floor plate with gasket cement (plastic type) and place gasket on plate. Raise front edge of plate until hooks on plate rest on hull floor. Raise rear end of plate and line up screw holes with drift punch. Install floor plate retaining screws.

Section XVI. TRANSFER UNIT

154. Description and Data

a. DESCRIPTION. The transfer unit is mounted on support brackets on the hull floor between and partially ahead of the two transmissions. The transfer unit performs two functions: it combines the power output of the two engines and it provides two speed ranges forward and one in reverse, all of which are manually selected. Each transfer unit input gear is connected to the input shaft by means of a sliding coupling, which in turn is connected to the input clutch lever on the outside of the case. This arrangement permits disconnecting either one of the engines if an engine is disabled, so that the vehicle can be driven by one engine without the drag of the disabled engine.

b. DATA.

Number of speeds forward	2
Number of speeds reverse	1
Gear ratio in "LOW"	2.34 to 1
Gear ratio in "HIGH"	1.03 to 1
Gear ratio in "REVERSE"	2.44 to 1
Type of gears	helical
Type of shift	manual, synchro mesh

155. Transfer Unit Control Levers

a. DESCRIPTION. The transfer unit shift hand lever is mounted in a bracket bolted to the hull floor at the right side of the driver's seat (fig. 7). When the lever is moved to the left and pushed forward, the lower end of the lever engages the reverse arm and lever assembly which is connected by rods to the reverse gear shifter shaft assembly. This slides the reverse gear clutch along the main shaft in the transfer unit, locking the reverse gear to the main shaft. When the lever is moved to the right and pushed forward or backward, the lower end of the lever engages the arm and lever assembly for the "HIGH" and "LOW" shifter shaft. This moves the high and low gear synchronizer clutch along the main shaft to engage the gear selected by the driver. Two throwout clutch handles one for each engine, are mounted on vertical shafts on each slide of the transfer unit. When a handle is in the rear position, the sliding clutch on the input gear is engaged with the input shaft. When the handle is moved forward, the sliding clutch is disconnected.

b. REMOVAL.

(1) *Remove periscope and spare head stowage box.* Remove four screws holding the box to the four supporting brackets. Push transfer unit shift hand lever to the "REVERSE" position and remove stowage box assembly.

(2) *Remove flare box.* Remove two screws holding flare box (next to driver's seat) to hull floor. Remove box.

(3) *Remove shift lever spring and guide.* Lift up on spring guide to compress spring until guide clears hole in shifting lever shaft. Remove spring and guide (fig. 113).

(4) *Remove shift lever shaft lock bolt.* Remove nut and washer from shift lever lock cap screw and remove screw (fig. 113).

(5) *Remove shift lever shaft.* Slide shift lever shaft out of support bracket toward driver's seat.

Caution: Arm and lever assemblies and spacer washers will drop down when shaft is removed.

Note position of parts before removing shaft so they will be installed in the proper sequence.

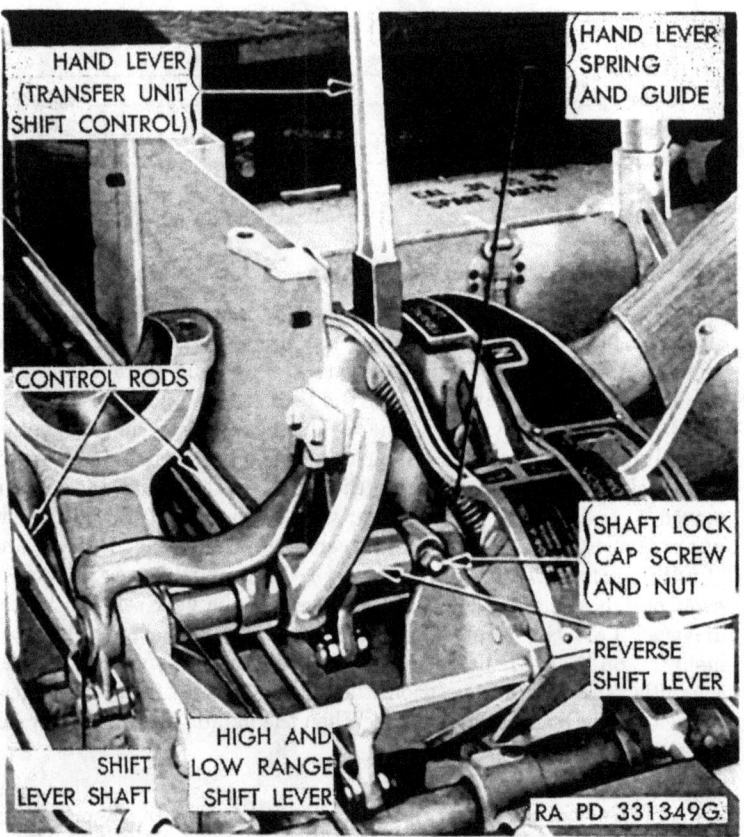

Figure 113. Transfer unit manual controls.

(6) *Remove shift lever.* Tilt top of lever toward right side of vehicle so that bottom part of lever comes out of side of support bracket. Raise driver's seat to allow room for removing lever.

(7) *Remove throwout clutch handle.* Remove throwout clutch handle clamp screw and lift off clutch handle (fig. 114).

c. INSTALLATION.

(1) *Install throwout clutch handle.* Place throwout clutch handle over clutch throwout arm shaft and install clamp screw.

(2) *Position shift lever in support bracket.* Slide top of shift lever through opening in left side of support bracket so that top of lever comes out through opening on quadrant.

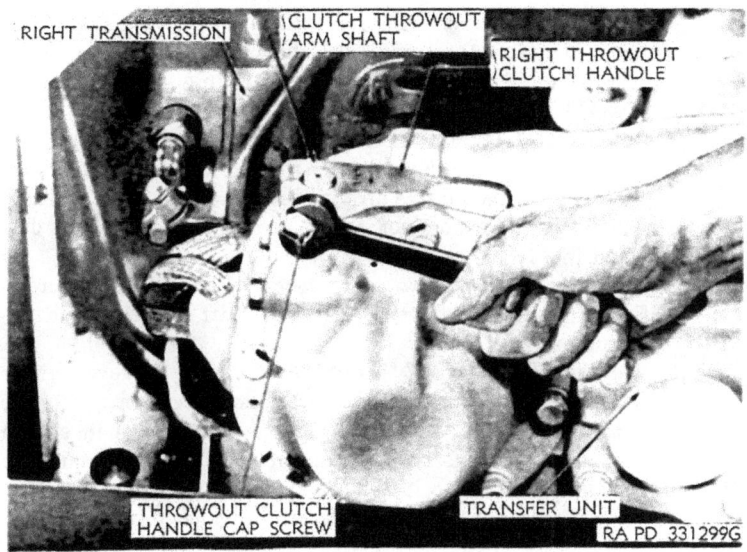

Figure 114. Removing engine input clutch lever.

(3) *Install shift lever shaft.* Slide shift lever shaft through opening in support bracket.

Note. Do not slide shaft all the way in.

Raise reverse shift lever and arm assembly and slide on end of shift lever shaft. Next, place one of the spacer washers on shaft and slide shaft through hole in shift lever. Install another spacer washer over shift lever shaft. Raise "HIGH" and "LOW" range shift lever and arm assembly up into position. Slide shaft through this arm also. Install the last spacer washer over the shift lever shaft and slide shaft into other end of support bracket (fig. 113). Line up hole in shaft and bracket and install lock bolt, nut, and washer.

(4) *Install shift lever spring and guide.* Slide spring and guide into opening on shift lever. Compress spring on guide so that guide will enter cutout on shift lever shaft (fig. 113).

(5) *Install flare box.* Position flare box on tapping blocks on hull floor next to driver's seat and install screws and washers.

(6) *Install periscope and spare head stowage box.* Place transfer unit shift hand lever in reverse position to allow room to install box. Slide box into position on the four supporting brackets and install four screws and washers.

d. LINKAGE ADJUSTMENT.

(1) *Check adjustment.* Move shift control hand lever from

"HIGH" to "LOW" and from "LOW" to "NEUTRAL" to locate by feel the middle of the neutral detent position. Lever should now be in line with gate in quadrant opposite "NEUTRAL." If it is not, adjust as outlined in (3) below. When "HIGH" and "LOW" shifter shaft is properly adjusted, move lever through gate toward driver. Slot in bottom of lever should line up with lug on reverse lever and arm assembly. If it does not, adjust reverse shifter shaft as outlined in (4) below.

(2) *Remove bulkhead extension cover.* Remove eight screws holding bulkhead extension cover. Lift out bulkhead extension cover.

(3) *Adjust "HIGH" and "LOW" shift rod.* Remove shifter shaft extension clamp bolt that locks extension to shifter shaft (fig. 115). Drive a wedge or small chisel in slot in extension until extension is free on shaft. Move hand lever toward "HIGH" position until shifter shaft extension slides off shifter shaft. Remove cotter pin and washer holding shift rod to shifter shaft extension. Loosen lock nut on shift rod and turn clevis in or out a sufficient amount so that, when the shifter shaft extension is assembled on the shifter shaft, the shift lever will be in the center of the gate in the quadrant. When adjustment is correct, install cotter pin and washer holding shift rod to shifter shaft extension and install clamp bolt holding shifter shaft extension to shifter shaft.

(4) *Adjust reverse shift rod.* After "HIGH" and "LOW" shift rod has been adjusted, adjust "REVERSE" shift rod in same

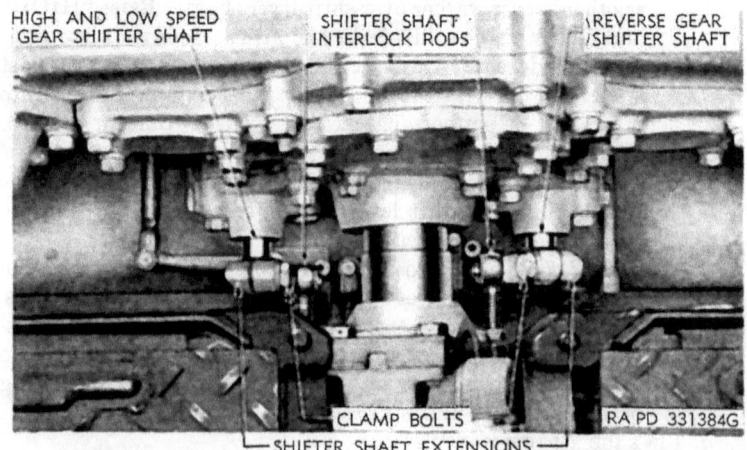

Figure 115. Transfer unit shift linkage.

manner so that the slot in the bottom of the shift lever will be in line with the lug on reverse lever and arm assembly. Shift lever should engage "HIGH" and "LOW" lever and "REVERSE" lever without interference.

Note. On late vehicles, the adjustment of the shift rods can be made without disconnecting the shift rod from the shifter shaft extension by merely loosening the lock nuts on the rod and turning the adjustment nut (which has left and right thread) in or out as required.

(5) *Install bulkhead extension cover.* Position bulkhead extension cover on bulkhead and subfloor and install attaching screws.

156. Transfer Unit Replacement

a. REMOVAL.
 (1) *Remove engines.* Refer to paragraph 97.
 (2) *Remove radiators.* Refer to paragraph 121.
 (3) *Remove lower radiator seal.* Remove four screws holding lower radiator seal to engine bulkhead. Lift out seal.
 (4) *Remove radiator center support.* Working from fighting compartment, remove two screws holding center radiator support to front of engine bulkhead. Remove four screws holding center radiator support to rear radiator support. Lift out support.
 (5) *Remove lower fire extinguisher lines.* Remove four clamps holding lines to hull floor. Disconnect lower center fire extinguisher lines at hull side wall connections and remove screws holding center bracket to hull floor.
 (6) *Remove throttle relay rods.* Disconnect and remove intermediate to rear throttle relay rod and rear relay as an assembly.
 (7) *Remove engine front support cushions.* Remove eight screws holding engine front support cushions to engine front support brackets. Remove cushions.
 (8) *Disconnect shifter shaft extensions.* Refer to paragraph 155*d*.
 (9) *Remove transfer unit.* Attach a rope sling around transfer unit as shown in figure 116. Attach chain hoist to rope sling and take up weight of transfer unit enough to clear rubber cushion supports. Remove the three transfer unit mounting screws. Raise transfer unit and slide toward rear of vehicle until propeller shaft yoke slides out of spline and shifter shaft extensions slide off shifter shafts. Raise transfer unit and remove from vehicle.

b. INSTALLATION.

(1) *Place transfer unit in vehicle.* Attach a rope sling around transfer unit, being sure that it is tied as short as possible (fig. 116), and connect chain hoist to unit. Lower unit into vehicle. Before lowering unit all the way into position, slide propeller shaft splined yoke into front end of transfer unit and also place the two shifter shaft extensions over the two shifter shafts. Slide transfer unit forward until it rests on the rubber cushion supports. Remove hoist and rope sling.

(2) *Install mounting screws.* Install two front and one rear transfer unit mounting screws. Tighten securely.

(3) *Connect and adjust linkage.* Refer to paragraph 155*d*.

(4) *Install lower fire extinguisher line.* Lower the fire extinguisher line into position at rear of transfer unit. Install two screws holding center bracket to hull floor and connect

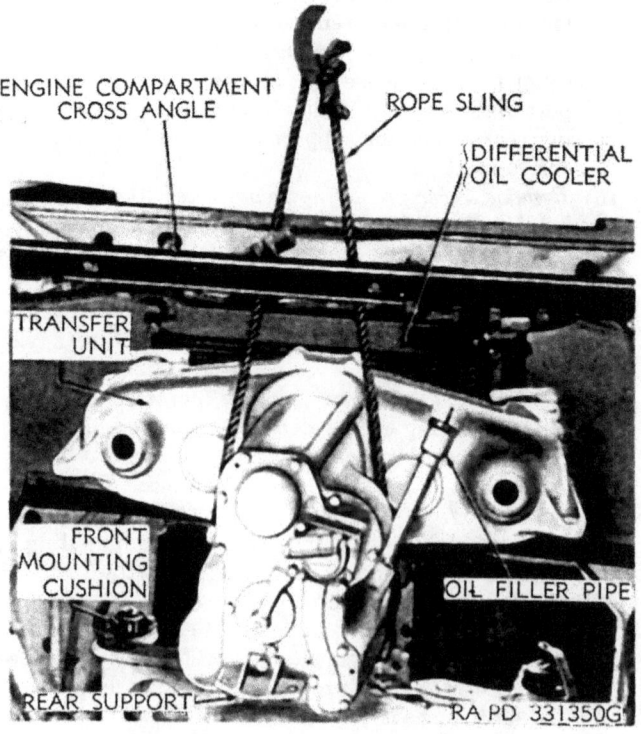

Figure 116. Removing transfer unit.

lines to couplings on hull side wall. Install four clamps and screws.

(5) *Install throttle relay rods.* Install intermediate to rear throttle relay rod and rear relay.

(6) *Install engine front mounting cushions.* Place the four front engine cushion rubber mounting pads on the front engine support mounting brackets and install eight screws and lock washers.

(7) *Install radiator center support.* Slide lower end of center radiator support through slot in bulkhead and install two mounting screws. Position upper end of support against rear radiator support and install four screws and washers.

(8) *Install lower radiator seal.* Place lower radiator seal in position in opening. Install four screws and washers holding seal to bulkhead.

(9) *Install radiators.* Refer to paragraph 121.

(10) *Install engines.* Refer to paragraph 98.

Section XVII. PROPELLER SHAFTS AND SUPPORT BEARINGS

157. Description and Data

a. DESCRIPTION. There are four propeller shafts used in this vehicle. Two of these shafts, called the main propeller shafts, extend from the transfer unit forward to the controlled differential. The rearmost shaft is the longest, extending up to the support bearing by the driver's seat. The forward shaft connects the rear shaft to the controlled differential. Two final drive propeller shafts extend from the controlled differential, one on each side, to the final drive housing. The propeller shafts are all of welded steel tube construction.

(1) *Main propeller shafts.* The rear main propeller shaft has a yoke for a universal joint welded to the rear end and has a short splined shaft welded to the front end (fig. 117). The second universal joint yoke is carried on this splined shaft and held in place with a nut, washer, and cotter pin. The front main propeller shaft is provided with a yoke for the universal joints welded to each end. The rear yoke of the rearmost universal joint is splined to provide a sliding joint with the transfer unit main shaft; and the front yoke of the foremost universal joint is splined to provide a sliding joint with the differential pinion shaft.

(2) *Final drive propeller shafts.* The final drive propeller shafts are of the same welded tube construction, but shorter and much larger in diameter, with larger universal joint yokes welded to each end (fig. 120). The yokes on each side of the controlled differential are splined to the output shafts and are a slip fit to compensate for movement of the unit and to provide clearance when removing the shafts. The yokes on the final drive input shafts are also splined, but are held on the shafts by a large nut and cotter pin.

(3) *Support bearing.* The main propeller shafts are mounted at their junction on a bearing support bolted to the hull floor just behind and to the right of the driver's seat. The rear propeller shaft is carried in this support on a double row ball bearing mounted in the bearing carrier (fig. 117). Lubricating fittings are provided in the carrier for lubrication of the bearing and bushing.

(4) *Tunnel and guard.* The main propeller shafts are mounted in a tunnel formed by stowage boxes on each side and sealed at the rear by a bolted plate (figs. 118 and 119). The top of the propeller shaft tunnel is covered at the rear by hinged covers which form part of the vehicle subfloor and in front by a periscope stowage box. The sides of the tunnel are supported by brackets attached to the hull floor. The final drive propeller shafts are protected by sheet metal covers which fit around the shafts and universal joints and are bolted together at mounting brackets on the controlled differential and final drive cases. Covers are mounted on the final drive housings. All covers are easily removable for service work.

(5) *Universal joints.* The connection between each transmission and the transfer unit is made by two universal joints and a coupling block. The transmission output shaft carries the rear yoke; the front yoke is integral with the short transfer unit input shaft.

b. DATA.

```
Propeller shafts_____  4
Type _____ welded tube
Universal joints_____  11
Type_____ needle bearing
Make_____ mechanic's
```

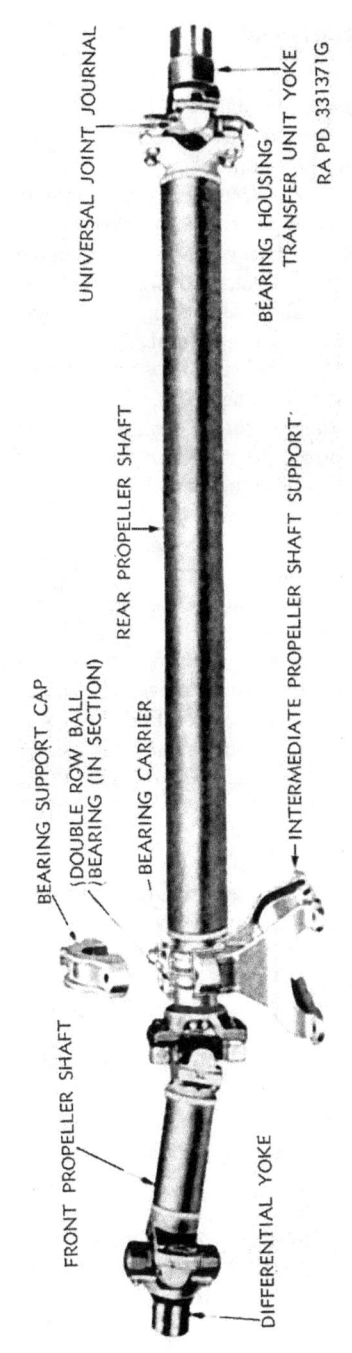

Figure 117. Main propeller shafts and mountings.

158. Main Propeller Shafts

a. REMOVAL.

(1) *Remove periscope stowage box.* Remove four cap screws holding periscope stowage box to support bracket and lift box from its position.

(2) *Remove left propeller shaft cover.* Remove one screw holding this cover in position and lift out cover.

(3) *Disconnect front universal joint.* Remove locking wires and four cap screws from front propeller shaft universal joint yoke, pry propeller shaft back to telescope rear yoke on transfer unit shaft and break joint.

Caution: Tie bearing caps to cross of universal joint to avoid losing needle bearings.

(4) *Remove collector ring assembly.* Remove two cap screws holding conduit guard cover. Remove four nuts and lock washers for periscope box and four nuts and lock washers for oddment box and remove both boxes. Lift collector ring and turret control box to one side.

(5) *Remove bulkhead extension seal.* Remove two screws holding seal in position at rear of tunnel and lift upper seal out of tunnel (fig. 119).

(6) *Remove bearing cap.* Remove two bolts holding bearing cap to bearing support and remove cap (fig. 118).

(7) *Remove propeller shafts.* Push the propeller shafts to the rear, telescoping them as much as possible. Pull assembly off transfer unit shaft, lift out of bearing support, and remove entire assembly from vehicle through the driver's door.

b. INSTALLATION.

(1) *Install propeller shafts.* Lower both the front and rear propeller shafts, as an assembly, into the vehicle through driver's door opening. Slide yoke on rear end of rear shaft on splines of transfer unit output shaft. Position front of rear shaft on bearing support.

(2) *Connect front universal joint.* Raise front end of front shaft and telescope shaft on transfer unit output shaft until sufficient room is provided to connect front universal joint. Remove wires holding bearing housings on crosses. Line up bearing housings with yoke on controlled differential pinion shaft. Install four cap screws and locking wires and tighten securely.

(3) *Install propeller shaft cover.* Install left propeller shaft cover and fasten in place with one screw.

(4) *Install bearing cap.* Position support bearing cap on support and attach with two bolts.

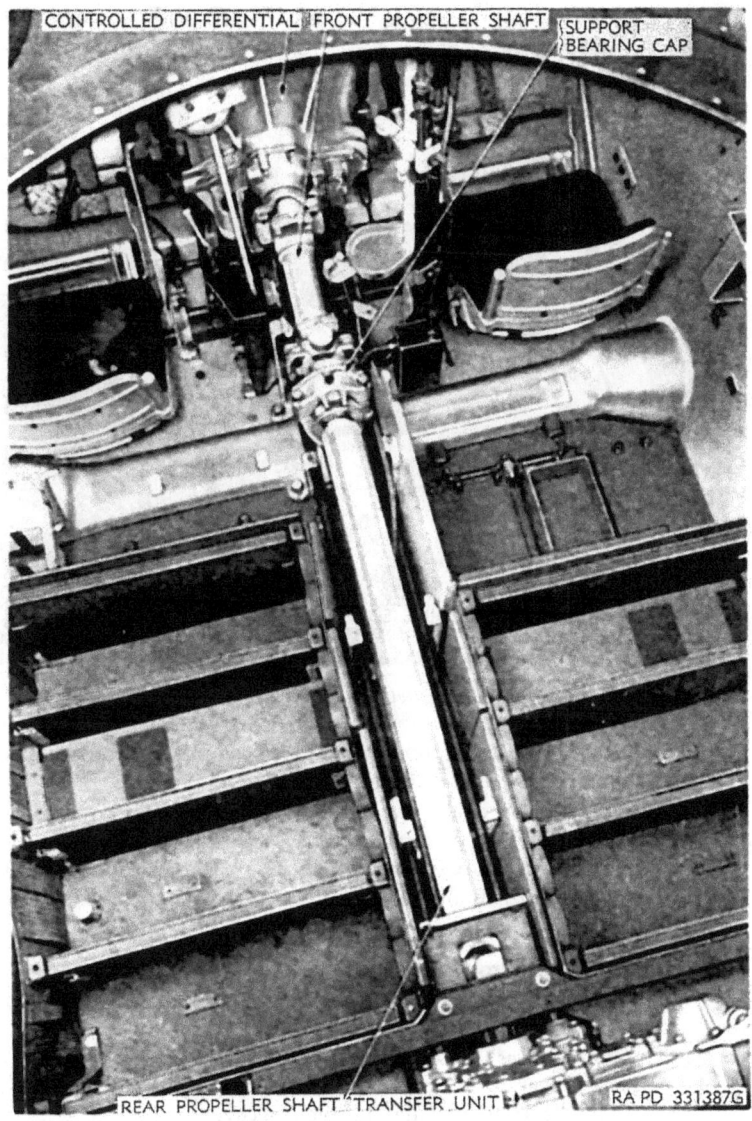

Figure 118. Main propeller shafts installed.

(5) *Install collector ring assembly.* Move collector ring and turret control box assembly back into position on mounting studs. Install oddment box at rear and periscope box at front, attaching each with four cap screws and lock washers. Install conduit guard cover and fasten with two cap screws.

235

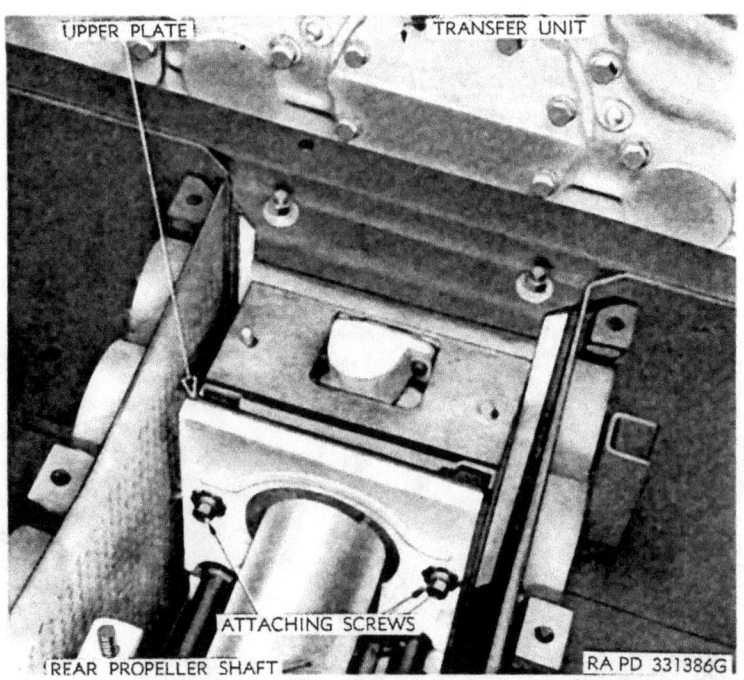

Figure 119. Bulkhead extension plate.

(6) *Install bulkhead extension seal.* Place upper seal in position around shaft and attach securely with two screws.

(7) *Close hinged covers.* Close bulkhead extension covers.

(8) *Install periscope stowage box.* Place periscope and spare head stowage box in position on support brackets and fasten securely with four cap screws.

159. Final Drive Propeller Shafts

a. REMOVAL.

(1) *Remove cover.* Remove four cap screws holding propeller shaft cover to mounting brackets on controlled differential and final drive housing and remove cover.

(2) *Disconnect outer universal joint.* Remove locking wire and four screws (fig. 120) holding universal joint bearing housing to yoke on final drive input shaft. Slide propeller shaft toward differential as far as it will go and disconnect universal joint.

(3) *Remove propeller shaft.* Slide propeller shaft toward side

Figure 120. Removing final drive propeller shaft.

of vehicle to pull yoke off differential output shaft and remove propeller shaft assembly from vehicle.

Caution: Run a piece of wire through bearing housing and mounting screw holes to prevent housing from falling off and losing needle bearing.

b. INSTALLATION (fig. 120).

(1) *Install propeller shaft.* Lower propeller shaft assembly into vehicle. Lift shaft into position. Engage yoke on differential output shaft spline and push yoke into differential as far as possible. Remove wire holding housings on cross of final drive universal joint. Line up holes in yoke and install four screws holding housings to yoke. Install locking wire through each two screw heads.

(2) *Install covers and panels.* Position propeller shaft covers and panels around shafts, and install four screws holding cover to brackets on controlled differential and final drive housing.

160. Propeller Shaft Support Bearing

a. REMOVAL.

(1) *Remove stowage boxes and collector ring.* Refer to paragraph 158a.

(2) *Separate front and rear propeller shafts.* Disconnect uni-

versal joint between front and rear propeller shaft at rear yoke and separate shafts.

(3) *Remove yoke.* Remove cotter pin and nut on front end of rear propeller shaft, holding yoke to propeller shaft, and remove yoke.

(4) *Remove support bearing cap.* Remove two bolts holding support bearing cap in position and remove cap (fig. 118).

(5) *Remove support bearing and carrier.* Lift the forward end of the rear propeller shaft until it is clear of the support bearing and with a piece of wood, lightly tap bearing and carrier off shaft.

b. INSTALLATION.

(1) *Install support bearing and carrier.* Position bearing and carrier on forward end of rear propeller shaft and drive into place with hammer and block of wood.

(2) *Install yoke.* Install yoke on front end of rear propeller shaft; tighten nut securely and lock with new cotter pin.

(3) *Connect front and rear shafts.* Push propeller shaft to the rear to allow enough clearance so that yoke will line up with universal joint. Install four screws that hold yoke to universal joint.

(4) *Install support bearing cap.* Place cap in position over bearing carrier bushing and secure with two bearing bolts.

(5) *Install stowage boxes and collector ring.* Refer to paragraph 158b.

Section XVIII. CONTROLLED DIFFERENTIAL

161. Description and Data

a. DESCRIPTION. The controlled differential, located at the front of the vehicle, transmits engine power to the final drive units and, in addition, contains the brake drums and bands that permit steering and stopping the vehicle.

b. DATA.

Brake rims	2
Size	15 x 4¼ in.
Brake linings, number per rim	3
Size	13 x 4 in.
Differential gearing, type	spur
Differential ratio	1.92 to 1 max
Drive gearing, type	spiral bevel
Drive gearing ratio	2.62 to 1

162. Steering and Brake Band Adjustment

a. GENERAL. The steering and brake band adjustment compensates for lining wear. The steering and brake bands should be adjusted whenever required by lining wear or whenever the controlled differential is replaced.

b. ADJUSTMENT PROCEDURE.

(1) *Remove adjusting hole plugs.* Remove band adjusting hole plug from each side of controlled differential.

(2) *Adjust band.* Insert a $1\frac{1}{16}$-inch socket wrench or wrench-41–W–642–200 through plug hole and engage adjusting nut (fig. 121). Turn adjusting nut clockwise to tighten band.

> *Note.* Brake band adjusting nut has a cylindrical surface on pressure side instead of the usual flat face. It is important that this adjustment be made by half turns only, so that this cylindrical surface will always be seated firmly against cross pin when adjustment is completed.

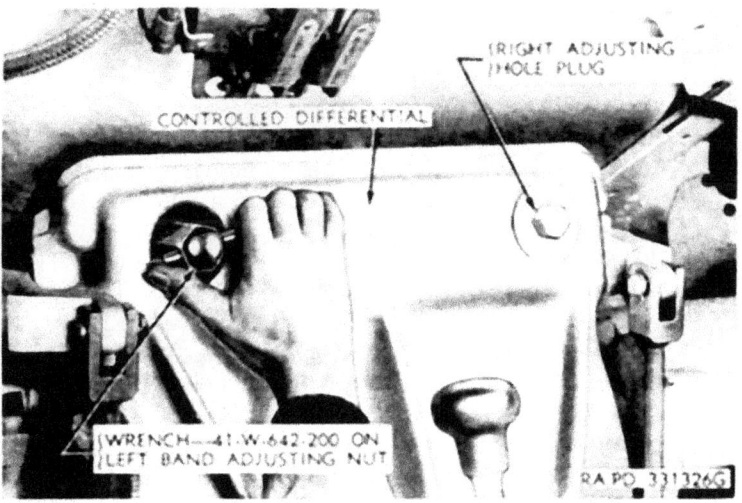

Figure 121. Adjusting steering brake band.

(3) *Check adjustment of steering and brake bands.* Check adjustment by pulling back on steering brake hand lever. The adjustment is correct when the following conditions prevail:

(*a*) Brake band must free when lever is in full forward position.

(*b*) To make second check, place a spring scale at the center of rubber grip on end of steering brake hand lever and pull back on lever until a force of 12–15 pounds torque is exerted. In this position, lever should be six notches back from forward position on quadrant.

Note. On new bands, lever should move back only five notches at a pressure of 12–15 pounds torque.

Repeat check for other brake band. If no spring scale is available, adjustment may be made by observing the following precautions: first, brake band must be free when lever is in full forward (released) position; second, levers should not pull back more than five or six notches under normal steering pressure; and third, levers should not pull back more than three additional notches under full steering or stopping pressure.

Note. Brakes adjusted on the "loose side" will not only last longer, but will prevent "glazing" of bands and provide easier steering.

(4) *Install adjusting hole plugs.* Install both band adjusting hole plugs in differential case.

163. Steering and Brake Shoes

a. REMOVAL.

(1) *Remove differential opening cover.* Remove 18 screws holding differential opening cover to hull front plate. Attach hoist to lifting handles on front cover. Remove cover and gasket and discard gasket.

(2) *Remove differential case cover.* Remove 18 screws that hold differential case cover to differential housing. Lift off cover and discard gasket.

(3) *Remove band adjusting nut and link pin.* Back off adjusting nut and remove nut. Slide adjusting brake band yoke out of operating shaft yoke pin and remove tension spring and washers; then remove link to operating shaft pin. Remove cotter pin and slide out pin holding link to brake shoe (fig. 122).

(4) *Remove brake band assembly.* Grasp brake band assembly with both hands and pull out of housing, permitting shoes to rotate around brake drum.

(5) *Remove other brake band assembly.* Repeat (3) and (4) above for removal of other brake band assembly.

b. INSTALLATION.

(1) *Install brake band assembly.* Install brake band assembly by rotating assembly around brake rim.

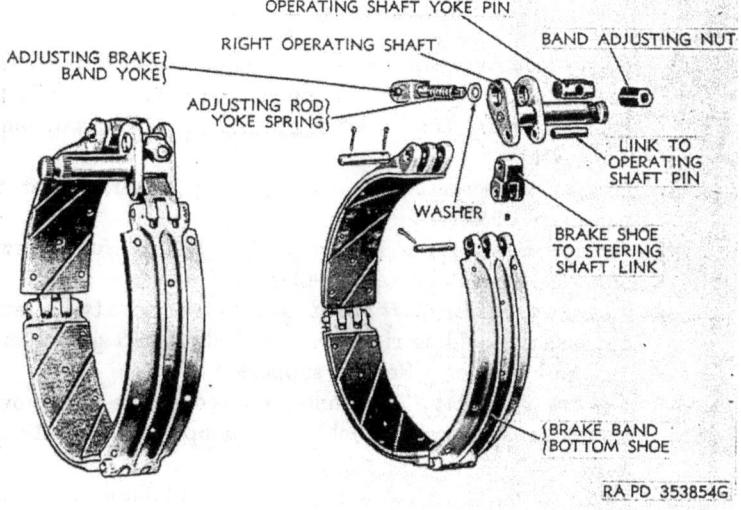

Figure 122. Steering brake band assembly.

(2) *Install link to shoe pin.* Line up pin holes in steering brake link and holes in shoes. Slide link to shoe pin into position and install cotter pin (fig. 122).

(3) *Install band adjusting nut.* Place steering yoke to shaft pin in position in operating shaft. Slide tension spring and washer over yoke and push yoke through operating shaft yoke pin. Install band adjusting nut on end of yoke.

(4) *Install other brake band assembly.* Repeat (1) through (3) above for other brake band assembly.

(5) *Adjust brakes.* Refer to paragraph 162.

(6) *Install differential case cover.* Place a new differential case cover gasket on differential case. Coat gasket with gasket cement. Place cover over gasket and install 18 screws and washers holding cover to case. Using a torque wrench, tighten the five ½-inch screws to 80–85 foot-pounds torque and the thirteen 7/16-inch screws to 45–50 foot-pounds torque.

(7) *Install differential opening cover.* Place a new gasket over edge of hull opening, attach hoist to cover, and lower into position on front of hull. Line up screw holes with drift punch and install 18 cover attaching screws. Remove hoist. Using a torque wrench, tighten screws to 80–85 foot-pounds torque.

164. Oil Pump and Screen

a. REMOVAL.

(1) *Drain oil.* Drain oil by removing differential drain cover under hull floor at front of vehicle and remove differential drain plug (fig. 158). Allow oil to drain and install plug and cover.

(2) *Remove periscope and spare head stowage box.* Refer to paragraph 155*b*.

(3) *Remove right propeller shaft guard.* Remove four screws and washers and remove propeller shaft guard.

(4) *Remove propeller shaft guard support.* Remove two screws and washers holding right front propeller shaft guard support to hull floor. Remove support.

(5) *Remove oil pump.* Disconnect oil cooler lines. Remove seven screws and washers holding oil pump to differential case (figure 123). Remove pump.

(6) *Remove oil pump strainer.* Remove screw holding oil pump strainer to inside of differential case; slide strainer down toward bottom of case and lift out of oil pump opening (fig. 123).

(7) *Clean screen.* Cut locking wire, remove wing nut from bottom of oil pump strainer assembly, and remove cover and strainer screen. Wash all parts in dry-cleaning solvent or volatile mineral spirits paint thinner and assemble.

b. INSTALLATION (fig. 123).

(1) *Install oil pump strainer.* Lower oil strainer through oil pump opening and slide pipe on strainer into differential case, making sure rubber gasket is on end of pipe. Install screw and washer holding oil pump strainer to case.

(2) *Install oil pump.* Place a new oil pump gasket on oil pump body and position pump on differential case. Install seven screws and washers holding pump to case. Connect oil cooler lines.

(3) *Install propeller shaft guard support.* Place propeller shaft guard support over mounting holes on hull floor and install two screws and washers.

(4) *Install propeller shaft guard.* Position right propeller shaft guard against supports and install four screws and washers.

(5) *Install periscope and spare head stowage box.* Refer to paragraph 155*c*.

(6) *Fill controlled differential.* Refer to paragraph 62.

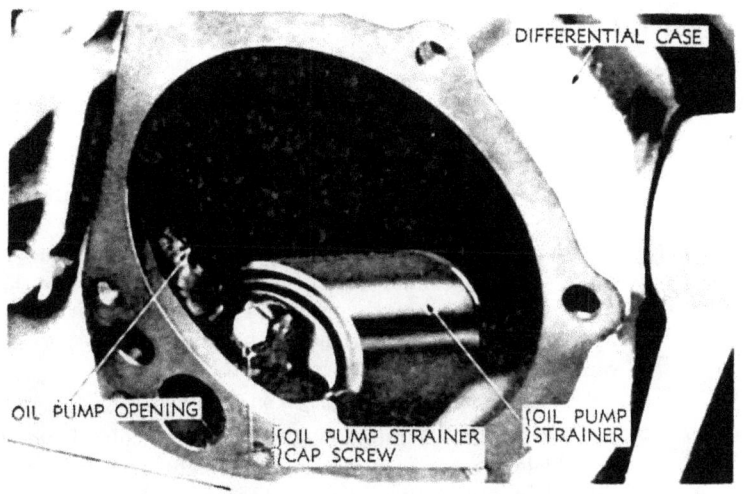

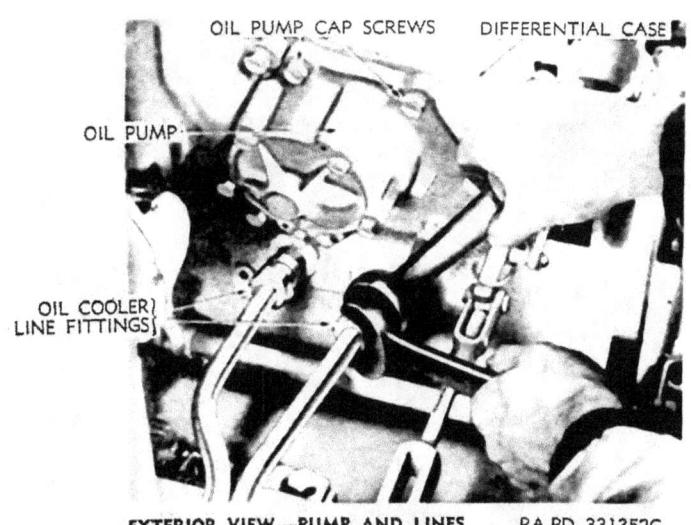

Figure 123. Removing differential oil lines.

165. Oil Cooler

a. REMOVAL.
  (1) *Traverse turret.* Turn turret 90° to left to clear radiator air inlet grille.
  (2) *Remove radiator air inlet grille.* Remove two screws from each end of radiator air inlet grille and lift out grille (fig. 82).
  (3) *Disconnect flexible oil cooler hoses.* Disconnect both oil cooler hoses at bulkhead connection by holding hose stationary with one wrench and turning coupling with another wrench (fig. 124). Plug ends of hoses immediately to prevent oil from running out over radiators.
  Note. Radiators will quickly plug if oil drips on them.
  (4) *Remove cooler.* While a helper supports cooler from top of vehicle, remove four cap screws on fighting compartment side of bulkhead that hold hanger to bulkhead. Lift out cooler.
  (5) *Remove cooler hoses.* After cooler has been removed from vehicle, turn on one end to drain oil from unit; and then

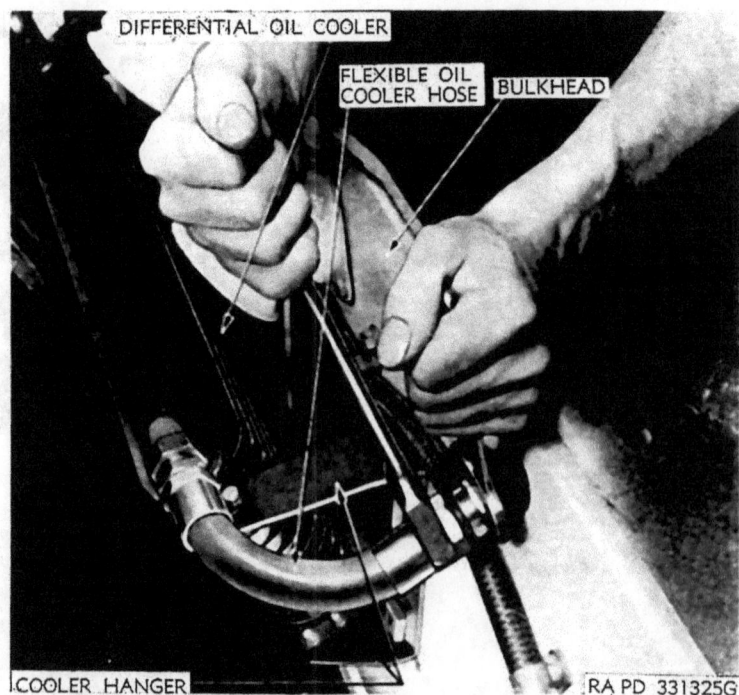

Figure 124. Disconnecting oil cooler hose.

remove both cooler hoses from ends of cooler by unscrewing hose connections.

b. INSTALLATION.

(1) *Install cooler hoses.* Coat threaded end of cooler hose with gasket cement and install in fitting on end of cooler.

(2) *Install cooler.* Lower cooler assembly into position over front of radiators until holes in cooler hanger line up with holes in bulkhead. While helper holds cooler in this position, install four oil cooler mounting screws from inside of vehicle.

(3) *Connect oil cooler hoses.* Connect both oil cooler hoses to connections on bulkhead.

> Note. Use two wrenches to connect coupling to keep hoses from twisting (fig. 124).

(4) *Install radiator air inlet grille.* Lower radiator air inlet grille into position in opening and install four attaching screws and washers.

(5) *Traverse turret.* Traverse turret to straight-ahead position.

166. Controlled Differential Replacement

a. REMOVAL.

(1) *Remove differential opening cover.* Refer to paragraph 163a(1).

(2) *Remove periscope and spare head stowage box.* Refer to paragraph 155b(1).

(3) *Disconnect steering brake levers.* Release brakes and push levers forward to the fully released position. Remove cotter pins and clevis pins, holding right and left steering brake cross shaft rods to lower end of steering brake cross shaft link rod relay lever on controlled differential housing (fig. 125).

(4) *Remove hand lever return springs.* Grasp end of spring with heavy pliers and pull back until spring is clear of relay lever (fig. 125). Release spring gradually until it is free. Leave front end of spring connected to hull front plate.

(5) *Remove left propeller shaft guard.* Remove screw holding left propeller shaft guard (fig. 125) to support brackets and remove guard.

(6) *Remove right propeller shaft leg guard.* Remove four screws holding right propeller shaft leg guard to support brackets and remove guard.

(7) *Remove stop light switches.* Remove cotter pin and slide stop light switch control rod off pin on relay arm (fig. 125). Remove two screws holding stop light switches to front differential mounting brackets.

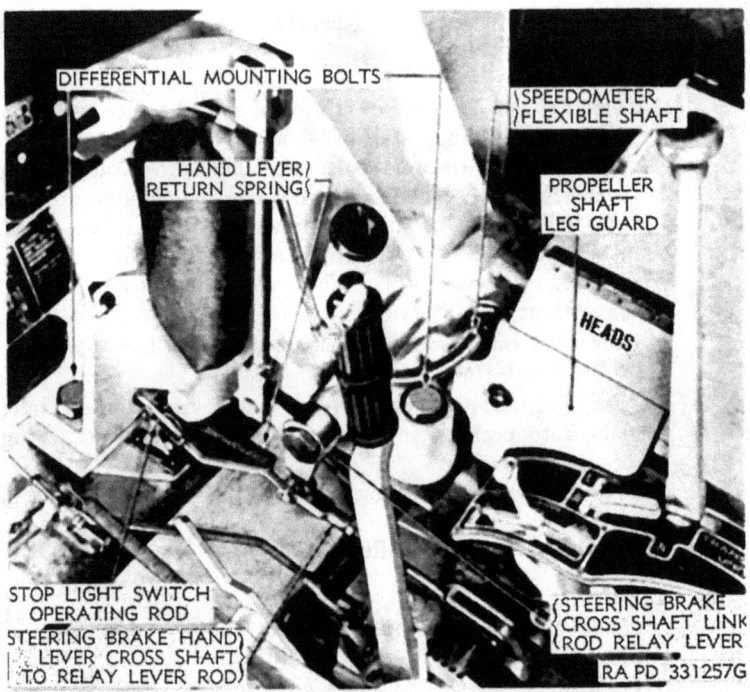

Figure 125. Steering brake linkage and differential mounting.

(8) *Drain oil.* Drain oil by removing differential drain cover under floor at front of vehicle (fig. 158) and remove differential drain plug. After oil has drained, install plug.

(9) *Disconnect oil cooler lines.* Disconnect oil cooler lines at right rear corner of differential by unscrewing coupling on line from fitting on housing (fig. 123).

Caution: Place rags or container under lines to catch any oil that may drain out of cooler lines.

(10) *Remove final drive propeller shaft shields.* Remove four screws holding each half of final drive propeller shaft shields to mounting brackets. Remove shields. Remove two screws holding shield brackets to sides of controlled differential housing.

(11) *Disconnect inner ends of final drive propeller shafts.* Remove locking wires from inner universal joint bearing cap screws. Remove four screws holding universal joint bearing caps to controlled differential output yoke. Slide yoke into controlled differential housing.

Caution: Tie bearing caps to cross of universal joint to avoid losing needle bearings.

(12) *Disconnect differential from front propeller shaft.* Remove locking wires from front universal joint bearing caps. Remove four screws holding bearing caps to differential yoke.

Caution: Tie bearing caps to cross of universal joint to avoid losing needle bearings.

(13) *Disconnect speedometer flexible shaft.* Disconnect speedometer flexible shaft (fig. 125) on differential by unscrewing knurled nut on speedometer shaft and pulling speedometer case. Remove shaft from clip on case and lay shaft to one side.

(14) *Remove differential mounting bolts.* Remove two front and one rear differential mounting bolts and nuts and remove bolts.

(15) *Remove differential.* Attach chain hoist to lifting bracket on differential cover. Take up weight of differential on hoist until it clears mounting brackets on hull floor. Raise differential very slowly, and see that it does not strike instrument panel (fig. 126). Remove differential and lower to ground. Remove chain hoist.

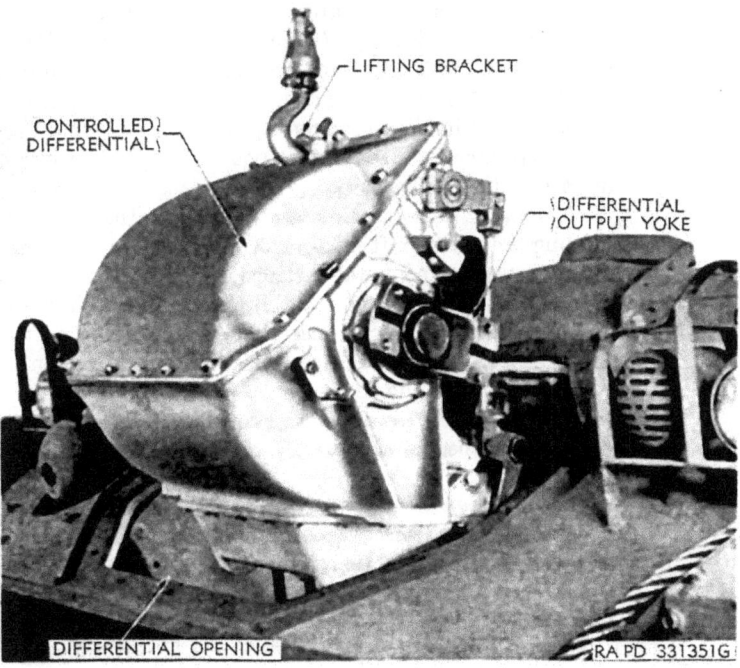

Figure 126. Removing controlled differential.

b. INSTALLATION.
   (1) *Place differential in vehicle.* Attach a chain hoist to lifting bracket on differential cover. Raise differential over opening in hull front deck. Lower assembly, being careful that it does not strike instrument panel. Before assembly is lowered completely into vehicle, raise propeller shaft universal joint and line up with yoke on differential. Lower assembly on support brackets. Remove chain hoist from cover.
   (2) *Install differential mounting bolts.* Install two front and one rear differential mounting bolts through boss on differential case and brackets on hull (fig. 125). Install nuts and washers, and tighten nuts to 200 foot-pounds torque using a torque wrench.
   (3) *Install speedometer flexible shaft.* Place speedometer flexible shaft in clip on differential case. Slide speedometer core into drive shaft sleeve and install knurled nut holding shaft to sleeve.
   (4) *Connect controlled differential to front propeller shaft.* Install four screws holding bearing caps to differential yoke. Using a torque wrench, tighten screws to 80 foot-pounds torque. Lock screws in pairs with locking wire.
   (5) *Connect inner ends of final drive propeller shafts.* Remove locking wires which prevent bearing caps from falling off. Position final drive bearing caps on differential output yokes. Install screws holding bearing caps to yoke. Using a torque wrench, tighten screws to 80 foot-pounds torque. Lock screws in pairs with locking wire.
   (6) *Install final drive propeller shaft shields.* Install brackets holding propeller shaft shields to side of differential housing. Position upper and lower section of propeller shaft shield over mounting bracket and install eight screws and washers.
   (7) *Connect oil cooler lines.* Insert oil cooler lines into fittings in controlled differential case and connect coupling on lines to fittings in case (fig. 123).
   (8) *Fill controlled differential.* Fill controlled differential with oil as prescribed by the lubrication order. Check level on gage.
   (9) *Install stop light switches.* Place stop light switches in position on front differential mounting brackets and install four screws and washers. Slide stop light switch control rod over pin on steering brake relay arm and install flat washer and cotter pin (fig. 125).
   (10) *Install right propeller shaft guard.* Place right propeller shaft guard against support brackets and install four screws and washers.

(11) *Install left propeller shaft leg guard.* Place left propeller shaft leg guard against support brackets and periscope stowage box and install screw and washer.

(12) *Install hand lever return springs.* Make sure front end of spring is hooked to bracket on hull front plate. Pull back on end of spring with large pliers, and hook end of spring over cutout in steering and braking lever relay (fig. 125).

(13) *Connect steering brake levers.* Line up hole in connecting rod yoke with hole in steering brake cross shaft. Install clevis pin and cotter pin.

(14) *Adjust linkage.* Adjust length of connecting rod to permit free assembly of pin at steering lever by loosening lock nut and rotating clevis. Turn clevis one-half turn tighter to assure taking out all slack and tighten lock nut. Connect rod to lever and install clevis pin and cotter pin. Adjust other lever in same manner.

(15) *Adjust steering brakes.* Refer to paragraph 162.

(16) *Install periscope and spare head stowage box.* Refer to paragraph 155c(6).

(17) *Install differential opening cover.* Refer to paragraph 163b(7).

Section XIX. FINAL DRIVES

167. Description and Data

a. DESCRIPTION. The final drive units are mounted in separate housings bolted to the front of the hull on each side (fig. 128). The final drives are designed so that they are interchangeable, right and left, and can be installed on either side of the vehicle. Power is transmitted to the final drives from the controlled differential through short propeller shafts which have universal joints at each end. The herringbone gear set provides a reduction ratio of 2.94–1. Power output is through the sprocket shaft which carries the track driving hub and sprockets.

b. DATA.

Bracket	cast steel
Cover	cast steel
Gear ratio	2.94 to 1
Number of sprocket teeth	13
Type of gearing	spur (herringbone)
Sprocket (used with steel track T72 or T72E1)	D76210
Sprocket (used with rubber track T85E1)	7049476

168. Hubs and Sprockets

a. REMOVAL.

(1) *Break track.* Break track below and to the rear of final drive and lift track over and to the rear of hub and sprockets (par. 172).

(2) *Remove hub and sprocket assembly* (fig. 127). Remove 10 nuts attaching hub to shaft and remove hub and sprocket as an assembly.

> *Note.* When steel track T72 or T72E1 is replaced by rubber track T85E1 sprocket—D76210 must be replaced by sprocket—7049476.

(3) *Remove sprockets.* Remove 13 bolts attaching each sprocket to hub and remove sprockets.

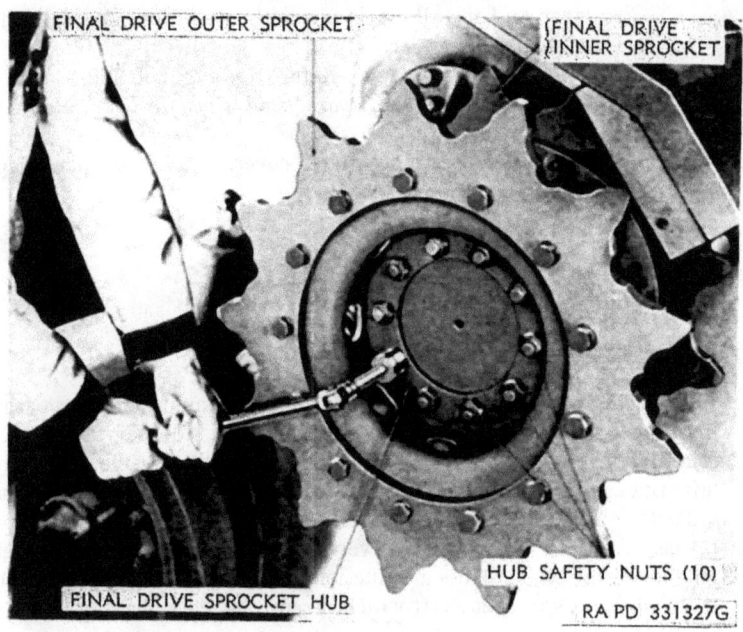

Figure 127. Removing track sprocket.

b. INSTALLATION.

(1) *Install sprockets.* Install sprockets on hub and tighten 13 bolts on each sprocket securely.

(2) *Install hub and sprocket assembly on shaft.* Position hub and sprocket assembly on shaft, line up holes, and install and tighten the 10 safety nuts securely.

(3) *Install track.* Refer to paragraph 172.

(4) *Adjust track.* Refer to paragraph 171.

c. REVERSING SPROCKETS. When the teeth on the sprockets become worn from use, the sprockets can be reversed on the hub or the complete hub and sprocket assembly can be changed from right to left and vice versa. This will present a new tooth surface to the track and provide longer life.

Figure 128. Disconnecting final drive from vehicle.

169. Final Drive Replacement

a. REMOVAL.
  (1) *Drain lubricant.* Remove final drive drain cover that protects final drive drain plug from bottom of hull beneath final drive (fig. 158). Remove drain plug. Install plug as soon as all oil has been drained.
  (2) *Remove guards.* Remove four screws holding halves of final drive propeller shaft guard to controlled differential and to final drive case and remove guard.
  (3) *Disconnect propeller shaft.* Remove locking wires and four screws holding the universal joint yoke on the final drive input shaft to the universal joint (fig. 120). Push the final drive propeller shaft sufficiently to allow universal joint to clear yoke. Lower outer end of propeller shaft after tying the universal joint bearing caps to the journal with locking wires to prevent accidental disassembly.

(4) *Break track.* Refer to paragraph 172.
(5) *Remove hub and sprocket assembly.* Remove 10 nuts attaching hub to shaft and lift hub and sprockets off as an assembly (fig. 127).
(6) *Remove final drive assembly.* Cut locking wires and loosen 18 mounting bolts (fig. 128) that hold final drive bracket flange to hull sidewall. Connect hoist to final drive lifting sling—41–S–3832–35 and install on final drive bracket. Remove attaching bolts, pry final drive assembly off dowel pins, and remove from vehicle (fig. 129).

Figure 129. Removing final drive assembly.

b. INSTALLATION.

(1) *Install final drive assembly.* Coat inner edges of bracket flange with joint sealing compound. Raise final drive assembly with hoist and sling—41–S–3832–35 and line up dowel

pins and bolt holes. Push the assembly into place and install the mounting bolt, tightening them carefully to 170–180 foot-pounds torque. Install locking wires. Remove sling.

(2) *Connect propeller shaft.* Push final drive propeller shaft and raise into position opposite yoke on final drive. Disconnect locking wires holding bearing housings on universal joint cross and line up holes in yoke with holes in universal joint housings. Install four screws holding bearing housings to yoke. Tighten screws to 130–140 foot-pounds torque, and install locking wire through each pair of screws.

(3) *Install guards.* Position halves of propeller shaft guards around shaft and install four screws holding guard to controlled differential and final drive cover.

(4) *Install hub and sprocket assembly.* Refer to paragraph 168.

(5) *Connect track.* Refer to paragraph 172.

(6) *Adjust track.* Refer to paragraph 171.

(7) *Add lubricant.* Make sure drain plug and protecting plate are installed. Remove filler plug and add lubricant as prescribed by the lubrication order. Install filler plug.

Section XX. TRACKS AND SUSPENSION

170. Description and Data

a. DESCRIPTION.

(1) Two individually driven types of tracks (steel, T72 and T72E1 or rubber, T85E1) provide the necessary traction to propel the vehicle (fig. 130). Each complete track is composed of separate track shoes of cast steel or rubber chevrons, with center guides. The steel track shoes are connected together with straight pins carried in rubber bushings (fig. 133). The rubber track shoes are held together by end connectors which lock with wedges on the shoe studs. The steel shoes are of interlocking design to eliminate vibration and wear which occur when track wheels pass over openings between track shoes. Two types of steel track shoes are used, the solid and drilled types. Grousers cannot be used with the solid type; however, the drilled type can be identified by three cast holes which are used when installing 28½-inch extended grousers. Two final drive sprockets, one on each side, pull the tracks forward over the supporting rollers and lay them down in the path of the advancing track wheels.

(2) Ten dual track wheels, five on each side, are carried on individual arms attached to independent torsion bars and mounted so as to be easily removed (fig. 130). The ten arms

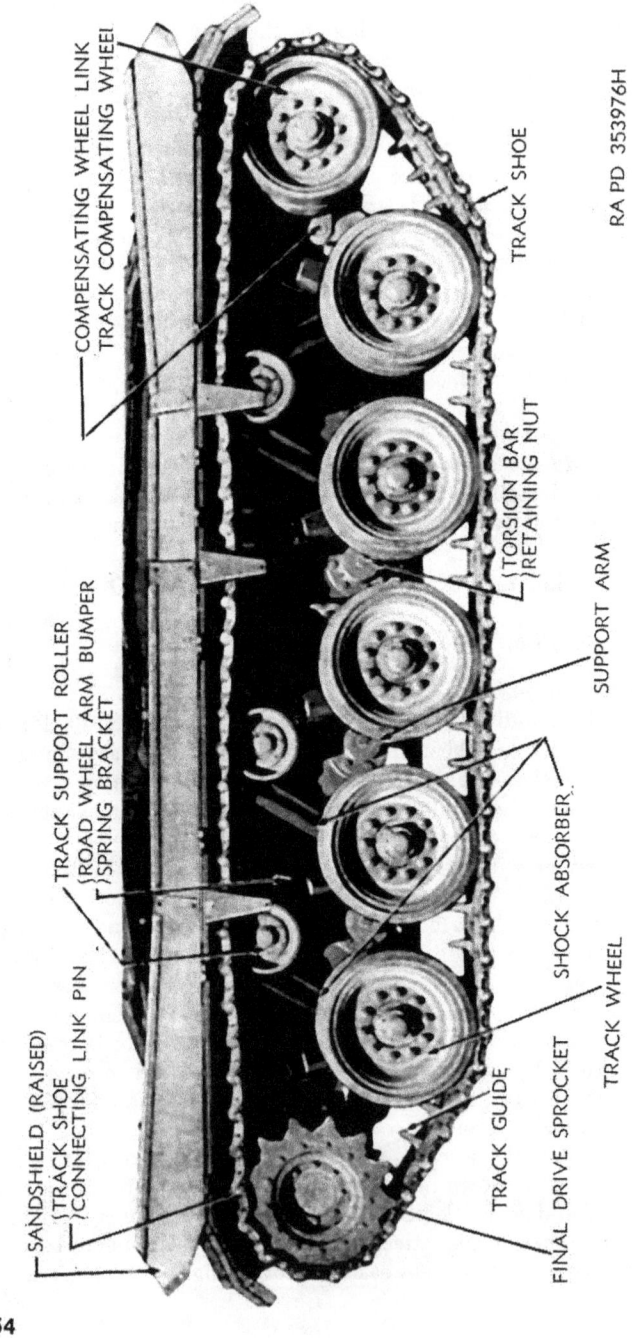

Figure 130. Track suspension system.

are each mounted on two roller bearings carried in housings which are bolted to the sides of the hull just above the floor level. The arms are splined to the torsion bars which extend through protective tunnels on the hull floor. The torsion bars for opposite wheels are staggered to permit carrying the wheels at the same level. Double-acting, airplane-type, hydraulic shock absorbers are provided for the two front and two rear suspension arms on each side. They are mounted on the rear suspension arms on each side and on the hull side walls (fig. 130).

(3) An adjustable compensating wheel for each track is mounted at the rear of the hull and is connected to the rear track wheel arm by a link (fig. 130), which is so arranged that any decrease in track tension caused by lifting of the rear track wheels will be offset by movement of the compensating wheel toward the rear and any increase in track tension caused by dropping of the rear track wheels will be offset by movement of the compensating wheel toward the front.

(4) Three dual, rubber-tired track support rollers are mounted on each of the upper hull sides (fig. 130) to support the track as it returns to the drive sprocket.

b. DATA.

(1) *Track shoes.*
Number per track_____ 75
Width (steel)_____ 16 in.
Width (rubber)_____ 16½ in.
Type_____ steel (with center guide and rubber bushings)
Type_____ rubber chevron (with center guide and rubber bushings)

(2) *Track.*
Pitch_____ 5½ in.

(3) *Support arms.*
Number (each side)_____ 5
Type_____ solid steel
Type of springing_____ torsion bar
Number of bearings_____ 2 (straight roller)

(4) *Track wheels.*
Number (each side)_____ 5
Type_____ dual, demountable, rubber-tired
Number of bearings (each wheel)_____ 2 (taper, roller)

(5) *Torsion bars.*
Number (each side)_____ 5
Type_____ solid steel

(6) *Track support rollers.*
  Number (each side) _____ 3
  Type _____ dual, demountable, rubber-tired
  Number of bearings _____ 2 (taper, roller)
(7) *Shock absorbers.*
  Number (each side) _____ 4
  Type _____ hydraulic, airplane-type
(8) *Road wheel arm bumpers.*
  Number (each side) _____ 5
  Type _____ volute spring

171. Track Adjustment

a. GENERAL. Extreme care must be taken at all times to see that the tracks are properly adjusted to prevent unnecessary wear and breakage.

b. TRACK ADJUSTMENT.

(1) To check the track adjustment, move vehicle to level ground to assure normal track tension. Raise sandshields, place a 4-foot pry bar between the track and hull at a point between the two rear track support rollers and push the track down with 200-pounds pressure, as exerted at track. Lay a straightedge along the top of the track between the two rear rollers and measure the sag midway between these rollers (fig. 131). This sag should be from 2 to 2¼ inches (steel track). Track blocks must be adjusted to above figure when sag becomes 3½ inches or over. The proper adjustment for rubber track (T85E1) is 3¼ inches.

Figure 131. Checking track tension—steel track.

(2) If the track adjustment is incorrect, loosen clamping bolt at rear of compensating wheel arm (fig. 132) and slide clamping bolt stop up from adjusting nut. Tighten or loosen track adjusting nut to obtain correct track tension using wrench—41-W-1315. Tighten clamping bolt after positioning stop on flat of adjusting nut.

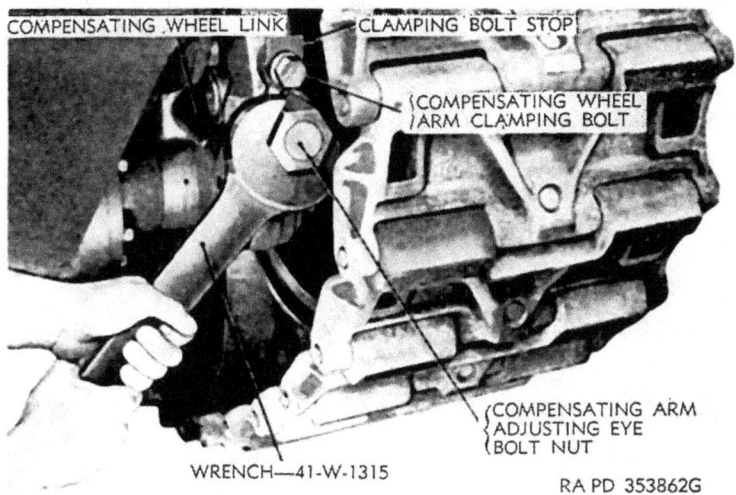

Figure 132. Adjusting track tension.

172. Track Replacement

a. REMOVAL.

(1) *Steel track.* Install track connecting fixtures—41-F-2997-86 on track, midway between compensating wheel and rear track wheel (fig. 133) and take up track tension. Remove track shoe pin wedge bolt, lock washer, and plain washer (fig. 132). Using a brass drift, drive wedge out of block. Tighten track connecting fixtures to relieve track tension and drive out track link pin, using knock-out pin—41-P-560-300 (fig. 133).

(2) *Rubber track.* To remove rubber track (T85E1) (fig. 136), loosen clamp bolt at rear of compensating wheel arm (fig. 134) and slide lock plate up from adjusting nut. Loosen adjusting nut until track is loose. Remove wedge nuts and wedges on both sides of tracks. Drive outer connector off first (fig. 137), then drive inner connector off.

(3) *Move vehicle.* Start engines, put transfer unit and transmission shift levers in "DRIVE" position, and run engines just fast enough to turn track drive sprocket slowly. Run upper

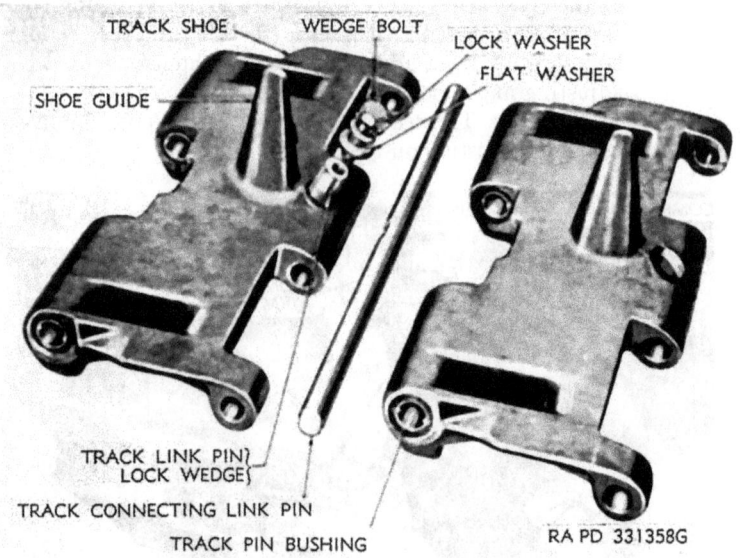

Figure 133. Steel track shoes—exploded view.

Figure 134. Removing track pin—steel track.

half of track forward off support rollers and drive sprockets. Tow vehicle off track. If new track is to be installed, place new track in front of old track, and tow vehicle onto new track.

Caution: If the steel track (T72 or T72E1) is replaced by the rubber track (T85E1), sprocket—7049476 must be used as replacement for sprocket—D76210 (par. 169).

b. INSTALLATION.

(1) *Steel track.* Lay track out on ground and tow vehicle on track so that rear end of track projects approximately 10 inches beyond rear track wheel. Pull front end of track up over drive sprocket and start engines. Put transfer unit in "REVERSE" and transmission in "DRIVE" and guide track back over support rollers and compensating wheel as rotation of drive sprocket moves upper half of track back. Mount track connecting fixtures—41–F–2997–86. Bring track ends together by tightening fixtures and drive in track link pin (fig. 135). Be sure that grooves in pin enter guides in pin bushing (fig. 133) and that pin enters bushings without damaging bushings ends. Install track link pin lock wedge in hole in shoe, with beveled end in, tap wedge lightly to its seat, and install plain washer, lock washer, and wedge bolt. Tighten bolt to 120 foot-pounds torque. Adjust track tension (par. 171).

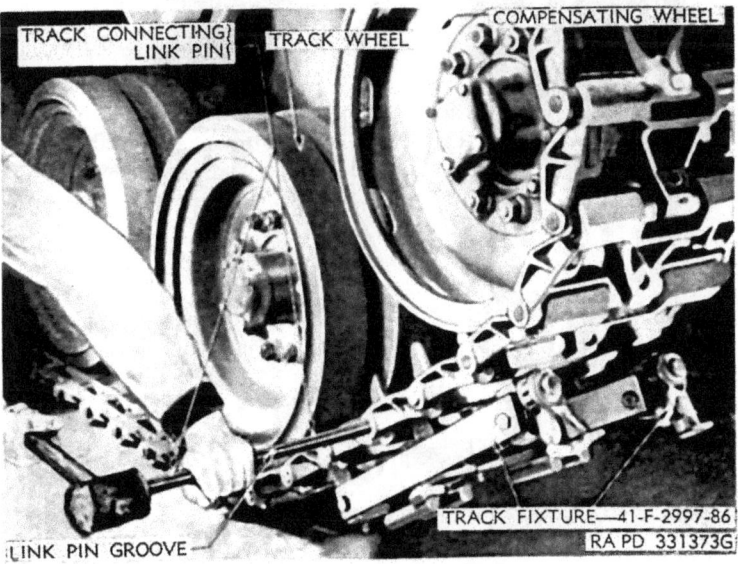

Figure 135. Connecting steel track.

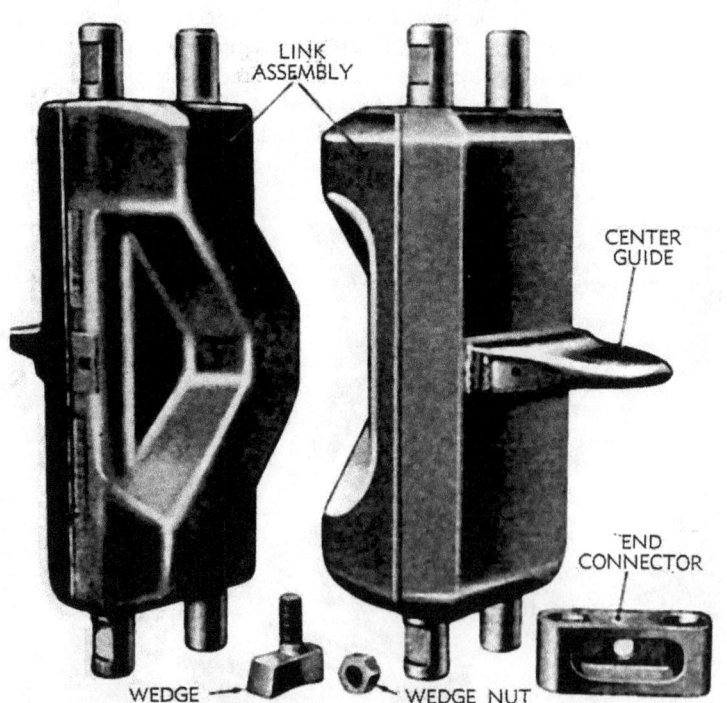

Figure 136. Rubber track shoes.

Figure 137. Removing track link end—rubber track.

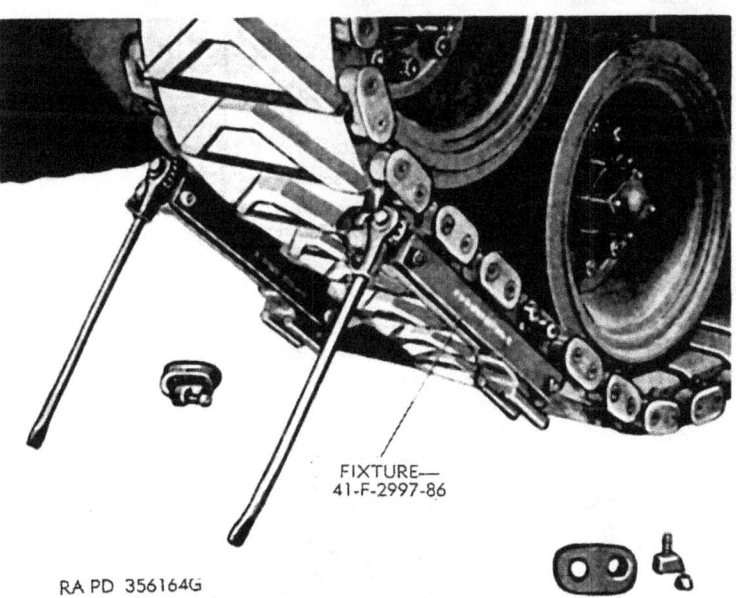

Figure 138. Connecting rubber track.

(2) *Rubber track.* To replace rubber track (T85E1) (fig. 138), mount track connecting fixture, draw track ends together, drive outer connector on track pins until near track fixture, remove outer track connecting fixture and continue to drive connector until against block; install wedge and wedge nut. Replace inner connector in same method. Install wedge and wedge nut. Drive wedges in and tighten nuts securely.

Caution: Be sure to install track with chevron pointing down over the sprocket.

173. Compensating wheel, Bearings, and Oil Seals

a. INTERCHANGEABILITY. The compensating wheel hub, hub cap, bearings, and oil seals are identical with respective track wheel parts.

b. REMOVAL. Break track at rear of vehicle (par. 172) and pull top half of track forward to clear compensating wheel. Remove 10 nuts and locks holding dual compensating wheel on studs and remove wheels. Remove four hub cap mounting screws and remove hub cap and gasket. Take out cotter pin, nut, and washer holding hub on spindle (fig. 139). Pull out on hub enough to loosen outer bearing. Remove outer bearing. Lift off hub and inner bearing. Pry grease seal off spindle bearing spacer with screwdriver.

c. INSTALLATION. Place new grease seal on spindle with featheredge toward arm and, using oil seal replacer—41–R–2383–950, tap over bearing spacer until oil seal contacts spacer (fig. 139). Place wheel hub on bench with wheel side down and pry grease retainer out of hub. Place new grease retainer in replacer—41–R–2390–450 (fig. 140) and position replacer on hub. Drive retainer into position carefully to prevent cocking in hub (fig. 141) and remove tool. Check retainer to see that it is square in the hub counterbore and that the flange of retainer is true. Grease inner roller bearing assembly thoroughly with lubricant as prescribed by lubrication order and push on spindle shoulder until bearing contacts spacer. Push hub assembly on spindle, making sure inner bearing enters race in hub. Pack outer roller bearing with prescribed grease, and install on spindle, holding wheel hub centered on spindle and pushing outer bearing into race. Install keyed washer on spindle and seat on bearing. Install nut on spindle.

d. ADJUSTMENT.

(1) Rotate hub and, at the same time, tighten nut to 200 foot-pounds torque. Back off nut until there is no torque, then tighten nut to 75 foot-pounds torque. Back off nut to first cotter pin hole with minimum travel of 15° and install cotter pin. If the first hole is less than 15° travel, back off nut to

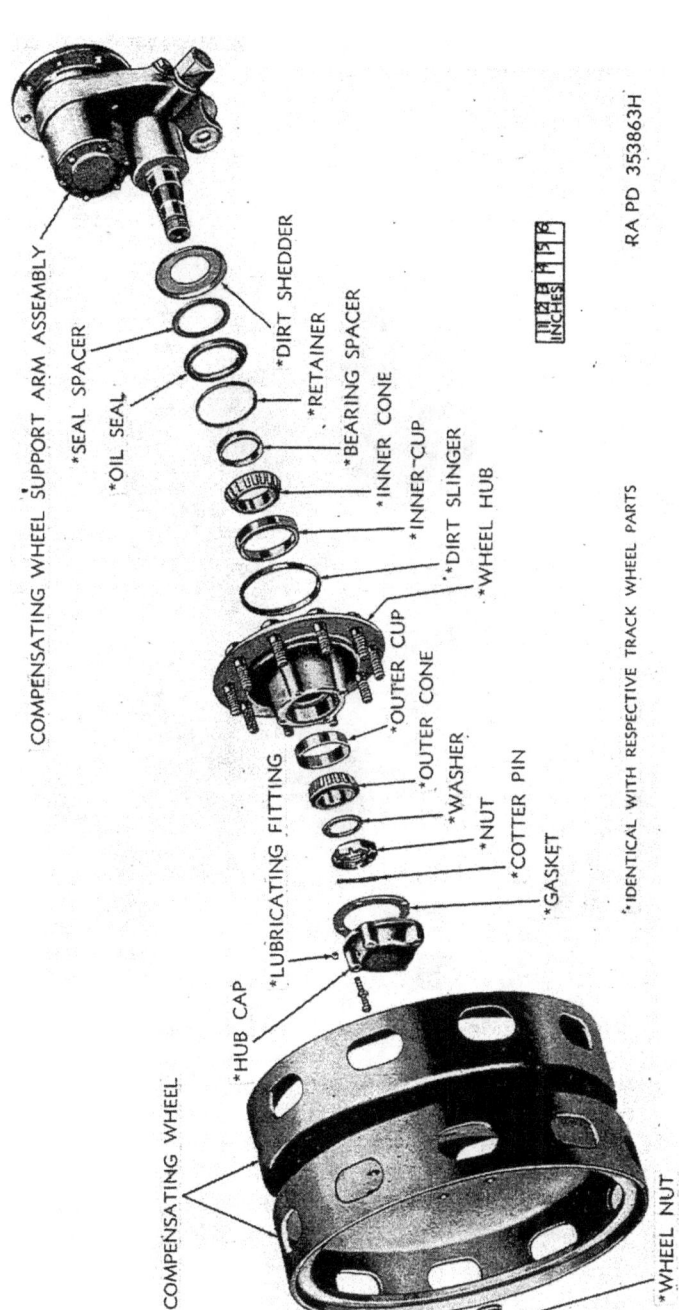

Figure 189. Compensating wheel, hub, and bearings.

next hole or 30° more. Position hub cap and install four mounting screws. Using pressure gun, force grease into hub until it starts coming out of grease shield at rear. Wipe off all excess grease. Lift wheels and install on hub studs; the first wheel with the concave side in the second with the concave side out. Install 10 nuts with locks, and tighten evenly to 250–300 foot-pounds torque. Install and adjust track (par. 171).

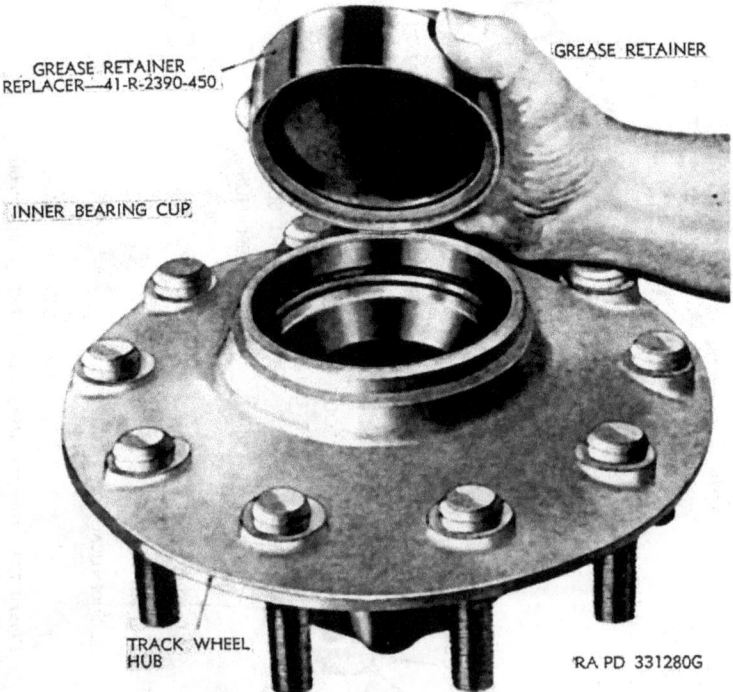

Figure 140. Wheel hub grease seal retainer in replacer.

(2) An alternate method of adjusting the wheel hub adjusting nut can be used if a torque wrench is not available. Using a wrench with an 18-inch handle, tighten the adjusting nut as tight as possible. Back off the adjusting nut no less than three-quarters of a flat and no more than one full flat and insert cotter pin.

Figure 141. Installing wheel hub grease seal retainer.

174. Compensating Wheel Support Arm Assembly

  a. REMOVAL. Remove compensating wheel and hub as previously outlined in paragraph 173 above. Remove cotter pin, nut, and washer from compensating link bolt. Install slide hammer adapter—41-A-18-245 on link bolt head, being sure that stud is screwed all the way in bolt and that hex shoulder of adapter seats on bolt head. Attach slide hammer puller—41-P-2957-33 and pull bolt (fig. 142). Push link end down and out of support arm. If necessary, hook hoist on rear suspension wheel and raise wheel. This will move link end to rear of vehicle so that drift and hammer can be used to drive out bolt. Remove six attaching screws, remove compensating arm cover (fig. 142), and discard gasket. Remove lock ring holding arm on support spindle as an assembly. Remove snap ring holding arm on support spindle and lift arm as an assembly off spindle. Remove screws holding spindle to tapping plate on hull and remove spindle.

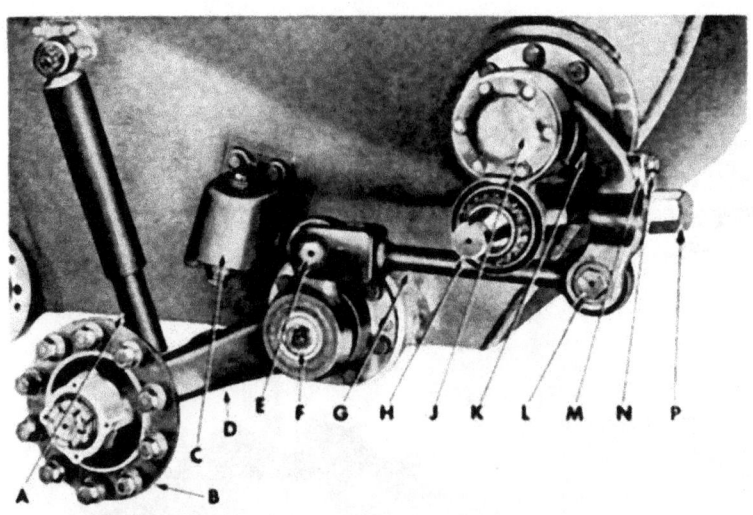

A—SHOCK ABSORBER
B—WHEEL HUB
C—ARM BUMPER SPRING BRACKET
D—SUPPORT ARM
E—COMPENSATING WHEEL LINK BOLT
F—TORSION BAR RETAINING NUT
G—COMPENSATING WHEEL LINK
H—COMPENSATING SUPPORT ARM SPINDLE
J—COMPENSATING WHEEL SUPPORT ARM COVER
K—INNER SUPPORT ARM
L—COMPENSATING WHEEL LINK BOLT
M—CLAMPING BOLT STOP
N—CLAMPING BOLT
P—ADJUSTING NUT

RA PD 331361H

Figure 142. Compensating wheel and track wheel support arm and linkage.

b. INSTALLATION. Coat flange of spindle with joint sealing compound, position spindle on hull tapping plate, and install eight attaching screws dipped in sealer. Tighten screws to 240–260 foot-pounds torque, using torque wrench. Install arm assembly on spindle, seeing that shoulder on spindle enters bearing races and that grease seal slips over bearing spacer. Install snap ring on spindle end to hold assembly. Using a new gasket, position support arm cover, and install six attaching screws. Connect compensating link to arm (fig. 143) and drive in bolt, using a soft hammer, then install washer, nut, and cotter pin. Tighten nut on connecting bolt to 180 foot-pounds torque (min). Install compensating wheel and hub (par. 173) and adjust (par. 173).

c. COMPENSATING ARM BEARING REPLACEMENT.
   (1) *Remove compensating arm and lever.* Refer to *a* above.
   (2) *Remove bearings.* Remove lever clamping bolt, clamping bolt stud, and washer (fig. 143). Remove adjusting nut and

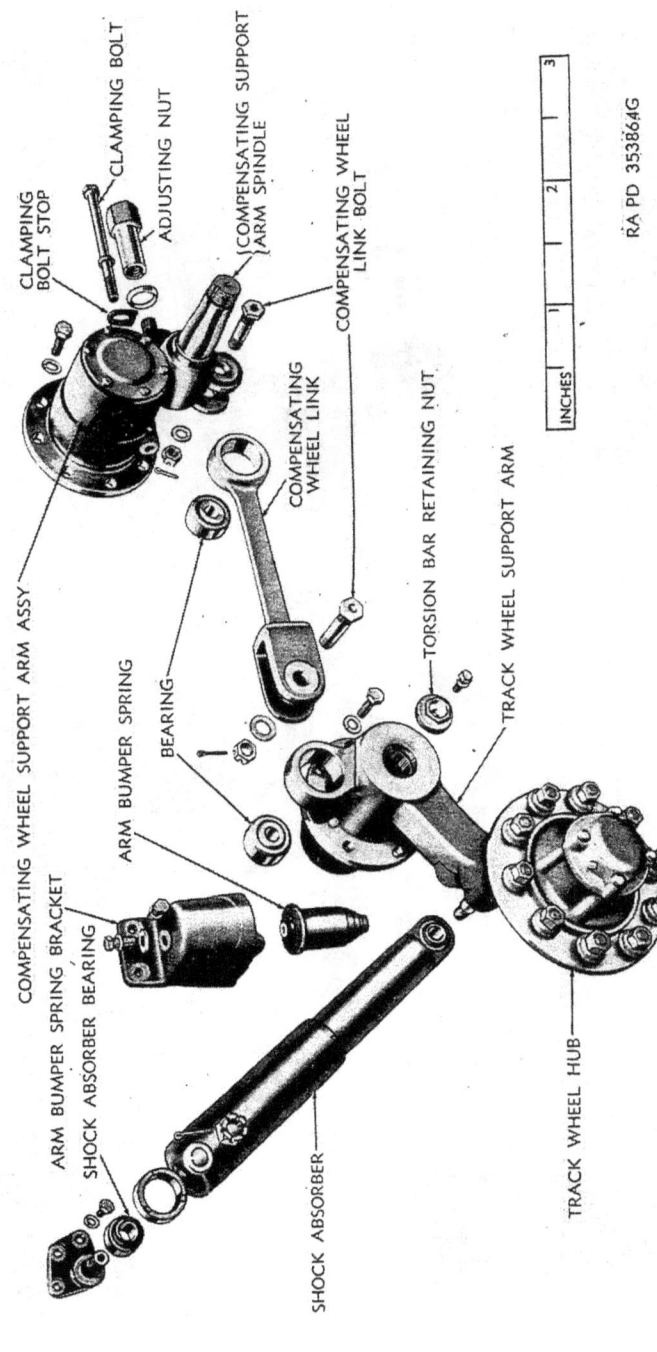

Figure 143. Compensating wheel and track wheel support arm and linkage—exploded view.

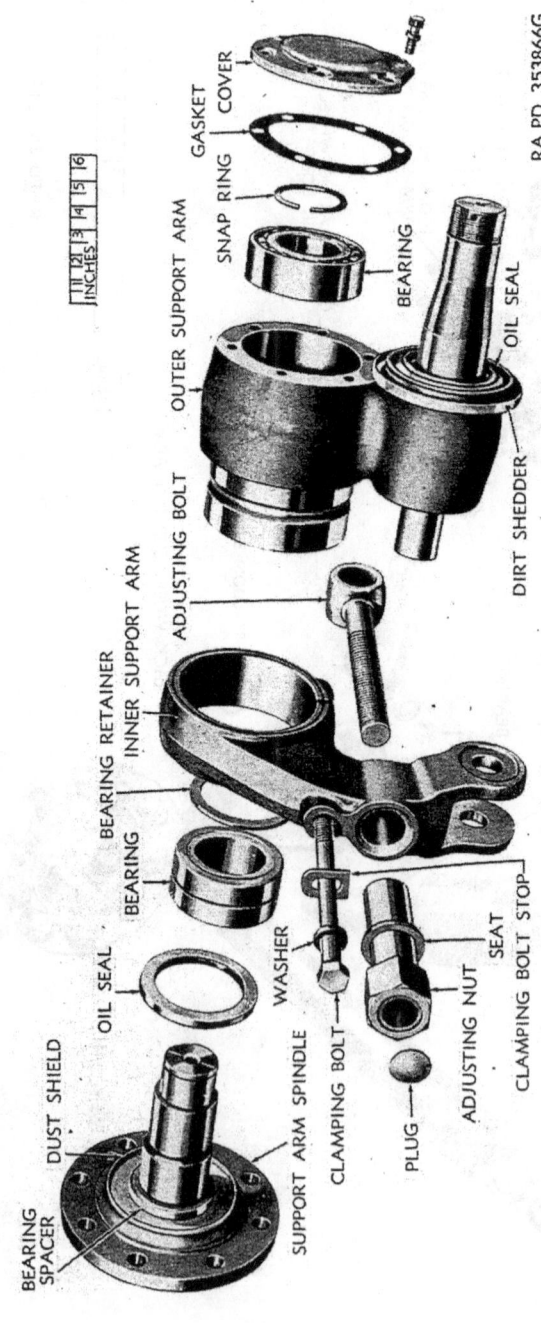

Figure 144. Compensating wheel support arm assembly—exploded view.

thrust washer. Remove inner support arm eye bolt from outer support arm. Tap inner roller bearing and grease retainer out of outer support arm. Tap out outer ball bearing.

(3) *Install bearings.* Grease bearings thoroughly with lubricant prescribed by lubrication order. Using a block of wood or brass drift, install bearings in outer support arm by tapping in until they seat on shoulders. Place new grease seal in inner side of arm, with featheredge toward vehicle, and tap in until it seats on shoulder of arm. Install adjusting eye bolt in inner support arm and place adjusting nut (thoroughly greased with lubricant) and thrust washer on eye bolt. Install inner support arm with eye bolt and nut on outer support arm. Install clamping bolt stud and lock washer on clamping bolt and turn bolt into inner support arm.

(4) *Install compensating arm and lever.* Refer to *b* above.

175. Compensating Link

a. REMOVAL. Remove compensating wheel and hub (par. 173). Remove nut, cotter pin, and washer from inside end of compensating link rear bolt. Install adapter—41–A–18–245 on link bolt, attach slide hammer puller—41–P–2957–33 to adapter, and pull bolt (fig. 145). Working at front end of compensating link, remove bolt holding link to track wheel support arm in same manner and remove link.

Figure 145. Removing compensating link bolt.

b. INSTALLATION. Lift compensating link into position on track wheel support arm (fig. 145) and line up connecting bolt holes. Place compensating link bolt in position and drive in, using a soft hammer. Install washer and nut on bolt and tighten to 180 foot-pounds torque (min). Connect rear end of link to compensating arm in same manner. Install compensating wheel and hub (par. 173), and adjust (par. 173). Connect track and adjust (par. 171).

176. Track Wheels, Bearings, and Oil Seals

a. INTERCHANGEABILITY. The track wheel hub, hub cap, bearings, and oil seals are identical with respective compensating wheel parts.

b. REMOVAL.

(1) *Remove wheel.* If wheel lifter—41–L–1400 is available, proceed as follows: Loosen track adjusting nut several turns to relieve track tension. Place wheel lift block over track guides between drive sprocket and No. 1 track wheel. Drive vehicle forward slowly until wheel to be removed rests on center of block (fig. 146). Place support arm hook under support arm and over top of arm bumper spring bracket, making sure that top of hook is between spring mounting screw and bracket mounting screws (fig. 146). Drive vehicle, slowly, either forward or backward until track wheel next to the one to be removed rests on top of the wheel lifting block. Set brakes and stop engines. Remove 10 nuts which holds wheels to hub and lift off wheel. If wheel lifter—41–L–1400 is not available, set brakes and place a block of wood (2 x 4 in or 2 x 6 in) against inside edge of track. Position hydraulic jack, with base of jack on track and block of wood under support arm to be lifted (fig. 147). Lift support arm until wheel clears track guides. Remove 10 nuts which hold dual wheels to hub and lift off wheels.

Caution: Exercise care in lifting arm to prevent arm end from slipping off jack, as a lift of approximately 9,000 pounds is required to raise arm sufficiently to remove wheels.

(2) *Remove wheel hub.* Remove four hub cap mounting screws, remove cap, and discard gasket. Remove cotter pin, nut, and keyed washer which hold hub assembly on spindle. Pull hub out enough to move outer roller bearing out of cup in hub and remove bearing. Pull hub assembly off spindle and remove inner bearing.

(3) *Remove oil seal.* Pry oil seal off bearing spacer on spindle and discard.

Note. Exercise care in prying off seal or dirt shedder in back of seal will be damaged.

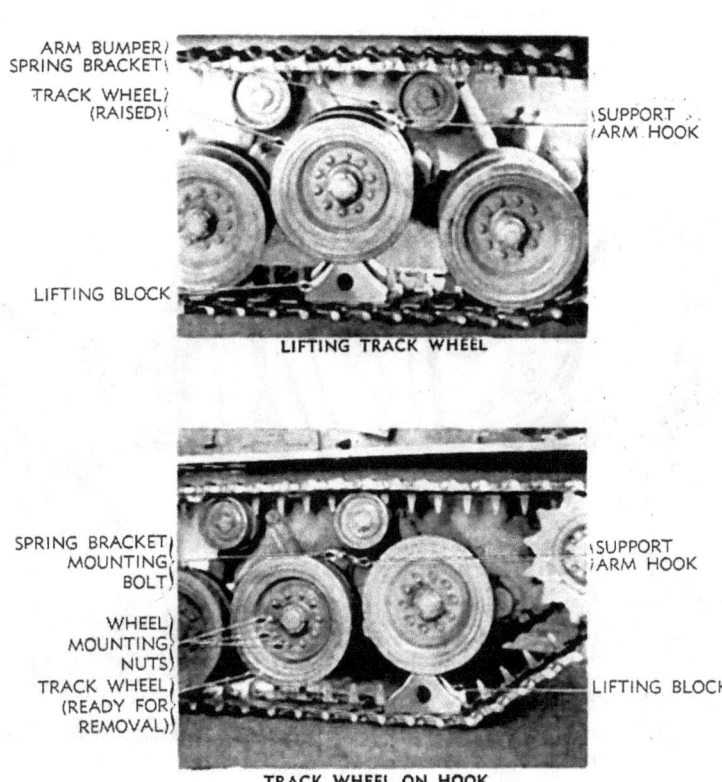

Figure 146. Removing track wheel, using lifter—41-L-1400.

Place hub assembly on bench, with wheel side down, and pry out grease retainer.

(4) *Remove bearing races.* Drive bearing races out of hub carefully, using a brass drift and hammer.

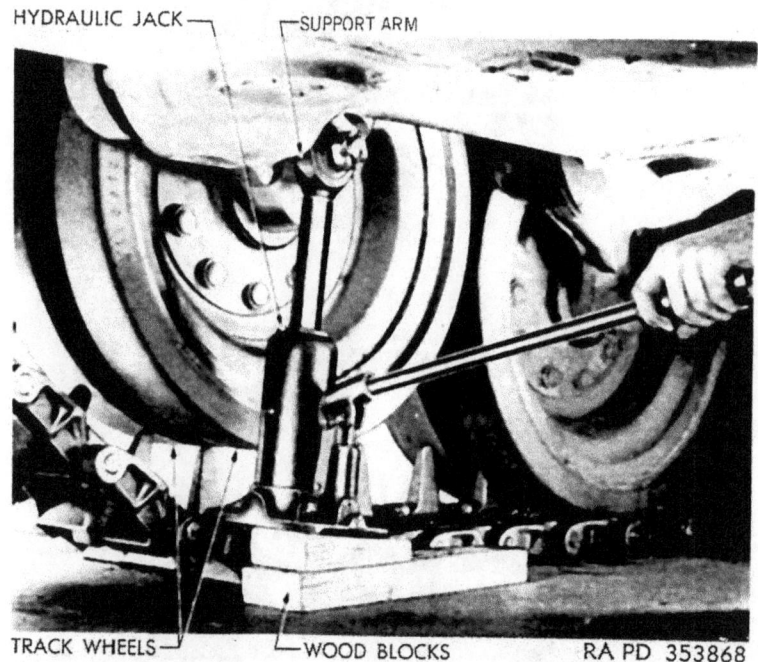

Figure 147. Lifting track wheel, using hydraulic jack.

c. INSTALLATION.

(1) *Install bearing races.* Drive bearing races into wheel hub, using a brass drift or block of wood and being careful to prevent cocking of race in hub.

(2) *Install oil seal.* Place a new oil seal over spindle, with featheredge of seal toward arm, and, using oil seal replacer—41–R–2383–950, drive seal into position on bearing spacer (fig. 148). Place new grease retainer in replacer—41–R–2390–450 (fig. 148) and position replacer on hub. Drive retainer into position, taking care to prevent cocking in hub, and remove tool. Check retainer to see that it is properly seated in hub counterbore and that the flange in retainer is true.

(3) *Install wheel hub.* Pack inner roller bearing thoroughly with the prescribed grease and push on spindle and over shoulder. Lift hub assembly onto spindle and over inner

bearing. Pack outer bearing with grease and push on spindle and into race in hub, centering hub on spindle. Install keyed washer and bearing retaining nut. Turn nut to 200 foot-pounds torque. Back off adjusting nut until it is free. Tighten the nut again to 75 foot-pounds torque. Turn back the adjusting nut to the first cotter pin hole, provided this means turning the nut 15° or more, otherwise, turn to second cotter pin hole. Install cotter pin.

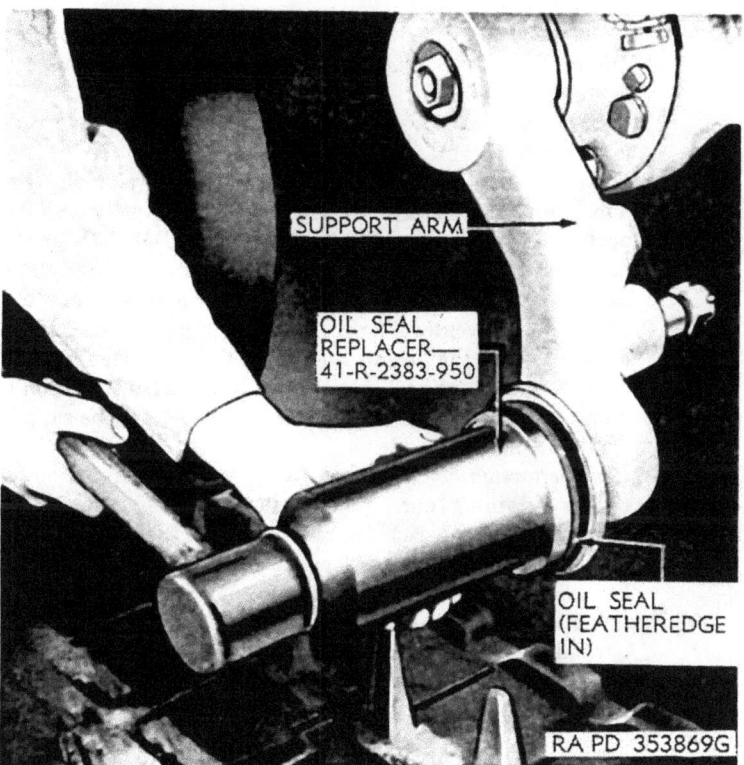

Figure 148. Installing wheel hub seal.

(4) *Install wheel.* Position dual track wheels on hub. Install 10 mounting nuts and tighten with torque indicating wrench—41-W-3634 to 250-300 foot-pounds torque. If wheel lift was used for removing wheels, drive vehicle forward or backward, slowly, until track wheel lifting block is under wheel supported by support arm hook. Remove hook from support arm and cushion stop bracket. Drive vehicle in reverse, slowly, until lifting block is between drive sprocket and No. 1

wheel. Remove lifting block. Adjust track (par. 171). If wheel lift was not available and wheels were removed by using a hydraulic jack, position wheels on hub, install mounting nuts, and tighten to 250–300 foot-pounds torque, with torque indication wrench—41–W–3634. Lower track wheels to track and remove hydraulic jack and blocking.

177. Torsion Bars

a. REMOVAL.

(1) *Disconnect shock absorber.* Disconnect shock absorber at lower end (par. 180).

(2) *Remove track wheel.* Remove track wheel by using hydraulic jack or lifter—41–L–1400 (par. 176).

(3) *Inspect torsion bars and support arm.* Remove track wheel hub (par. 176). Check backlash or free movement of support arms at the arc of wheel spindle. Backlash should not exceed three-eighths of an inch. Raise support arm by hand to take up backlash and check the distance from center of spindle to center line of track support rollers. If the distance is less than shown in figure 149, the torsion bar may be distorted or support arm may be bent. If installation of a new torsion bar does not correct the condition, the support arm must be replaced.

(4) *Remove torsion bar.* Remove screw which holds torsion bar to bar retaining nut. Remove bar retaining nut. Screw adapter—41–A–18–245 for slide hammer into torsion bar, being sure that stud is screwed all the way into bolt and that hex shoulder of adapter seats on bolt head. Attach slide hammer puller—41–P–2957–33 to adapter and pull torsion bar out of support arm (fig. 150).

b. INSTALLATION.

(1) *Indentifying torsion bars.* The torsion bars have designating arrows stamped on the arm end. The arrows indicate the rotation of the bars when the wheel and arm are raised or, in other words, the direction of bar "spring." The four front torsion bars on the right side of the vehicle have designating arrows in clockwise rotation. The four front torsion bars on the left side of the vehicle have arrows pointing in the counterclockwise direction. The rear torsion bar on the right side of vehicle has an arrow indicating counterclockwise and the rear torsion bar on the left side of the vehicle has a clockwise arrow. Be sure to install torsion bars in the correct locations.

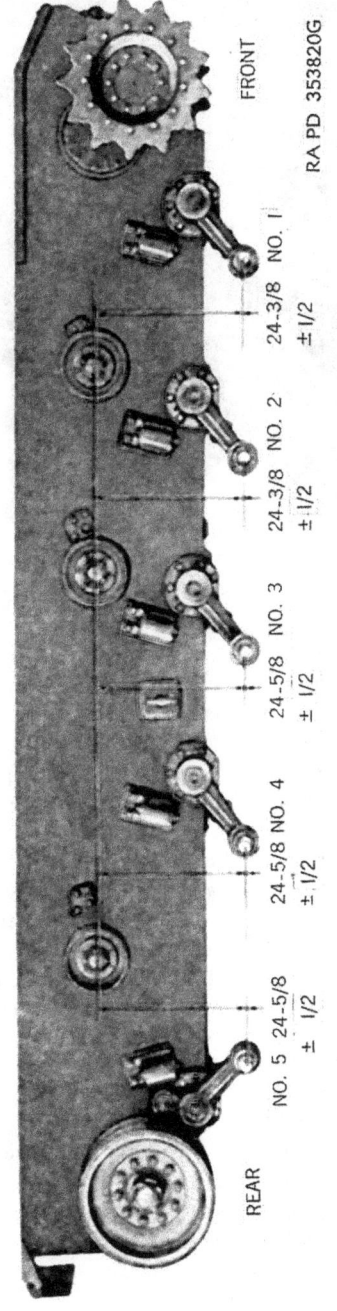

Figure 149. Torsion bar and support arm serviceability limits.

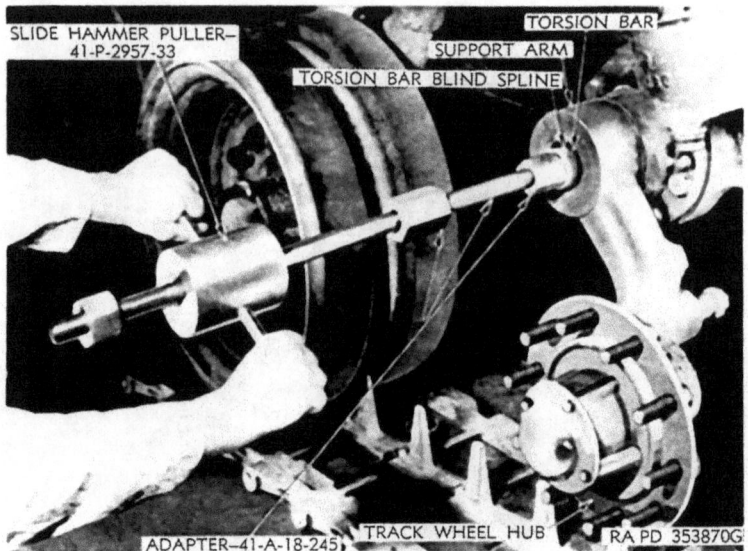

Figure 150. Removing torsion bar.

(2) *Install torsion bars.* Slide the correct torsion bar into the tunnel until the inner end of the bar contacts the anchor. Examine the outer end of the torsion bar and rotate it until the small slot that is cut in the chamfer is straight up. This will insure the blind spline at the anchor end of the bar also being straight up, a requirement for proper entry of the bar into its anchor. Gently tap the bar into the anchor with a soft hammer. If the bar will not enter readily, loosen the cap screws which hold the anchor supporting plate and make certain that the four cap screws are loosened approximately the same amount. The two cap screws which hold the anchor to the supporting plate do not have to be loosened. After the torsion bar has entered the anchor, raise the support arm until the small slot in the outer edge of the arm lines up with the blind spline in the outer end of the torsion bar. The torsion bar can then be tapped all the way into place. Install the torsion bar retaining nut and tighten it to 75 foot-pounds torque. Install the cap screw in the retaining nut and tighten it to 65 foot-pounds torque. Tighten the torsion bar anchor plate screws to 35 foot-pounds torque.

(3) *Inspect torsion bars and support arm.* Check backlash and/or torsion bar set (*a* above) (fig. 149). Install track wheel hub.

(4) *Install track wheel.* Install track wheel using hydraulic jack or lifter—41–L–1400 (par. 176).

(5) *Connect shock absorber.* Connect shock absorber lower end to support arm (par. 180).

178. Track Wheel Support Arm

a. REMOVAL. Remove track wheel and torsion bar (par. 177). Remove seven bolts that hold support arm assembly to hull side and lift or hoist arm out of hull opening.

b. INSTALLATION. Coat support arm mounting plate flange on hull with joint sealing compound. Lift or hoist arm assembly into position on hull side and line up mounting bolt holes with drift punch. Dip seven arm mounting bolts in joint sealing compound, install in arm flange, and tighten to 240–260 foot-pounds torque. Install torsion bar and track wheel (par. 177).

179. Track Support Rollers, Bearings, and Seals

a. REMOVAL. Place block of wood on track wheel nearest roller to be removed. Set jack on wood block and raise track until it clears roller (fig. 151). Remove five bolts that hold roller mounting bracket to hull side wall and remove roller.

b. INSTALLATION. Coat flange of roller mounting bracket with joint sealing compound. Lift support roller assembly up until track guides enter space between rollers and position mounting bracket on hull side wall. Line up mounting screw holes in bracket flange and hull side wall. Dip mounting bolts in joint sealing compound and install, drawing up evenly.

c. BEARING AND SEAL REPLACEMENT.

(1) *Removal.* Remove track support roller as outlined in *a* above. Remove four screws that hold hub cap to wheel hub, remove cap, and discard gasket. Remove cotter pin and nut that hold bearings on roller spindle. Pull out on roller to move outer bearing and keyed washer out of roller hub. Remove outer bearing and washer. Lift roller and hub assembly off mounting bracket spindle. Remove inner bearing and grease retainer. Pry grease seal off bearing spacer and discard.

Note. Remove grease seal carefully to prevent damage to dust shedder.

Remove bearing races out of hub, using adapter—41–A–18–296 for outer bearing race (fig. 152) and adapter—41–A–12–550 for inner bearing race, together with screw—41–S–1047–200.

277

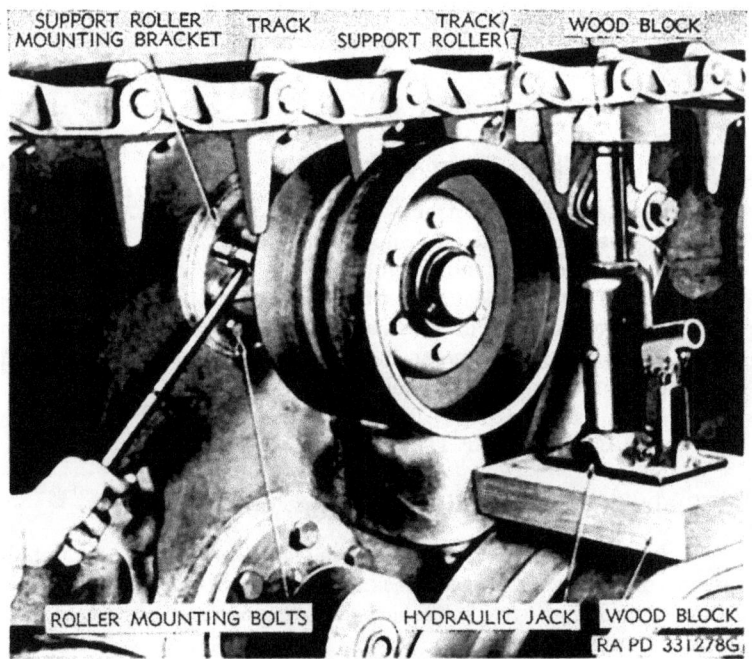

Figure 151. Removing track support roller.

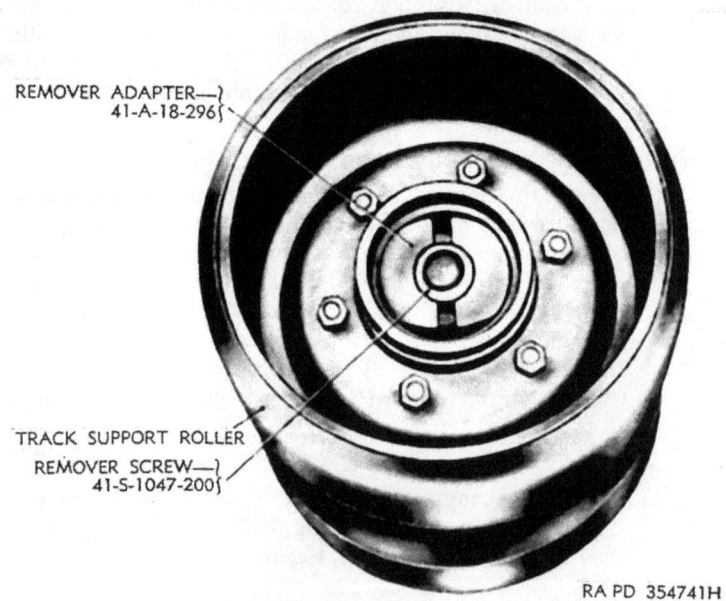

Figure 152. Bearing race remover adapter.

(2) *Repair.* When the track support roller brackets became loose and cannot be tightened with existing bolts, proceed as follows:
  (*a*) Block up track at loose support roller bracket and remove bracket.
  (*b*) Drill and tap existing holes in side of hull for 5/8–18NF–2 x 1 5/8-inch bolts.
  (*c*) Ream existing holes in support roller bracket to fit new bolts.
  (*d*) Position support roller bracket in place and attach to hull, using lock washers—121574 and 5/8–18NF–2 x 1 5/8-inch bolts—224252.
(3) *Installation.* Install bearing races in roller hub by driving carefully into place with brass drift or block of wood until they seat in shoulders of hub. Place new grease seal on support spindle, with featheredge toward mounting bracket and drive in position on bearing spacer, using seal replacer—41–R–2397–875 (fig. 153). Place new grease retainer in position on replacer—41–R–2396–375 with flanged edge against shoulder of tool and drive into support roller (fig. 154), until it seats on hub shoulder. Pack inner bearing with grease, as prescribed by lubrication order, and position on spindle. In-

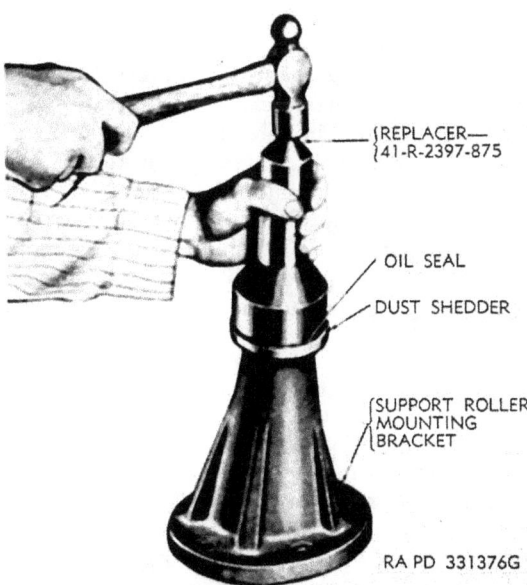

Figure 153. Installing track roller grease seal.

Figure 154. Installing track roller hub grease retainer.

stall roller and hub assembly on spindle and over grease seal and inner bearing, being careful to prevent buckling grease seal leather. Pack outer bearing with grease and install on spindle and into race in hub. Install keyed washer and nut on spindle and tighten adjusting nut to 150 foot-pounds torque. Back off nut until there is no torque. Tighten nut to 75 foot-pounds torque. Back off nut to first cotter pin hole, with minimum travel of 15°, and install cotter pin. If first hole is less than 15° travel, back nut off to next hole. Using a new gasket, position hub cap and install four screws. Install support roller assembly (*b* above).

180. Shock Absorbers

a. REMOVAL. Working at upper end of shock absorber, remove cotter pin, nut, and washer that hold end of shock absorber to mounting bracket on hull side. Screw puller—41-P-2907-196 on bushing flange of shock absorber. Hold body of puller with large crescent wrench and turn down puller bolt, pulling shock absorber off pin (fig. 155). Working at lower end of shock absorber, remove cotter pin, nut, and washer that hold shock absorber to mounting pin on support arm.

Using puller—41-P-2907-196 remove lower end of shock absorber from mounting pin in same manner.

Note. Use of a longer ¾-16NF-2 bolt in shock puller will permit pulling lower end of shock absorber completely off pin.

b. INSTALLATION. Lift shock absorber into position and enter upper end bearing over mounting bracket pin. Drive bearing into position on mounting bracket carefully with a soft hammer, install washer and nut, and tighten nut to 180 foot-pounds torque, using a torque wrench. Install cotter pin. Mount lower end of shock absorber on pin in support arm in the same manner.

181. Arm Bumper Spring Bracket

a. REMOVAL. Remove bolt and spacing washer which hold volute spring in bracket and remove spring. Remove two bolts inside of bracket which hold assembly to hull side wall. Remove two mounting bolts which hold upper end of bracket on hull side (fig. 146) and remove stop bracket.

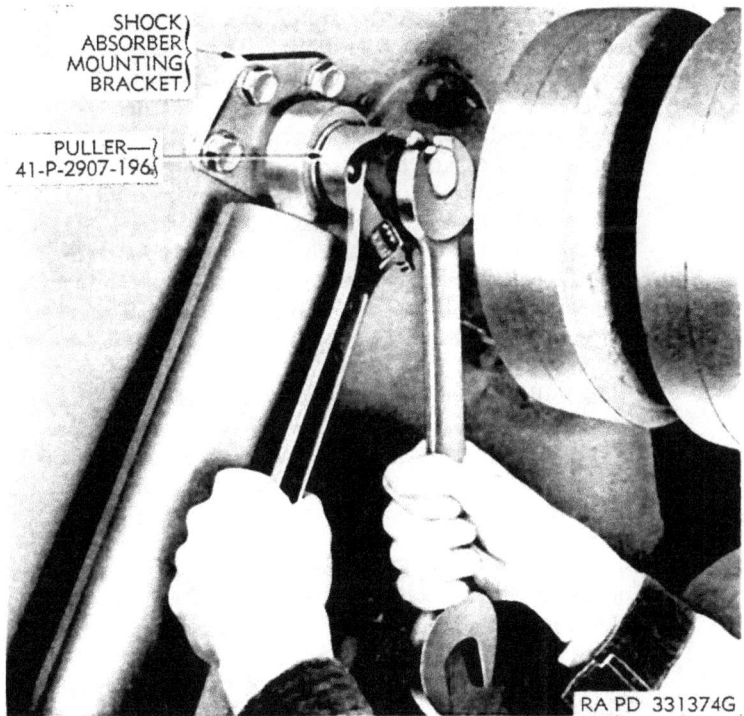

Figure 155. Removing shock absorber.

Note. Early production model vehicles have had this bracket welded to the hull, for front torsion bars, to prevent shearing under certain driving conditions. If this is the case, it will be impossible to remove the bracket without first breaking the welds.

b. INSTALLATION. Identify proper bracket for right and left side. When bracket is properly installed, the boss on the bottom of the bracket should point forward toward the wheel end of support arm. Position bracket on hull side wall and install two upper screws and washers. Threads of bolts should be well coated with joint sealing compound before installation. Working under bracket, install two lower mounting bolts and washers. Tighten the four attaching bolts to 80–85 foot-pounds torque.

Section XXI. HULL

182. Description and Data

a. DESCRIPTION.

(1) The hull of the vehicle is a completely welded structure except for portions of the front, top, and floor, which are removable for service operations. These removable portions consist of a plate above the controlled differential at the front of the vehicle, two drivers' doors over the drivers' seats, a large hinged door over the engine compartment, an air inlet grille ahead of, and an air outlet grille in back of the engine compartment door, and removable covers (one on each side) over each fuel tank and over each pair of batteries. Openings in the bottom of the hull include the escape hatch door, two large inspection plates (one under each engine and transmission) and the small covers just beneath the drain plugs for the engines, hydramatic transmissions, transfer unit, controlled differential, and final drives.

(2) The hull floor carries the mounting brackets for the engine and transmission supports, the transfer unit supports, and the differential supports. It also incorporates the tunnels in which the torsion bars for the track suspension wheels are carried. The hull is divided into two compartments: the fighting compartment at the front and an engine compartment at the rear. These compartments are separated by a bulkhead that extends from side to side and from the roof down to the bulkhead extensions, which in turn extend forward to cover the transfer unit.

(3) Seats for the driver and assistant driver are mounted in the front of the hull. These seats have both forward and backward and up and down adjustments (par. 28). Protective

pads for driver and assistant driver are provided around the final drive propeller shafts, at the side of the controlled differential, and on the periscope head rests.

(4) The various stowage items carried in the hull are mounted in sheet metal containers, which are bolted or latched to the hull floor and side walls. Ammunition stowage is provided under hinged covers which, when closed, serve as a subfloor. The 75-mm shells are carried in double-walled containers.

b. DATA. For detailed data on the hull refer to paragraph 5.

183. Sealing Hull Parts

a. GENERAL. Joint sealing compound should be applied to detachable hull parts when they are assembled to the vehicle to prevent water, dust, or fine sand from seeping through the joints and entering the hull.

b. PARTS TO BE SEALED.
  (1) Differential opening cover.
  (2) Final drive housings.
  (3) Drain valve flanges.
  (4) All floor pans and drain plug covers.
  (5) All bolts extending through the side of hull or through floor of hull.

184. Drivers' Doors

a. REMOVAL. Release door latch lever and allow spring pressure to raise door (fig. 10). With a helper to support door in this position, remove eight screws and washers holding hinge and lever assembly to hull roof. Remove door and hinge assembly by sliding lever out of hinge opening. Remove door.

Note. Do not lose shims from under hinge flange.

b. INSTALLATION. Place shims over hinge flange and coat edge of flange with joint sealing compound. Lift door and hinge assembly into position above hinge opening. Slide latch lever and hinge through hinge opening and place in position on hull roof.

Note. Flat spot on hinge flange should be toward door opening.

With helper holding up end of door to keep flange on hinge even with hull roof, install eight screws and washers that hold hinge flange to hull roof.

185. Bulkhead Doors

The bulkhead doors (fig. 156) are made in two sections. The bottom or lower half of the door may be slid up or down by releasing

the handle on the front of the door. This sliding door allows the
engine fans to draw foul air or smoke from the fighting compartment.
This door can be locked in any position by opening door the desired
amount and then pushing down on latch. The top half or complete
door is removable for the purpose of cleaning out leaves, twigs, or
other foreign matter from the front of the radiator cores. To remove doors, raise the two spring levers at the top and allow doors
to tilt forward. Raise levers to clear stops on doors, swing door to
approximately 90° from bulkhead, and remove from bottom retainer.

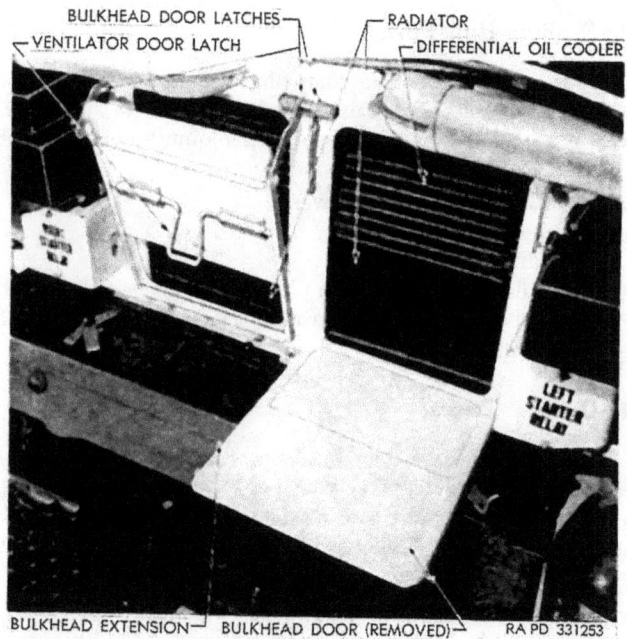

Figure 156. Bulkhead doors.

186. Protective Pads

a. GENERAL. The protective pads which can be replaced by the
using organization personnel are the periscope pads, final drive propeller shaft knee pads, controlled differential hand guard pads, and
the crash pads around the door openings.

b. PERISCOPE PADS.

    (1) Working in driver's seat, remove two screws and washers
holding periscope pad to holder and remove pad.

    (2) To install pad, place it on holder and install two mounting
screws and lock washers.

c. FINAL DRIVE PROPELLER SHAFT GUARD KNEE PADS.
   (1) Remove four screws and lock washers holding both halves of guards to brackets mounted on controlled differential and final drive housing. Remove three nuts and washers holding knee pad to lower guard and remove pad.
   (2) To install pad, place three studs on knee pad through slots in guard and install nuts and washers. Place upper or lower half of guards on mounting brackets and install screws and washers.

d. CONTROLLED DIFFERENTIAL HAND GUARD PAD.
   (1) Remove screw and washer holding top of guard to final drive propeller shaft guard mounting boss on controlled differential. Remove bottom screw holding guard to differential case. Slide guard out from between mounting boss on differential case and final drive propeller shaft guard mounting bracket. Remove four nuts and washers holding hand guard pad to support and remove pad.
   (2) To install pad, place studs on pad through holes in support and install nuts and washer. Place support over mounting holes on controlled differential and install screws and washers.

e. DRIVER'S DOOR OPENING PAD AND/OR SEAL.
   (1) Working in driver's seat, remove 12 screws holding driver's door opening pad retainer to under side of hull roof. Push pad downward into vehicle. This pad is pressed into position so that it will hold the weather seal in position on the deflector around door opening. To remove weather seal, pull from retainer.
   (2) To install, coat edges of rubber weather seal with joint sealing compound and place around door opening deflector. Place seal in position and raise pad tightly up against weather seal. Install 12 screws and washers holding pad and seal in position in door opening.

187. Seats

a. REMOVAL. Remove seat back by pulling straight up until frame post of back slides out of seat assembly. Remove seat cushion. Raise seat to the "UP" position by releasing control lever on right side and, at same time, take weight off seat. Remove four screws and washers holding seat assembly to hull floor. Remove seat through driver's door opening.

b. INSTALLATION. Lower seat assembly through driver's door opening and position over tapped holes in hull floor. Loosely install four screws and washers holding seat assembly to hull floor. Adjust seat in its slotted holes and tighten screws. Place seat cushion on seat and slide seat back into holes on seat frame.

188. Propeller Shaft Tunnel

The propeller shaft tunnel is formed by the ends of the ammunition boxes on each side and sealed at the rear by a baffle bolted to the hull floor. The top of the propeller shaft tunnel is covered by hinged doors which form part of the vehicle subfloor and cover stowage boxes. The removal of the propeller shaft and tunnel covers is outlined in paragraph 158.

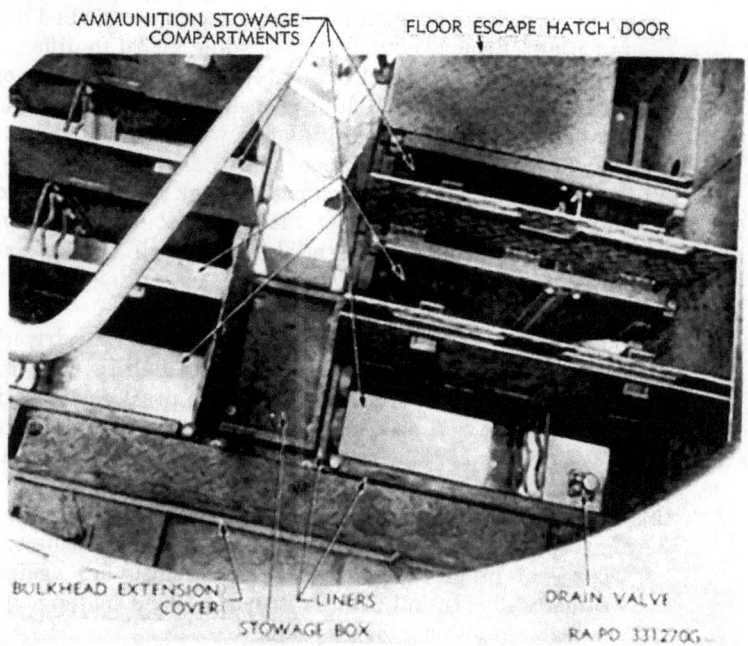

Figure 157. Ammunition boxes.

189. Ammunition Boxes

The rounds of 75-mm ammunition are stored beneath the hull subfloor.

Note. The use of liquid in containers in ammunition racks is discontinued (par. 54).

190. Subfloor

The hull subfloor consists of the covers of the ammunition boxes, which are bolted to the hull floor. The front hinged cover over the propeller shafts is for lubricating the two oil cups at the front propeller shaft support bearing.

191. Drain Valves

a. DESCRIPTION. There are five hull floor drain valves located at various places on the hull floor (fig. 158). To allow water to drain from the hull floor, push down on knob to compress spring and turn knob one-quarter turn to right. This will lock valve in the open position. After water has drained, turn knob to left to release lock and close valve.

b. REMOVAL. With helper holding heads of bolts from underneath the vehicle, remove four safety nuts and allow valve assembly to drop down.

c. INSTALLATION. Coat edges of valve cage with joint sealing compound and hold in position from bottom of hull. Coat four bolts with sealing compound and push in from bottom of hull. With helper holding heads of bolts, install four safety nuts. Wipe off surplus sealing compound.

192. Periscopes

The driving periscopes in the hull are in the driver's doors. Operation of these periscopes is outlined in paragraph 32.

Section XXII. TURRET

193. Description

The turret is of curved armor plate, all-welded, approximately 60 inches inside diameter. The sides of the turret are 1 inch thick; the roof is one-half inch thick. The turret rotates 360° on a continuous ball-bearing mount. This bearing is completely enclosed for protection from direct hits and from dirt or water. The turret is traversed either by hand or by the hydraulic traversing mechanism. The turret is circular in shape, except that a bulge extends to the rear (opposite the gun). The radio is mounted in this bulge. There is no turret basket, but seats for the commander, gunner, and loader are attached to the turret and rotate with it. Two hinged doors provide access to the turret. One door is in the top of the commander's cupola and the other is on the right side of the turret. The commander has vision through a periscope in the top of the cupola door and six vision blocks at the base of the cupola. A periscope for the gunner is located forward of the cupola. A pistol port is located on the right side of the turret to permit discarding empty 75-mm shells with turret closed.

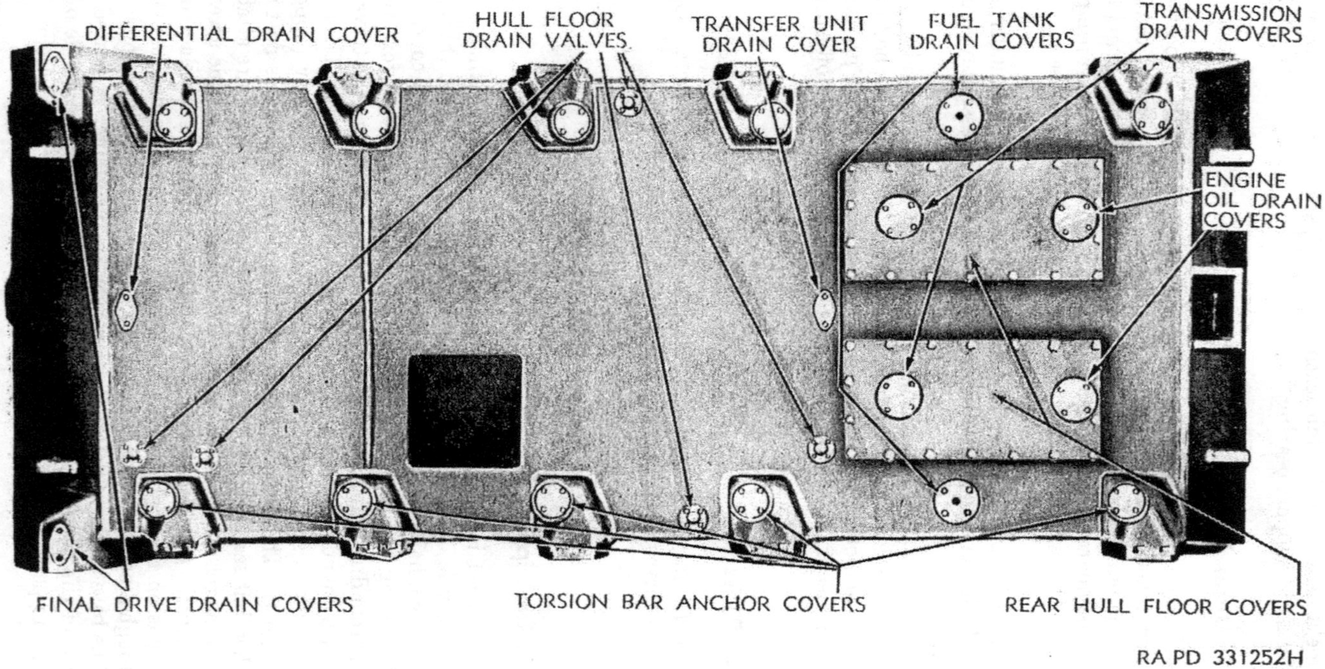

Figure 158. Drain openings in hull.

194. Doors and Latches

a. TURRET DOOR.

(1) Remove hairpin locks from each end of turret door hinge torsion rod. Remove screw holding dust cap to outer door hinge and remove cap (fig. 159). Raise door to vertical position to relieve tension from torsion rod and slide torsion rod hinge cap off end of rod. Pull torsion rod out of anchor on turret roof and out of hinges. Remove door.

(2) To install, place door in vertical position over hinges on turret roof. Slide torsion rod through hinges and into anchor welded to turret roof. Place hinge cap over outer end of torsion rod so that cap fits into cutout hinge. Lower door to closed position, and install dust cap and screw. Install retaining clips on each end of torsion rod.

b. CUPOLA DOOR (fig. 160).

(1) Open cupola door to the vertical position and hold in this position. Remove eight screws holding cap assemblies to outer ends of hinge. Remove caps. Torsion springs are now in the released position. Pull all six springs out of tube. While supporting door, push tube out of hinge bushings. Lift out door.

(2) To install door, place door in vertical position next to hinges on cupola. While holding door vertical, install tube through bushings in hinges. Next, place six torsion springs in tube. Position cap assemblies over ends of hinges, making sure that torsion springs fit into cutouts in caps. Install eight screws holding caps to hinges. Close door.

c. DOOR LATCH.

(1) *Cupola door.*

(*a*) Remove two screws which fasten cupola door latch spring and handle to bottom side of cupola door. Slide out spacer and remove spring and handle from door (fig. 160).

(*b*) To install, place spring and latch handle in bracket on door, place spacer in bracket, and install two screws clamping spacer to bracket.

(2) *Turret door.*

(*a*) Remove nut and external tooth washer from end of latch lever shaft. Lift off inside latch from lever shaft and remove spacers (fig. 17). Slide latch lever shaft out of door.

(*b*) To install, place two spring washers under head of shaft and insert latch lever shaft through turret door. Place shim spacers over shaft and install inside latch, lock washer, and nut.

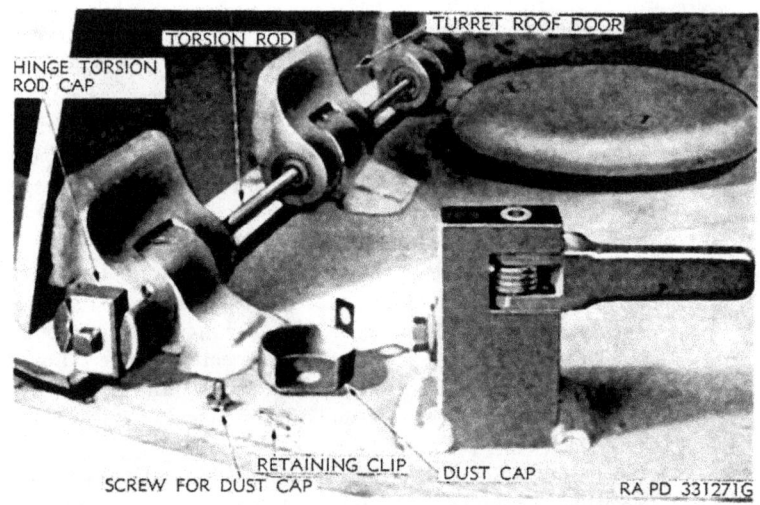

Figure 159. Turret door hinge.

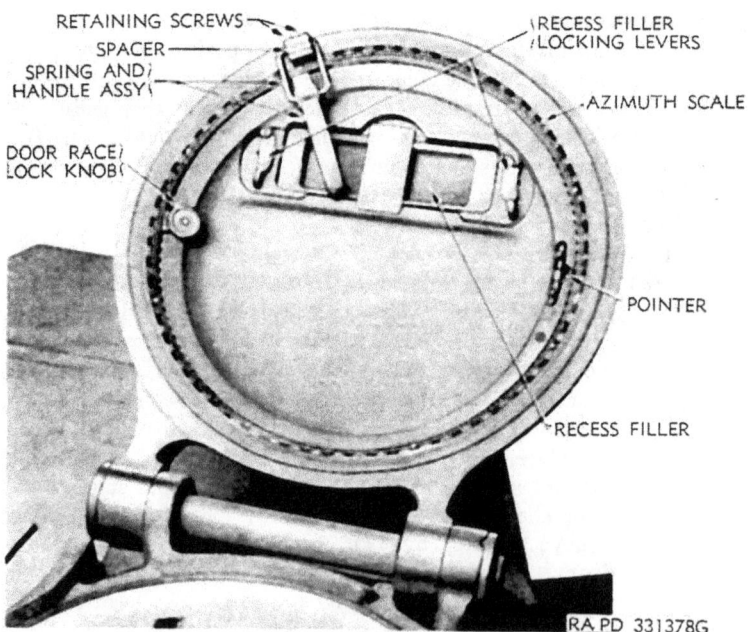

Figure 160. Cupola door and latch.

195. Seats

a. COMMANDER'S SEAT. The commander's seat, which is attached to the rear of the turret, can be raised or lowered in 1-inch variations by releasing a lever at the left lower side of the seat and raising or lowering seat to the desired position. The seat may be raised to the top position and folded back for additional height while riding with cupola door open.

   (1) To remove the seat assembly, remove the top stop screw in the center of the seat track (fig. 161) and slide out seat assembly.
   (2) To install seat, slide seat into seat track and install top stop screw.

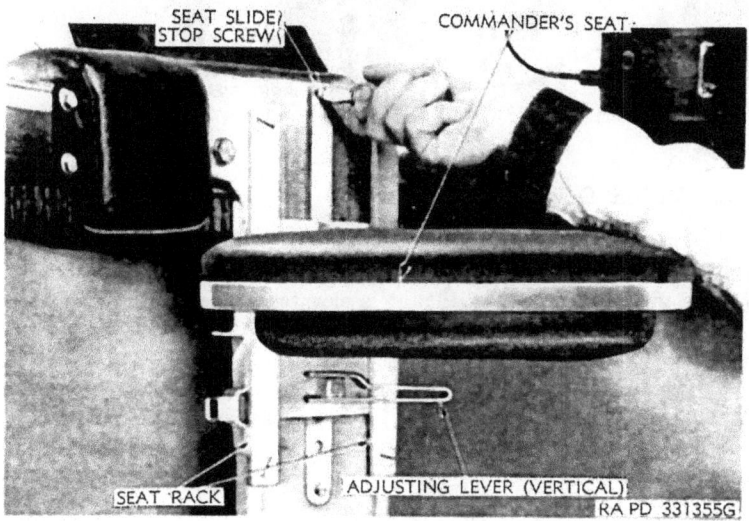

Figure 161. Removing commander's seat.

b. GUNNER'S SEAT. The gunner's seat may be adjusted to three positions by removing the locking pin at the bottom of the seat, raising or lowering seat to the desired position and installing adjusting pin (fig. 162).

Note. Early production model vehicles were not equipped with a foot rest for the gunner's seat, but have since been modified to include this foot rest.

c. LOADER'S SEAT. The loader's seat is attached to a hinged bracket at the rear of the turret and can be folded down for sitting, or folded back for storing.

   (1) To remove seat completely, release latch on bottom of seat and slide seat out of bracket.
   (2) To install, slide seat into bracket and engage latch lever.

291

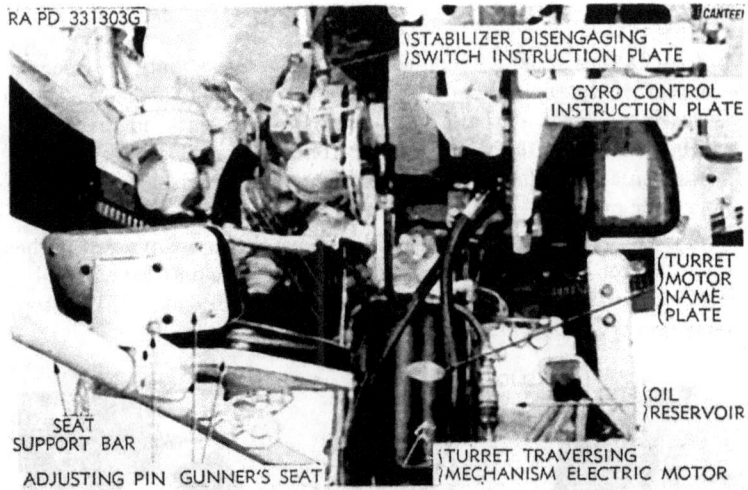

Figure 162. Gunner's seat.

196. Vision Devices

a. GUN-SIGHTING PERISCOPE. A periscope is mounted in the turret roof, forward of the commander's cupola. This unit is linked to the 75-mm gun in such a way that it is elevated or depressed with the gun. Periscope heads should be replaced in the event of damage (par. 238).

b. CUPOLA PERISCOPE. An emergency periscope (where periscope M15A1, par. 231 is not available) can be used in the commander's cupola door in the event of damage to the direct vision blocks in the base of the cupola.

(1) To remove, refer to paragraph 236*b*.

(2) To install the periscope in the door, remove the periscope recess filler by releasing the two latches at the ends of the plug and pulling down on filler. Insert the periscope holder into the opening and lock in position with the two latch levers. Insert the periscope in the holder and lock in position by sliding back the latch and tightening the knurled knob. The cupola door race plate and periscope can be rotated 360° by loosening the knurled lock knob at the base of the plate next to the azimuth scale and turning the plate (fig. 160).

c. DIRECT VISION BLOCK REPLACEMENT.

(1) Remove the two lock wedge screws and the jack screw locking block in position (fig. 163). Remove lock wedge and allow block to drop down.

(2) To install, slide block into position and install jack screw and two lock wedge screws.

197. Traversing Mechanism

a. REMOVAL OF PUMP. To replace the traversing pump, it is necessary to remove the complete pump and motor assembly as a unit to permit removal of the pump from the motor.

(1) *Disconnect oil lines and conduits.* Remove firing control handle from elevating handwheel by removing screw and washer from wheel (fig. 18). Remove two screws holding conduit guard cover to stabilizer pump. Disconnect three cables from terminal. Remove two clips holding conduit to motor support bracket. Place conduit to one side. Disconnect lower ends of both stabilizer hose connections at stabilizer pump.

Caution: Place rags or a container under pump to prevent oil from draining onto hull floor.

Remove two screws holding cal. .30 firing switch to hydraulic control handle. Remove one screw holding handle to lever. Remove two screws clamping conduit to lever. Disconnect cables from switch and pull conduit out of lever. Remove two screws holding traversing remote control cable to front of turret and elevating gear box. Remove two screws holding remote control handle to turret roof. Dis-

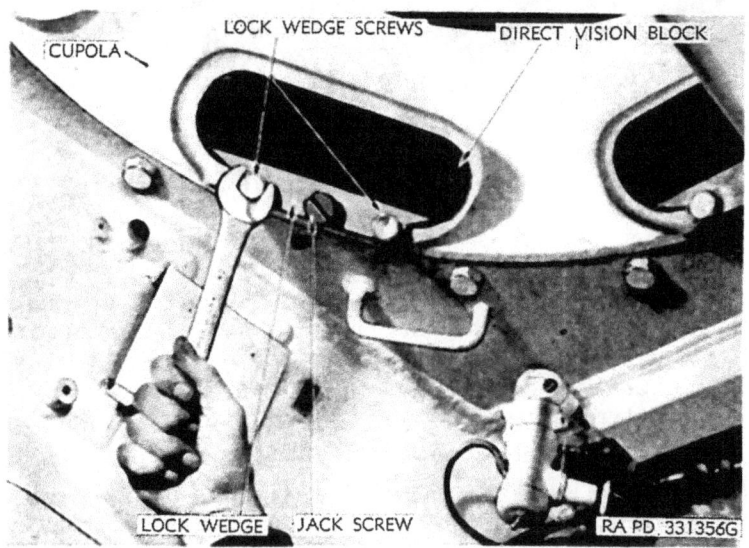

Figure 163. Removing cupola vision block.

connect oil feed line from traversing pump and bottom of oil reservoir. Remove line. Disconnect two oil feed lines from top of traversing pump and top of oil reservoir. Disconnect two oil feed lines from front end of traversing pump. Disconnect oil feed line from stabilizer pump to stabilizer reservoir.

(2) *Remove pump and pump motor assembly from turret.* Remove right lower screw holding turret motor to turret motor mounting bracket (fig. 164). Pull bracket to one side and

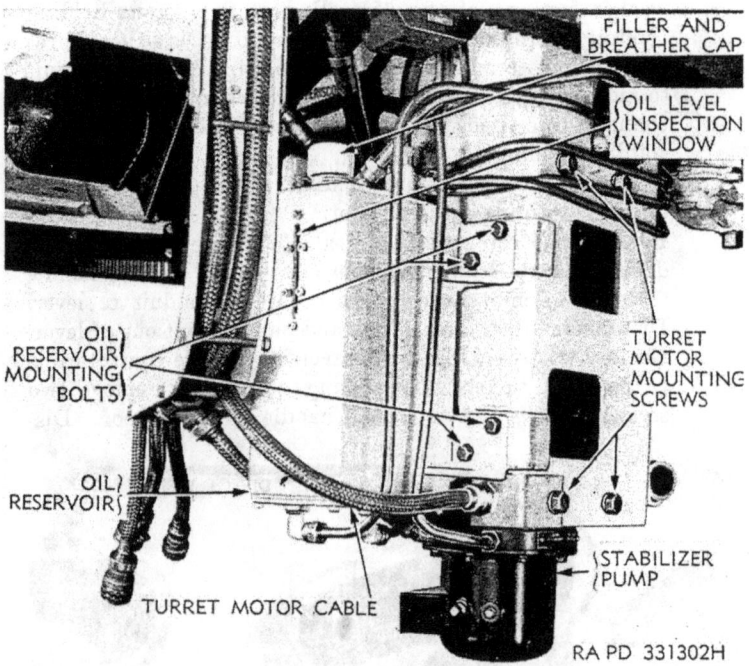

Figure 164. Stabilizer pump and turret motor.

disconnect turret motor feed cable from terminal on turret motor. Remove three remaining screws holding turret motor to turret motor mounting bracket.

Caution: Support weight of motor and pump assembly while removing last screw.

Remove assembly from vehicle and place on work bench.

(3) *Remove pump from pump motor.* Remove four screws and washers holding pump to pump motor. Lift off pump and discard gasket.

b. INSTALLATION OF PUMP.
   (1) *Install pump.* Place a new pump gasket over pump motor and install the four attaching screws and lock washers.
   (2) *Install pump and pump motor assembly in turret.* Lower the pump and pump motor assembly into the turret and bring up into position on the motor mounting bracket on front of turret. Install two upper and lower left turret motor mounting screws. Connect turret motor cable to terminal on turret motor. Place turret motor cable mounting bracket over the motor support bracket and install the last motor mounting screw.
   (3) *Connect conduits and oil lines.* Connect oil feed line from stabilizer reservoir to front of stabilizer pump. Connect two oil feed lines to front of traversing pump. Connect two oil feed lines from top of stabilizer pump to top of oil reservoir. Connect oil feed line from bottom of stabilizer pump to bottom of oil reservoir. Connect two stabilizer hoses to stabilizer pump. Position remote control cable against front of turret elevating gear box and install mounting screws through clips. Install two screws holding remote control handle to turret roof. Slide the firing control cable through the hydraulic control lever and assemble firing switch to ends of terminals. Position handgrip on end of operating lever and install attaching screw. Place switch in top of handgrip and install two mounting screws. Lock cable to lever at bottom by installing two screws through locking plate. Place stabilizer pump cables over terminals on stabilizer pump and install retaining nuts and washers. Position cover over junction box and install two attaching screws. Line up hole in firing switch with hole on elevating handwheel, and install screw and lock washer. Fill stabilizer reservoir and oil reservoir with hydraulic oil and bleed stabilizer.
c. OIL RESERVOIR REMOVAL AND INSTALLATION.
   (1) Place a container under the oil reservoir and disconnect oil feed lines at bottom of pot. Remove two bolts holding conduit to front of oil reservoir. Disconnect six oil feed line connections at top of oil reservoir. Remove four bolts holding oil reservoir to traversing and stabilizer pump motor mounting bracket (fig. 164). Remove reservoir.
   (2) Install oil reservoir over mounting holes on traversing and stabilizer pump motor mounting bracket. Install four attaching bolts and lock washers. Connect six oil feed lines to fittings on top of oil pot. Connect two oil feed lines to fittings on bottom of oil pot. Place cable mounting clips over mount-

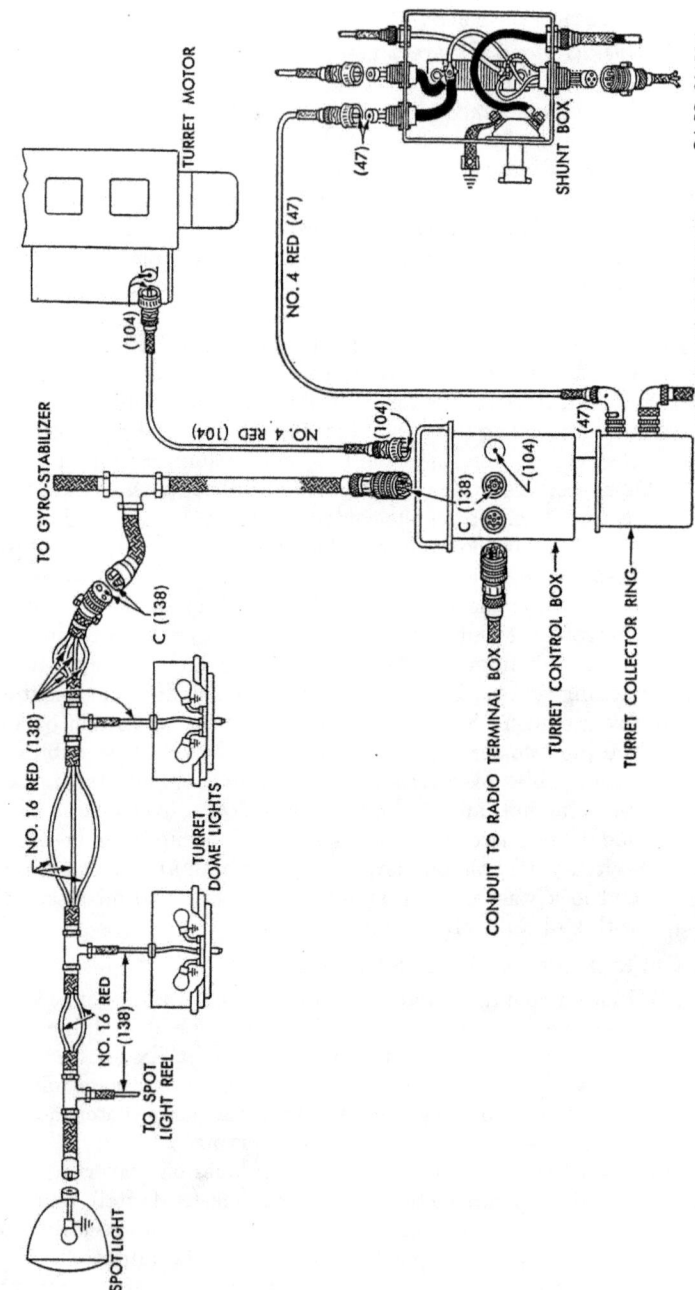

Refer to paragraph 101 for list of electrical circuit numbers.

Figure 165 — Turret feed circuit wiring diagram.

ing holes on front of oil reservoir and install two attaching screws and washers. Fill oil reservoir with hydraulic oil.
d. ELECTRIC MOTOR REMOVAL AND INSTALLATION.
   (1) Remove pump motor and separate from traversing pump as outlined in *a* above. Remove four safety screws holding pump to traversing and pump motor and remove pump. Discard gasket.
   (2) Install a new stabilizer pump gasket on bottom of pump motor and install four safety screws and lock washers. Install traversing pump on hydraulic traversing motor and install assembly as outlined in *b* above.

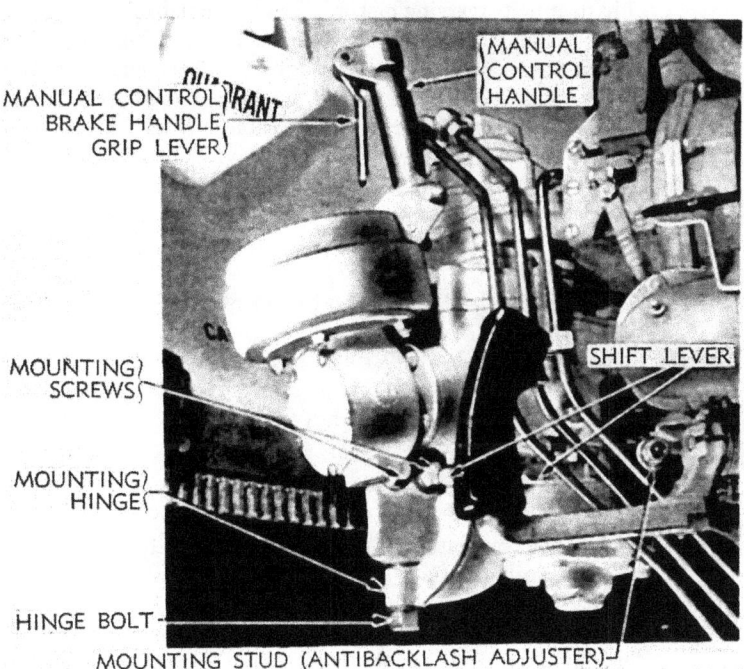

Figure 166. Traversing gear box mountings.

e. TRAVERSING GEAR BOX REMOVAL AND INSTALLATION.
   (1) Disconnect three oil feed lines at top of hydraulic traversing motor. Remove two screws with 3/4-inch heads holding gear box mounting bracket hinge to side of turret (fig. 166). Remove lock nut, nut, washer, and spring from antibacklash adjustment stud. Install both nuts, and lock together to remove stud. Remove stud. Remove screw with 1-inch head in front of gear box. Remove two screws with 3/4-inch heads

in front of manual control wheel on top of base ring and remove traversing gear box assembly.

Caution: Support assembly while removing last screw.

(2) Position traversing gear box over mounting holes on turret. Install holes on turret. Install two screws in front of manual control operating handle holding gear box hinge to turret. Install antibacklash stud through mounting bracket and into turret ring. Remove two nuts used for installing studs. Install two screws holding gear box mounting bracket hinge to turret. Connect three oil feed lines to top of hydraulic traversing motor. Position antibacklash spring over stud and install spacer, flat washer, nut, and lock nut. Rotate turret 360° to locate high spot between gear and turret ring. When high spot is located, tighten nut on spring until a slight drag is noticeable. Back off nut one-quarter turn and lock with lock nut.

f. DRAINING AND FLUSHING HYDRAULIC TRAVERSING SYSTEM. Drain, clean, and flush the entire system as follows:

(1) Turn off main battery switch and lower the gun barrel as far as possible.

(2) Disconnect oil feed lines at bottom of oil reservoir (fig. 164) and drain oil into a container.

(3) Remove bottom plate. With a clean rag soaked in dry-cleaning solvent or volatile mineral spirits paint thinner, clean out any sludge which may have accumulated on inside of oil reservoir.

Note. Avoid the use of waste whenever possible for this operation. Be sure that no pieces of lint are clinging to parts before assembling.

(4) Install bottom plate and oil feed lines and fill oil pot with hydraulic oil.

Caution: Under no circumstances is dry-cleaning solvent or volatile mineral spirits paint thinner to be used for this flushing operation. To do so will cause damage to the mechanism by "seizing" various close-fitting parts.

Run the mechanism for 15 to 20 minutes.

(5) Drain oil and fill with clean hydraulic oil specified on lubrication order (par. 62).

198. Turret Lock

a. REMOVAL. Disengage turret lock by turning handle to the vertical position; pull the handle outward and turn the handle until it is in the horizontal position. Remove two screws and washers holding turret lock to turret and remove lock.

b. INSTALLATION. Position turret lock over mounting holes on front of turret and install two mounting screws and washers. Turn handle to vertical position; push in handle and turn handle until it is in the horizontal position. Turret is now locked against rotation.

199. Antiaircraft Machine Gun Mount

a. GENERAL. The cal. .50 antiaircraft machine gun is stowed on the right side of the turret when the vehicle is operating in a heavily wooded area. In this position, the gun is below the top of the turret and is protected from low-hanging branches. When the vehicle is in combat, the gun is mounted on the tripod on top of the turret (fig. 3). To remove the gun from the stowed position, loosen the knob in the bracket enough to allow the pintle to be lifted out. Position the gun and gun mount in the tripod and tighten the knob in the head of the tripod to hold mount in place.

b. TRIPOD REMOVAL. Remove gun from tripod by unscrewing knob enough to allow pintle to be lifted out of tripod. Remove six nuts and washers holding tripod to turret roof. Remove tripod.

c. TRIPOD INSTALLATION. Place tripod over mounting holes in turret roof. Slide six bolts through tripod and turret roof. While one man holds heads of bolts on top of turret, another man from inside of turret installs six nuts and lock washers. Install gun in tripod.

Section XXIII. FIRE EXTINGUISHERS

200. Removal and Installation

a. REMOVAL. The fixed fire extinguisher, with valve assembly, can be replaced in the following manner:
  (1) Remove subfloor section surrounding extinguisher.
  (2) Back off union nut that holds main tube to extinguisher (fig. 167).
  (3) Back off nut holding remote control assembly to extinguisher valve.
  (4) Remove eye bolt nuts holding clamps in position on extinguisher.
  (5) Lift extinguisher and valve assembly out of vehicle.

b. INSTALLATION.
  (1) Position extinguisher and valve assembly on mounting bracket.
  (2) Line up control assembly to extinguisher valve and tighten nut securely.
    Caution: Before installing control assembly, be sure plunger pin is up in the set position. If it is down when installing head, extinguisher may be set off.

(3) Position outlet line on extinguisher and tighten union nut.
(4) Pull holding clamp ends together and draw up eye bolt nuts securely.
(5) Install subfloor surrounding extinguisher in position flush to subfloor of vehicle. See that rubber seal is in position around outlet line.

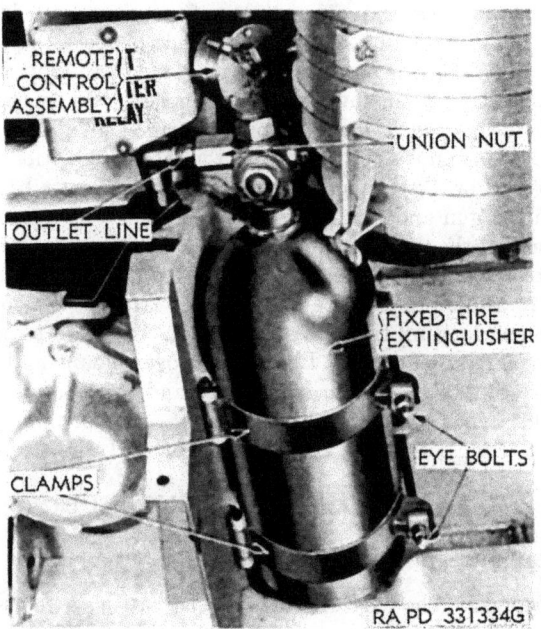

Figure 167. Fixed fire extinguisher installed.

201. Maintenance

a. HANDLING.
   (1) Any cylinder containing gas under high pressure is as dangerous as a loaded shell. The extinguisher cylinders should never be dropped, handled roughly, or exposed to unnecessary heat.
   (2) Red safety blow-off seal on valve head indicates if cylinder has been discharged due to high temperature. This should be examined regularly; if it is missing, the cylinder should be replaced.

b. MAINTENANCE.
   (1) Every 4 months, the control head must be removed from the cylinder. Weigh the cylinder to determine the weight of the carbon dioxide. Contents can be determined only by weight,

do not use a pressure gage. The empty weight and full weight is stamped on the valve of each cylinder. This does not include the control head and other attachments or the discharge horn on the portable cylinder. Weigh the cylinder with content but without the control head, other attachments, or horn. From this weight subtract the empty weight stamped on the valve. The net weight must be within 10 percent of the full weight (9 pounds for the 10-pound unit; 3.6 pounds for the 4-pound unit). Cylinders which do not meet these weights must be removed for recharging and a full cylinder installed. The portable extinguisher must be weighed in the same manner and replaced if not up to weight.

(2) While the control head is disconnected from the cylinder valve, test the operation of the pull cable and the pull lever on the control head to make sure the cam and plunger pin work freely. Connect the control head to the valve and install the locking pin and seal wire.

(3) After long, rough operations, examine the cylinders in general for bad dents or breaks. Check condition of valve, control heads, clamps, and, on the portable unit, the discharge horn. On fixed units, the cylinder connections, tubing, nozzles, and horns must be checked. Tighten all connections and replace any broken or damaged parts.

Section XXIV. COMBINATION GUN MOUNT M64

202. Maintenance

a. Check path of recoil and remove all obstructions.

b. Check quantity of oil in recoil cylinder as outlined in paragraph 214. Examine the projection of the replenisher piston rod. If projection is less than one-quarter to one-third of an inch, fill replenisher.

c. When filling replenisher, use oil specified on lubrication order (par. 62).

d. Manually elevate and traverse the gun through the full limits to see that no binding occurs. Check for any binding or lost motion in elevating gears, handwheel, or traversing gear. If binding exists, notify ordnance maintenance personnel.

e. Check all bolts and nuts for tight fit, also retaining screws and pivot pins.

f. Check recoil guide to see that it is free of paint and caked grease and properly lubricated.

g. Operate breech block and check operation of cam and cam retainer.

h. Check the handwheel for looseness on the shaft and proper lubrication. Check the elevating arc and pinion for broken or chipped teeth and check tooth contact by passing a piece of 0.0015-inch brass shim stock, 2 inches wide, between pinion and arc teeth. This should pass without binding. Not more than two thickness of shim stock are required to secure a good tooth contact impression on the shim stock.

i. Check for obstructions between the breech ring and the rear portion of the mount and remove if found.

203. Filling and Draining Recoil Mechanism

a. Check quantity of oil in recoil cylinder as follows:
  (1) Observe distance the replenisher piston rod projects beyond indicator (fig. 168). A projection of $3/16$–2 inches is normal.
  (2) Oil specified on lubrication order (par. 62) is added or removed as required to keep piston rod in normal position.

b. To add oil to replenisher, remove plug from filling valve (fig. 168) and screw oil gun hose into opening. Operate gun until piston rod projects at least one-quarter inch. Remove filler hose; insert and tighten filler plug.

c. To drain oil from replenisher remove plug from the filling valve and insert spout of oil gun into opening and depress the ball check, drain enough oil to allow the piston rod to return to normal position. Insert and tighten filler plug.

d. Oil leakage around the filling valve may be corrected by replacing the plug gasket. Other leakage from replenisher or recoil cylinder will be reported to ordnance maintenance personnel. Emergency firing of the gun may be continued as long as the replenisher piston rod is in sight.

Section XXV. MACHINE GUN MOUNTS

204. Cal. .30 Machine Gun Mount (Bow)

a. INSTALLATION. Place the cal. .30 Browning machine gun M1914A4 (flexible) in the cal. .30 mount D76459 so that the front mounting holes in the machine gun and bow mount aline. Secure the gun by inserting the front locking pin.

b. REMOVAL. Remove the front locking pin and withdraw the machine gun from the mount.

c. MAINTENANCE. Make sure that the traveling lock (fig. 168) seats securely in the rear mounting holes of the machine gun and that the lock spring holds the hook without play. When not in traveling position, the arm spring should keep the arm against the hull roof.

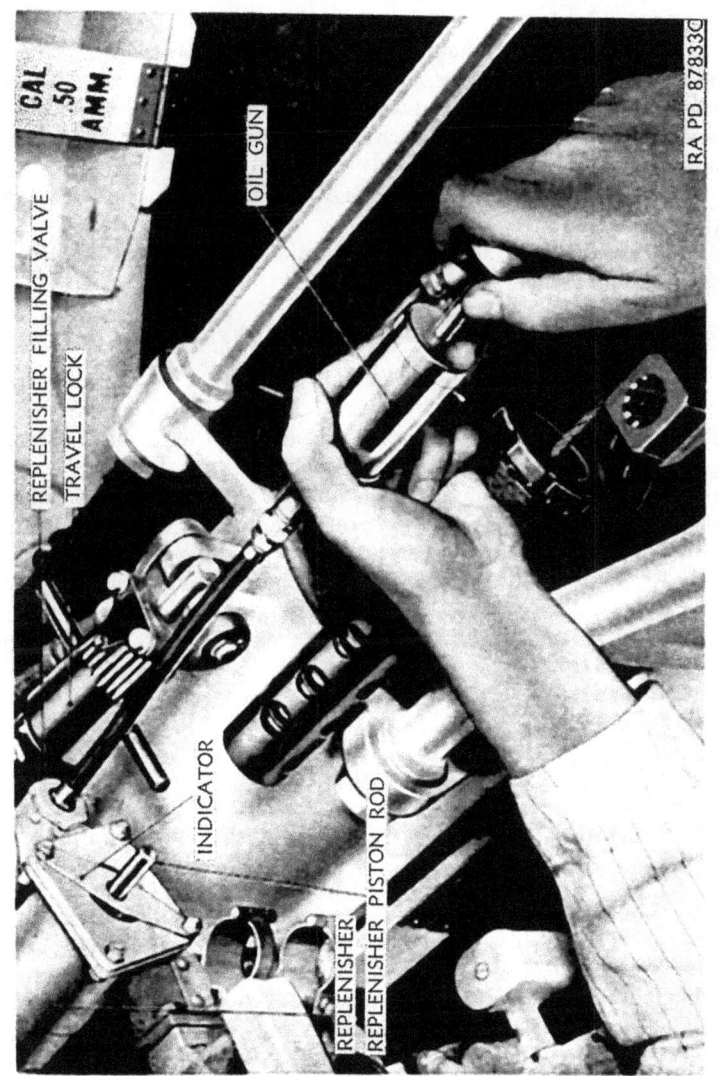

Figure 168. Filling replenisher.

205. Cal. .30 Machine Gun Mount (Combination Gun Mount M64)

a. INSTALLATION. Place the coaxial cal. .30 Browning machine gun M1914A4 (fixed) in its seat in the combination gun mount M64 (fig. 178). Aline the front mounting holes in the machine gun and mount and insert the front locking pin held by its chain. Swing up the elevating and traversing mechanism to aline its mounting holes with the rear mounting hole of the machine gun and insert the rear chained locking pin.

b. REMOVAL. Remove the front and rear locking pins and withdraw the machine gun from the mount.

c. MAINTENANCE. Make sure that the locking hooks and their springs hold the gun securely and rigidly.

206. Cal. .50 Machine Gun Mount (AA)

a. INSTALLATION. Release the mount pintle lock of the antiaircraft gun support on the turret, by loosening the locking knob (fig. 179). Install the cal. .50 machine gun mount pintle in the socket of the support. Tighten the knob to lock the mount against loosening. Place the cal. .50 HB Browning machine gun M2 in the cradle assembly of the mount. Aline the front and rear mounting holes in the machine gun and mount. Insert the front and rear locking pins.

b. REMOVAL. Withdraw the front and rear locking pin (fig. 179) and remove the machine gun from the mount. Loosen the locking knob and pull the mount from its seat in the support. Tighten the lock knob to avoid loss.

c. MAINTENANCE. The mount should pivot in the support. The travel lock should aline and engage smoothly.

Section XXVI. STABILIZER

207. Testing Stabilizer Electric Circuit and Operation of Oil Pump

a. GENERAL. One stabilizer electric circuit tester per tank is supplied (fig. 169). It is used by operating personnel in testing the continuity of the electrical circuit and the operation of the stabilizer oil pump with the control out of the circuit. A quick diagnosis can thus be made and the source of trouble can be located by a process of elimination. See paragraph 210 for detailed information. Operation of the stabilizer is described in paragraph 220.

b. PRELIMINARY OPERATIONS. Before installing the tester, make sure that the tank storage battery is fully charged.

(1) Turn the master battery switch to the "ON" position. Move

the gun master switch, in the turret control box, to "ON" position (fig. 19).
(2) Move the turret electric motor switch in the turret control box to "ON" position.
(3) Move the stabilizer switch, in the turret control box to "ON" position. The switch indicator light should glow.
(4) Squeeze the elevating shifter lever trigger (fig. 181) and move the lever to the right.

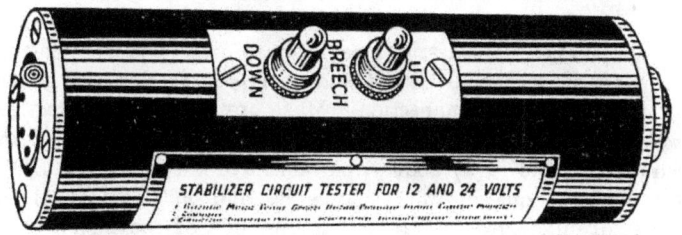

Figure 169. Stabilizer electric circuit tester—17–T–5505–750.

(5) Rotate the elevating handwheel clockwise and then counterclockwise to depress and elevate the gun. If the gun does not respond to the movement of the handwheel, check and make sure that gyro control (fig. 184) follows the movement of the handwheel, before making tests outlined in *c* below. If the gyro control does not follow the movement of the handwheel, examine the flexible shaft for breaks in its core.
(6) If the breech of the gun goes up and remains there while the stabilizer switch is turned on, check the mounting and operation of the recoil switch. This switch must be mounted so that the button does not protrude more than one-sixteenth inch from the body of the switch.

c. ELECTRICAL TEST. Be sure that operations outlined in paragraph 216 have been completed to place the stabilizer in normal operation.
(1) Rotate the stiffness rheostat dial to the number "10."
(2) Press release lever on receptacle of gyro control and withdraw multiprong connector, insert it in receptacle in base of tester. The pilot light in tester should now glow brightly for 24 volts supply, dimly for 12 volts, otherwise an open circuit exists.
(3) To locate open circuit—
  (*a*) Check the position of the switch plunger on the disengaging switch. The plunger should be fully released and protruding approximately one-quarter inch from the body

of the switch. Depress plunger manually. If plunger does not move freely, remove cover of switch box and inspect for broken contacts and loose or broken wiring.

(b) Remove cover from turret control box and inspect wiring for loose connections. Inspect stiffness rheostat and adjustable slide resistors for burned or broken wiring, loose contacts, corrosion, and foreign matter. Rotate the stiffness rheostat and inspect for poor contact. Temporarily bridge the rheostat with a jumper to check for open circuits. Tighten any loose contacts, remove any corrosion or foreign particles, and replace any defective components.

(c) Inspect multiprong connector plug for loose or grounded electrical connection. Make any necessary corrections.

d. OIL PUMP TEST. With the multiprong connector inserted in the tester receptacle (c(2) above), movement of the "UP" or "DOWN" switch attracts one side of the magnet bar and shuts off that oil orifice. The hydraulic power is applied to one side of the piston through the remaining orifice (fig. 170).

Caution: When testing as outlined in (1) and (2) below, press the button for an instant only to prevent damage from too rapid travel of the gun.

(1) Press the button marked "Breech Up"; the breech of the gun should rise rapidly.

(2) Press the button marked "Breech Down"; the breech of the gun should drop rapidly.

(3) If the gun does not respond to button in either (1) or (2) above, replace the oil pump assembly (par. 209a(7)).

(4) If the movement of the breech is opposite to that stated in either (1) or (2) above, interchange the yellow and green leads in the terminal box of the oil pump assembly (par. 209a(7)) (fig. 177).

(5) If the gun responds to either (1) or (2) above, but not *both*, check as follows:

(a) If it fails to respond to the "Up" test, an open circuit exists in the yellow lead circuit (fig. 171). Inspect circuit for broken or frayed wiring or poor contact at the multiprong connector or oil pump terminals.

(b) If it fails to respond to the "Down" test, an open circuit exists in the green lead circuit. Inspect circuit for broken or frayed wire or poor contact at the multiprong connector or oil pump terminals.

e. REPLACEMENT.

(1) If the tests described in c and d above show proper operation, the gyro control is probably at fault.

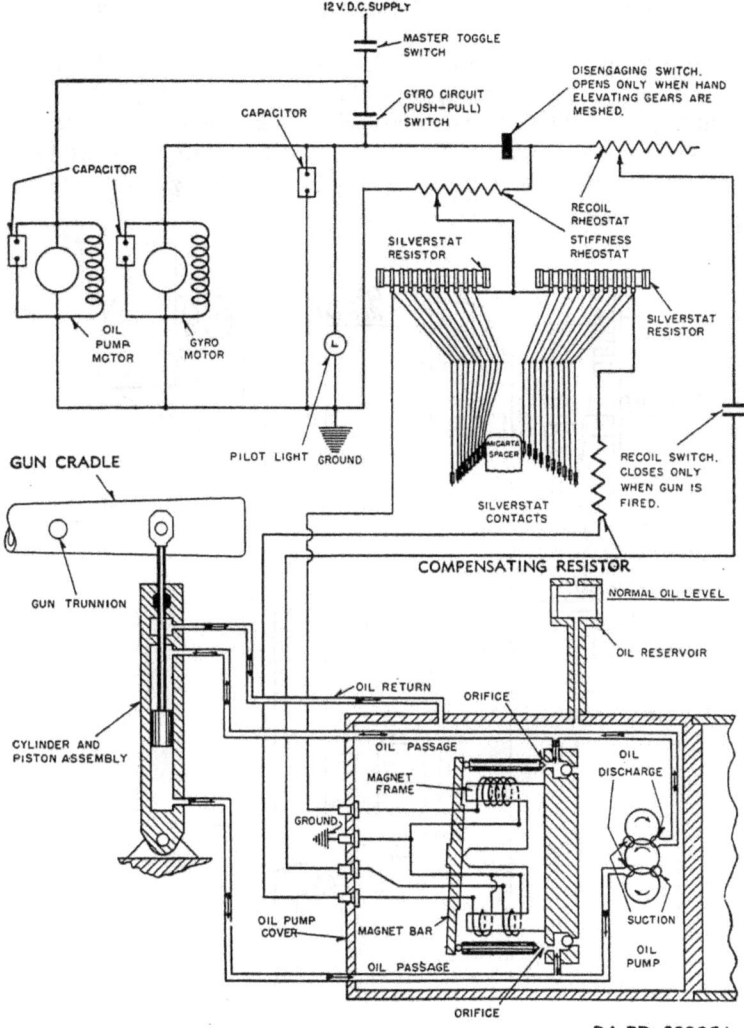

Figure 170. Stabilizer electric and oil circuits.

(2) Replace the gyro control (par. 209a(5)), insert the multiprong connector in the gyro control receptacle, check oil reservoir level (par. 220b(1)), and again check the stablizer.

208. Charging the Stabilizer with Oil and Purging the System

a. Move the turret electric motor switch in control box (fig. 19) to the "OFF" position.

b. Remove reservoir filler plug (fig. 173) from elbow on top of oil reservoir.

c. Pump oil specified on lubrication order (par. 62) into reservoir until full, using a pump type oil can or suction type oil gun. Install filler plug. When temperatures below 0° F. are reached, precautions must be taken to warm the oil in order to obtain satisfactory performance of the stabilizer. Running the heater generator to warm

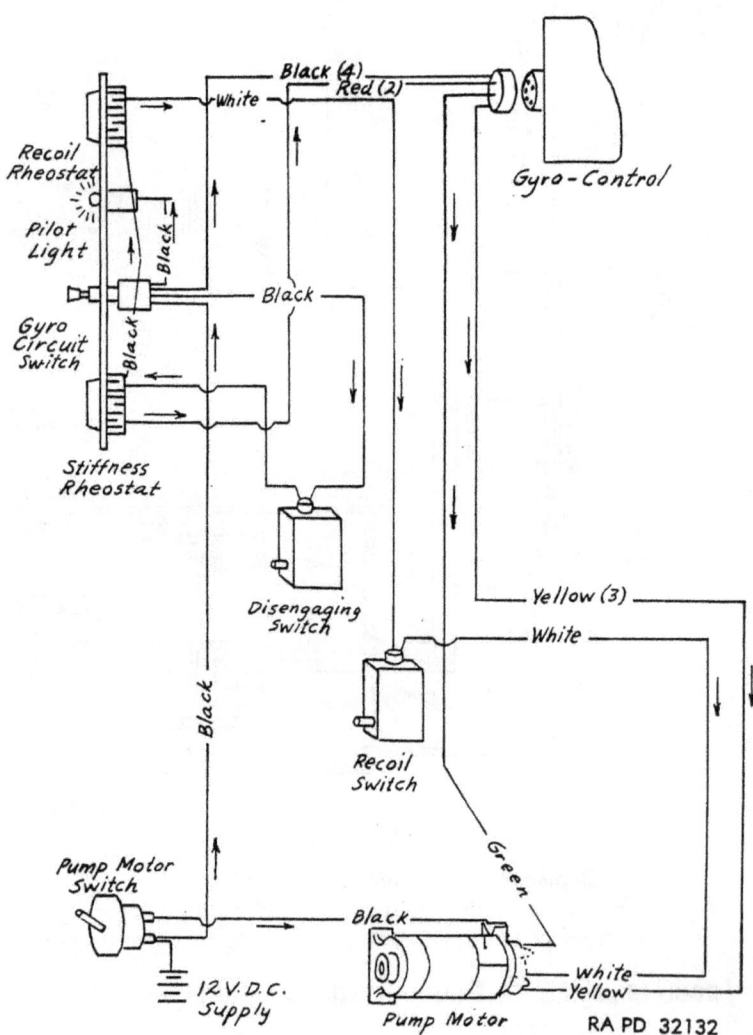

Figure 171. Stabilizer wiring diagram.

the crew compartment of the tank will help warm the stabilizer oil. If oil action becomes sluggish below 0° F., stabilizer should be drained and filled with recoil oil (special). At extremely low temperatures, recoil oil (special) may be diluted with recoil oil (light) as necessary to produce satisfactory operation. When the temperature rises above 0° F. the system will not be drained, only hydraulic oil will be added, when necessary, as the oils are miscible.

d. Squeeze the elevating shifter lever trigger (fig. 181) and rotate the lever to the right to disengage the hand elevating gears.

e. With the gun positioned over the front of the tank, slowly push the breach end up, remove the small hex-head plug from the front bleeder valve (fig. 175), loosen the fitting, connect rubber hose to valve (fig. 172), and drain oil from the valve until a solid flow of oil is obtained. Tighten valve and install plug.

Figure 172. Bleeding the stabilizer cylinder.

f. Push the breech down and repeat *e* above on rear bleeder valve.

g. Repeat *e* and *f* above, if necessary, until a solid flow of bubble-free oil is obtained from both valves. Tighten valves with wrench.

h. Check level of oil in reservoir and if not at least two-thirds full, remove plug and fill as outlined in *b* and *c* above. Install and tighten plug.

i. Squeeze the elevating shifter lever trigger and rotate the lever to the left to engage the hand elevating gears.

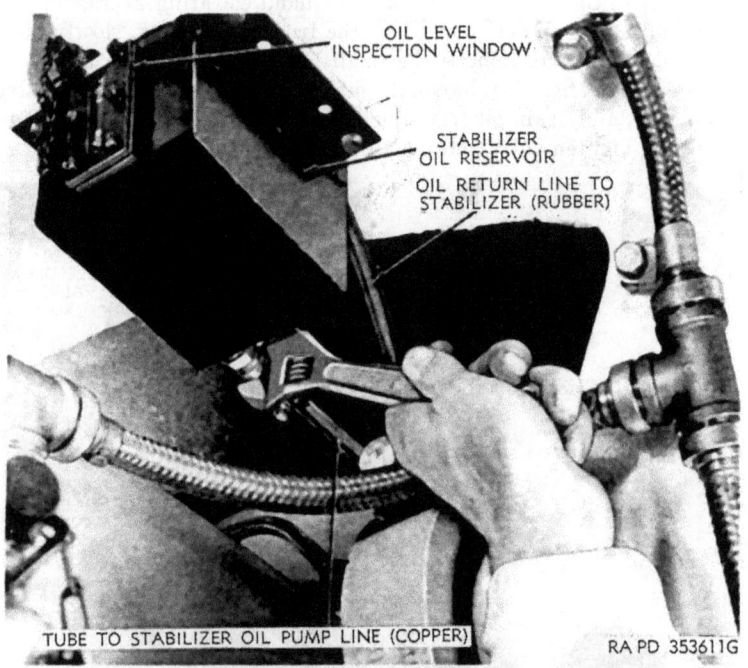

Figure 173. Removing oil line from stabilizer oil reservoir.

j. Move the turret motor switch (fig. 22) to "ON" position and note oil level in reservoir. If the oil level drops, there is air in the system and purging, as outlined in *e*, *f*, and *g* above, must be continued until all air is removed.

k. Operate stabilizer and check as outlined in paragraph 220*e*.

209. Maintenance of Stabilizer

a. REMOVAL. Move turret motor switch and stabilizer switch to "OFF" position. Remove filler plug from top of stabilizer oil reservoir, also drain plug from stabilizer oil pump, and drain oil from the system into a suitable container.

(1) *Oil pump oil lines.* Disconnect the four oil lines from the oil pump and collect surplus oil in the container. Tag lines for identification.

(2) *Cylinder and piston and oil reservoir oil lines.* Disconnect the three oil lines from the cylinder and piston assembly and the one oil line (fig. 173) from the oil reservoir and remove lines.

(3) *Cylinder and piston assembly.*

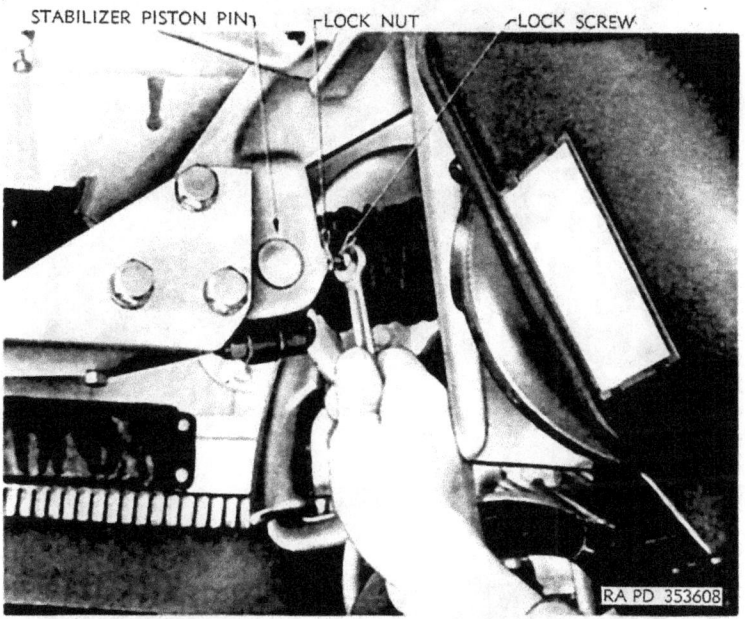

Figure 174. Loosening lock nut and screw on stabilizer cylinder and piston pivot pin.

(a) To disconnect the cylinder and piston assembly from the gun cradle, loosen the two lock nuts (fig. 174) and lock screws and drift out stabilizer pivot piston pin. Withdraw pin.

(b) To remove cylinder and piston assembly and mounting bracket (fig. 175), disconnect the four cap screws holding the bracket to the turret front and remove screws and washers, also cylinder and piston and bracket as an assembly.

Caution: Cover cylinder openings with masking tape to prevent dirt from entering system.

311

Figure 175. Removing stabilizer cylinder and piston assembly and bracket.

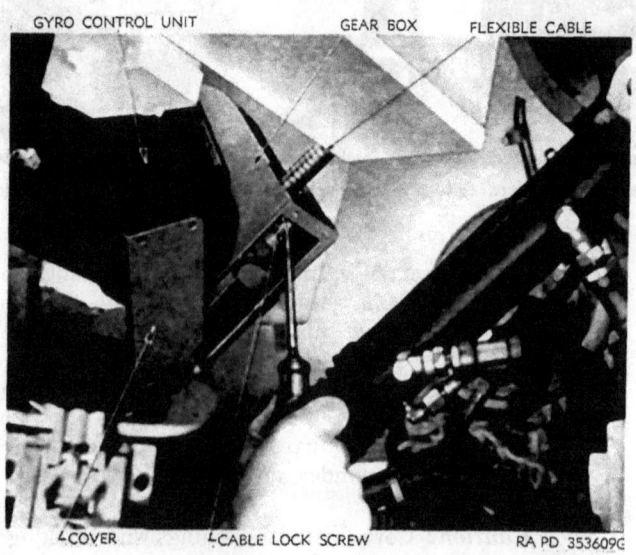

Figure 176. Removing cable from gyro control gear box.

(4) *Oil reservoir.* Loosen and remove the four nuts, bolts, and washers attaching the oil reservoir to its hangers in the turret roof and remove reservoir (fig. 173).

(5) *Gyro control and gear box.* Remove gyro control and gear box as follows:

  (*a*) Pull multiprong plug out of socket in gyro control unit.

  (*b*) Remove four screws from gear box cover (fig. 176) and remove cover.

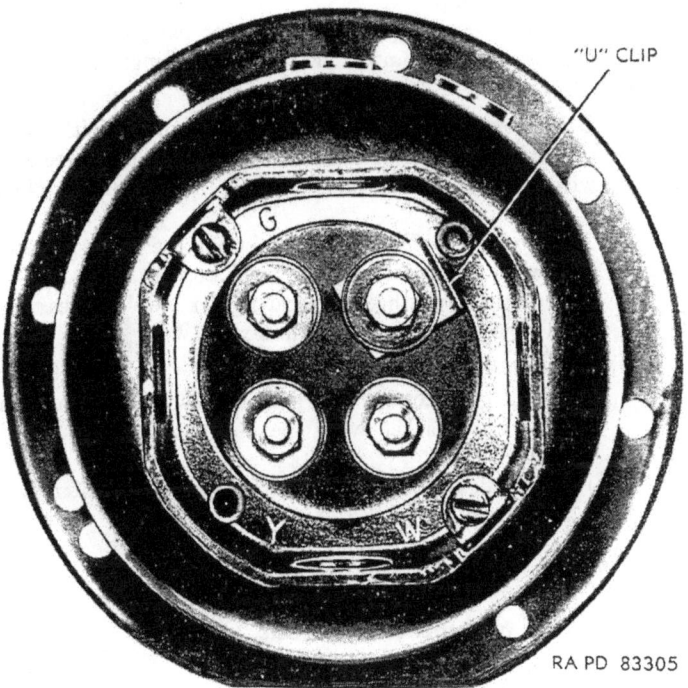

Figure 177. Stabilizer oil pump terminals.

  (*c*) Loosen flexible shaft lock screw and withdraw flexible shaft and felt washer (fig. 176).

  (*d*) Remove four cap screws and lock washers from gyro control gear box mounting bracket, support stabilizer control unit securely when removing screw, and remove gear box and gyro control as a unit.

(6) *Flexible cable and miter gear.* To remove flexible cable from handwheel remove six screws attaching flexible shaft gear box to handwheel shaft housing, and remove gear box and flexi-

ble shaft. A miter gear is attached to the handwheel shaft with elastic stop nut and key. Remove stop nut and drift out key and remove miter gear and key.

(7) *Oil pump.* To remove stabilizer oil pump from turret electric drive motor (**fig. 184**), remove the two screws which attach cover to pump terminal box. Remove colored wires from three terminals (marked with initial letter of color) and withdraw conduit and wires. Remove four socket-head screws or cap screws which attach pump flange to turret motor. Remove oil pump.

(8) *Interior parts.* No work on interior parts of gyro control, gear box, or oil pump is performed by using organization.

b. INSTALLATION.

(1) *Oil pump.* Attach oil pump to turret motor with the four socket-head screws or cap screws. Connect the three colored wires to the three terminals. Attach the cover to the oil pump with two screws.

(2) *Miter gear and flexible cable.* Attach the miter gear to the handwheel by means of the key and stop nut. Attach flexible cable to the handwheel by securing flexible shaft gear box to handwheel shaft housing with six screws.

(3) *Gyro control and gear box.*

(*a*) Install gyro control and gear box as a unit by attaching mounting bracket with the four cap screws and lock washers.

(*b*) Install flexible shaft and felt washer and tighten flexible shaft lock screw.

(*c*) Install gear box cover with four screws.

(*d*) Plug multiprong plug into receptacle in gyro control unit.

210. Isolation of Functional Defects

a. SYSTEM WILL NOT OPERATE.
 (1) Check for defective wiring (par. 207*c*).
 (2) The stabilizer oil pump may be frozen. Install the oil pump assembly (par. 209*a*(7)).
 (3) The disengaging switch may be open or inoperative. If the switch button is in the correct position for stabilizer operation (par. 220), notify ordnance maintenance personnel for replacement of the switch.

b. GUN FLUCTUATES ABOVE AND BELOW TARGET.
 (1) Air in the stabilizer system may cause this malfunction. Purge air from system (par. 208) and add oil if necessary (par. 208).

(2) If lubrication of the system is required, lubricate according to instructions on lubrication order (par. 62).

(3) Inadequate supply of oil in the stabilizer system would cause this malfunction. Charge the system with oil (par. 208). If leaks are found in the oil line, notify ordnance maintenance personnel.

(4) The recoil switch may be defective. If the switch plunger is in the correct position (par. 207b(6)) when the gun is in battery position and is not stuck, notify ordnance maintenance personnel.

(5) The gun mount trunnion bearings may be loose. Notify ordnance maintenance personnel.

(6) Loose cylinder or piston rod pivot pins are indicated by chattering. Notify ordnance maintenance personnel.

(7) Lost motion between gyro control and elevating handwheel will require adjustment of the gear box. Notify ordnance maintenance personnel.

(8) Check operation of oil pump (par. 207).

c. RAPID GUN VIBRATION KNOWN AS "HUNTING."

(1) Air in the stabilizer system may cause this malfunction. Purge air from system (par. 208) and add oil if necessary (par. 208).

(2) Loose cylinder or piston rod pivot pins or bearings are indicated by chattering. Notify ordnance maintenance personnel.

(3) Tighten mounting bolts on the gyro control if necessary.

(4) Lost motion between gyro control and elevating handwheel will require adjustment of the gear box. Notify ordnance maintenance personnel.

(5) Perform all tests indicated in paragraph 207.

d. RISING BREECH OR UNEVEN STIFFNESS.

(1) The recoil switch may be defective. Check the switch plunger. If it is in the correct position (par. 207) when the gun is in battery position and is not stuck, notify ordnance maintenance personnel.

(2) Perform the electrical test and the oil pump test (par. 207).

e. GUN DOES NOT RESPOND TO ELEVATING HANDWHEEL.

(1) Check flexible shaft to see that it is properly attached. Flexible shaft casing may be too long or too short. Notify ordnance maintenance personnel.

(2) Check condition of stiffness rheostat (par. 220).

(3) Check oil level in the stabilizer (par. 208). If necessary, charge the system with oil (par. 208). If leaks are found in the oil line, notify ordnance maintenance personnel.

(4) Loose cylinder or piston rod pivot pins or bearings are indicated by chattering. Notify ordnance maintenance personnel.//
(5) Perform the oil pump test (par. 207) to determine whether the gyro control or oil pump is at fault.//
(6) The circuit breaker in the turret control box (fig. 19) may have kicked out. Press circuit breaker reset button.

f. ELEVATING HANDWHEEL HARD TO OPERATE.
  (1) Check flexible shaft to see that it is properly attached. Flexible shaft casing may be too long or too short. Notify ordnance maintenance personnel.
  (2) Loose cylinder or piston rod pivot pins or bearings are indicated by chattering. Notify ordnance maintenance personnel.

g. BREECH DIPS EXCESSIVELY WHEN GUN IS FIRED, AT ALL RECOIL RHEOSTAT SETTINGS.
  (1) The recoil switch may be defective. Check if the switch plunger is in the correct position (par. 207) when the gun is in battery position and is not stuck, notify ordnance maintenance personnel.
  (2) Perform the electrical test (par. 207) to check for defective wiring.

h. GYRO CONTROL RUNS WHEN ELEVATING GEARS ARE ENGAGED FOR MANUAL OPERATION. Inadequate supply of oil in the stabilizer system would cause this malfunction. Charge the system with oil (par. 208). If leaks are found in the oil line, notify ordnance maintenance personnel.

i. OIL LEAKS AROUND PISTON ROD. Oil seal may be defective. Notify ordnance maintenance personnel.

j. LEAKS AROUND OIL LINE CONNECTIONS. These may be caused by defective or loose flares or connections. Notify ordnance maintenance personnel.

k. DROP IN STABILIZER OIL RESERVOIR OIL LEVEL.
  (1) If oil connections leak, notify ordnance maintenance personnel.
  (2) If oil leaks around piston rod, notify ordnance maintenance personnel.

l. BLEEDER VALVE LEAKS. Valve may be pulled down too tight, damaging the orifice, or the valve may be defective for other reasons. Notify ordnance maintenance personnel.

m. OIL PUMP VIBRATES.
  (1) Check oil pump (par. 207). Replace if necessary (par. 209).
  (2) The oil line may be constricted. Notify ordnance maintenance personnel.

Section XXVII. MAINTENANCE UNDER UNUSUAL CONDITIONS

211. Extreme-Cold Weather Maintenance

Refer to TM 9-2855 and TB ORD 193 for a general discussion of maintenance problems, the application of antifreeze compounds and arctic-type lubrication, handling of storage batteries in extreme cold, and dewinterization procedures.

212. Extreme-Hot Weather Maintenance

a. COOLING SYSTEM. Formation of scale and rust in the cooling system occurs more often during operation in extremely high temperatures, therefore, corrosion inhibitor compound (par. 118) should always be added to the cooling liquid.

b. BATTERIES.
  (1) *Electrolyte level.* In torrid zones, check level of electrolyte in cells daily and replenish, if necessary, with pure, distilled water. If this is not available, rain or drinking water may be used. However, continuous use of water with high mineral content will eventually cause damage to batteries and should be avoided.
  (2) *Specific gravity.* Batteries operating in torrid climates should have a weaker electrolyte than for temperate climates. Instead of 1.280 specific gravity as issued, the electrolyte (acid, sulfuric) should be diluted with pure, distilled water to readings of 1.200 to 1.240 specific gravity as specified in TM 9-2857. This is the correct reading for a fully-charged battery. This procedure will prolong the life of the negative plates and separators. Under this condition a battery should be recharged if below 1.160 specific gravity.
  (3) *Self-discharge.* A battery will self-discharge at a greater rate at high temperatures if standing for long periods. This must be considered when operating in torrid zones. If necessary to park for several days, remove batteries and store in a cool place. If this is not practicable operate engine at regular intervals to maintain charge.

c. HULL.
  (1) In hot damp climates corrosive action will occur on all parts of the vehicle and will be accelerated during the rainy season. Evidences will appear in the form of rust and paint blisters on metal surfaces and mildew, mold or fungus growth on fabrics, leather, and glass.

(2) Protect exterior surfaces from atmosphere by touch-up painting and keeping a film of engine lubricating oil (OE–10) on unfinished exposed metal surfaces. Cables and terminals should be protected by ignition insulation compound.

(3) Make frequent inspections of idle, inactive vehicles. Remove corrosion from exterior metal surfaces with abrasive paper or cloth and apply a protective coating of paint, oil, or suitable rust preventive.

213. Maintenance After Fording—Normal Fording

Vehicles which have not exceeded the normal fording depth should be treated as indicated below.

a. Suspension parts, track wheels, shock absorbers, tracks, etc., should be cleaned and lubricated.

b. Check engine compartment for indication of water in engine parts, transmission, distributor starter, or generator. Drain, clean, and lubricate as required. Replace soaked ignition cables.

c. Replace instruments containing water.

214 Maintenance After Operation on Unusual Terrain

a. MUD. Clean thoroughly all exposed parts as soon as possible after operation in mud, particularly if a sea of liquid mud has been traversed. Clean tracks, wheels, suspensions, and shock absorbers. Lubricate where possible to force out contaminated lubricant.

b. SAND OR DUST. Repaint surfaces blasted by sand. Clean engine, engine compartment, and grille screens. Lubricate completely to force out contaminated lubricants. Clean air cleaners, and fuel and oil filters.

CHAPTER 4

MATÉRIEL USED IN CONJUNCTION WITH MAJOR ITEM

Section I. ARMAMENT

215. General

a. This section contains essential information necessary for the crew to properly identify, connect, operate, and protect the armament while being used or transported with the vehicle.

b. Complete detailed coverage for service upon receipt of matériel, operation, and maintenance of the armament is contained in separate technical and field manuals (app. 4). Coverage for the combination gun mount is contained in TM 9-313.

c. Cleaning and lubricating instructions are contained in the lubrication order (figs. 27 through 34) and preventive maintenance services in tables II and III together with the instructions pertaining to the vehicle.

216. Description

a. The 75-mm gun M6 (fig. 178) was especially designed for use in the light tank. It is lighter in weight and smaller in outside diameter than other 75-mm tank and field artillery guns. The breech mechanism has a semiautomatic, horizontal sliding-wedge type of breechblock and a continuous-pull, self-cocking type firing lock.

b. The combination gun mount M64 mounts the 75-mm gun M6 as the primary armament. In addition, a cal. .30 machine gun M1919A4 (flexible) is mounted on the right side of the cradle (fig. 178). The recoil mechanism is a concentric hydrospring type mechanism that uses the cradle as the outside cylinder. The gunner's position is on the left side of the armament in this vehicle.

c. The gun stabilizer system, referred to merely as the stabilizer, is a hydraulically operated system which prevents the gun from being moved out of a preestablished vertical position by the "pitching" movement of the tank.

Note. This system should not be referred to as the "gyro stabilizer" because the word "gyro" is correctly used in reference to the stabilizer control unit only.

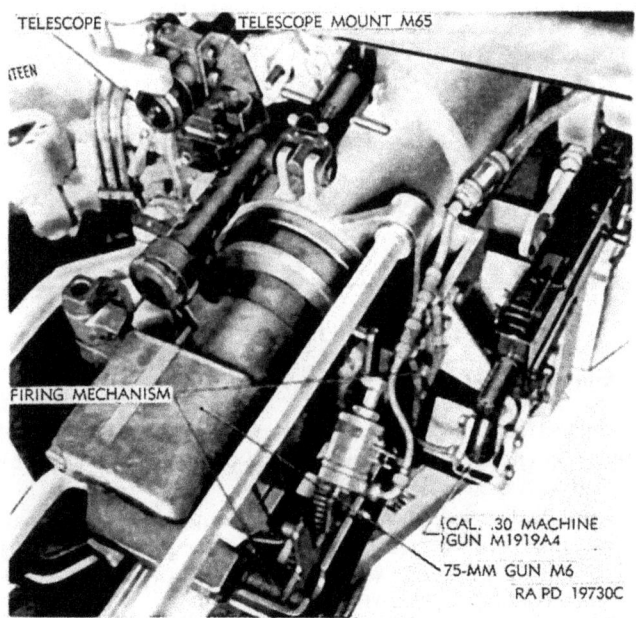

Figure 178. 75-mm gun M6 and combination gun mount M64 installed in light tank M24.

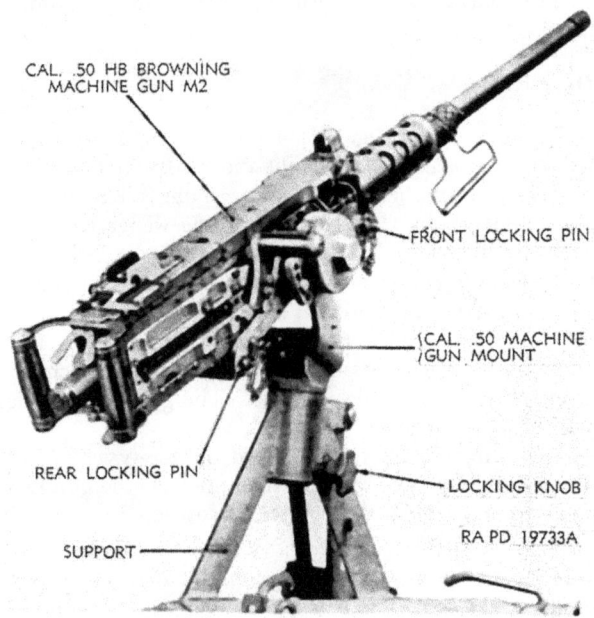

Figure 179. Cal. .50 machine gun in pintle mount on turret roof.

The main components of this unit are: gyro control, gear box (fig. 184), cylinder and piston (fig. 175), oil pump, turret electric motor (fig. 184), control-box switches, recoil and disengaging switches (fig. 181), oil lines and oil reservoir (fig. 173), flexible shaft, wiring and attached parts.

d. The coaxially mounted machine gun (fig. 178) is recoil-operated, belt-fed, and air-cooled. Another cal. .30 M1919A4 machine gun (fig. 180) is mounted in a ball mount in the bow of the tank.

e. A Browning machine gun, cal. .50, M2 heavy barrel (flexible) (fig. 180), recoil-operated, belt-fed, and air-cooled, is installed in a pintle mount on the outside of the turret.

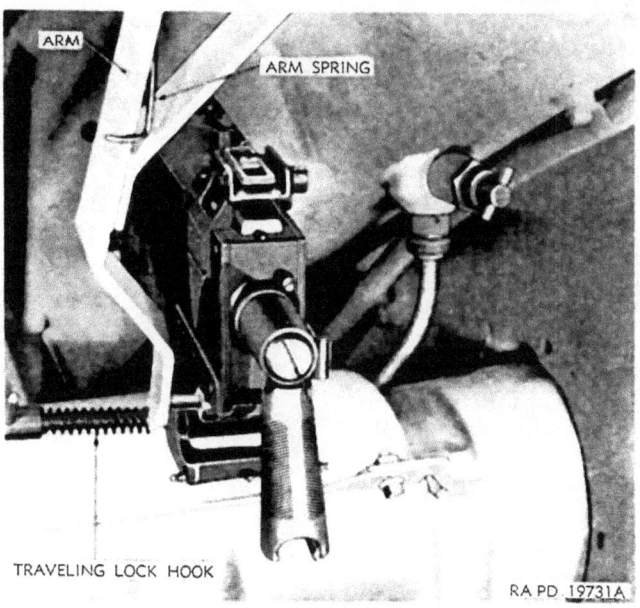

Figure 180. Cal. .30 machine gun installed in bow mount.

217. Tabulated Data

a. GUN, 75-MM, M6.
```
Weight_____ 406 lb
Length_____ 116.375 in.
Breechblock, type_____ horizontal, sliding-wedge
Firing lock, type_____ continuous-pull, self-cocking
Rate of fire (max)_____ 20 rd per min
```

b. MOUNT, COMBINATION GUN, M64:
Elevation (max) ---------------------------------- +267 mils
Depression (max) --------------------------------- −178 mils
Traverse (hydraulic and manual, continuous with turret) --- 6,400 mils
Mechanism, recoil, M22 -------------- concentric, hydro-spring
Recoil, normal ----------------------------------- 11½ in.
Recoil, maximum allowable ------------------------ 12½ in.
Recoil mechanism oil capacity -------------------- 9 qt (aprx)
Weight --- 1,174 lb

c. GUN MACHINE:
2—cal. .30, M1919A4 (flexible).
1—cal. .50, M2, HB (flexible).

d. MOUNT, MACHINE GUN:
1—cal. .30.
1—cal. .50 (AA).
1—tripod, cal. .30, M2.

e. STABILIZER:
1—stabilizer.

f. WEAPONS INSTALLED BY USING TROOPS:
5—gun, submachine, cal. .45, M3
    or
4—gun, submachine, cal. .45, M3
    and
1—carbine, cal. .30, M1.
    with
1—launcher, grenade, M3.

218. Preparation for Firing

a. PLACING THE GUN IN FIRING POSITION.
  (1) Remove all muzzle and breech covers and stow in place provided.
  (2) Unhook the gun travel lock (fig. 168) by turning its adjustable "T" shaped handle. Depress the gun slightly and swing the lock down until it engages in latched position.
  (3) To traverse the gun, first see that the driver's and assistant driver's doors are closed, then disengage the turret traversing lock (par. 42).
       Caution: If the traversing lock is disengaged when the tank is on an incline, the turret will rotate rapidly if the shift lever is set for power operation.
       Refer to chapter 2, section IV for traversing controls and turret operation.
  (4) Install the sighting and fire control instruments as described in chapter 4, section II.

(5) Tactical instructions for laying the piece are contained in FM 17–75.

b. BEFORE-OPERATION INSPECTION.

(1) Before operation, inspect the gun tube to see that gun is unloaded and that no dirt or foreign matter is in the chamber and bore. Remove preservative from bore and chamber with burlap wrapped on cleaning brush or staff.

(2) Turn on gun master switch (fig. 19) in turret control box, pull safety knob to "FIRE" position (fig. 182), and trip electric firing trigger in elevating handwheel handle (fig. 181). Note the pronounced "click" of "dry" firing. If the click is not heard, the firing solenoid plunger (fig. 182) is not releasing the latch lever of the firing mechanism (fig. 178). Push solenoid plunger in by hand to see that it is not stuck and see that plunger contacts the trigger hammer latch lever when solenoid is energized. Adjust solenoid so that clearance between plunger and lever is approximately one thirty-second of an inch. After completing check, turn switches to "OFF" position and place safety knob on "SAFE" position.

219. Operating the 75-MM Gun in Elevation

a. The gun is always elevated or depressed by rotating the elevating handwheel (fig. 181).

b. Rotate the handwheel clockwise to depress the gun and counterclockwise to elevate it.

c. Refer to paragraph 220 for stabilizer-controlled operation.

220. Operating the Stabilizer

a. GENERAL. The stabilizer holds the 75-mm gun and the cal. .30 machine gun in the combination gun mount at a predetermined elevation while the tank is in motion. The stabilizer controls the gun movement in a vertical plane within limits regardless of the pitching of the tank. The stabilizer has no control of the movement in the horizontal plane due to the roll of the tank.

b. STARTING THE STABILIZER.

(1) Check oil level in reservoir and make sure that it is at least two-thirds full. Add oil, if necessary, as directed in paragraph 208.

(2) Make sure the vehicle battery switch is turned "ON."

(3) Turn "ON" the turret motor switch and stabilizer switch located in the turret control box and warm up the stabilizer system until the oil pump located under the turret motor feels

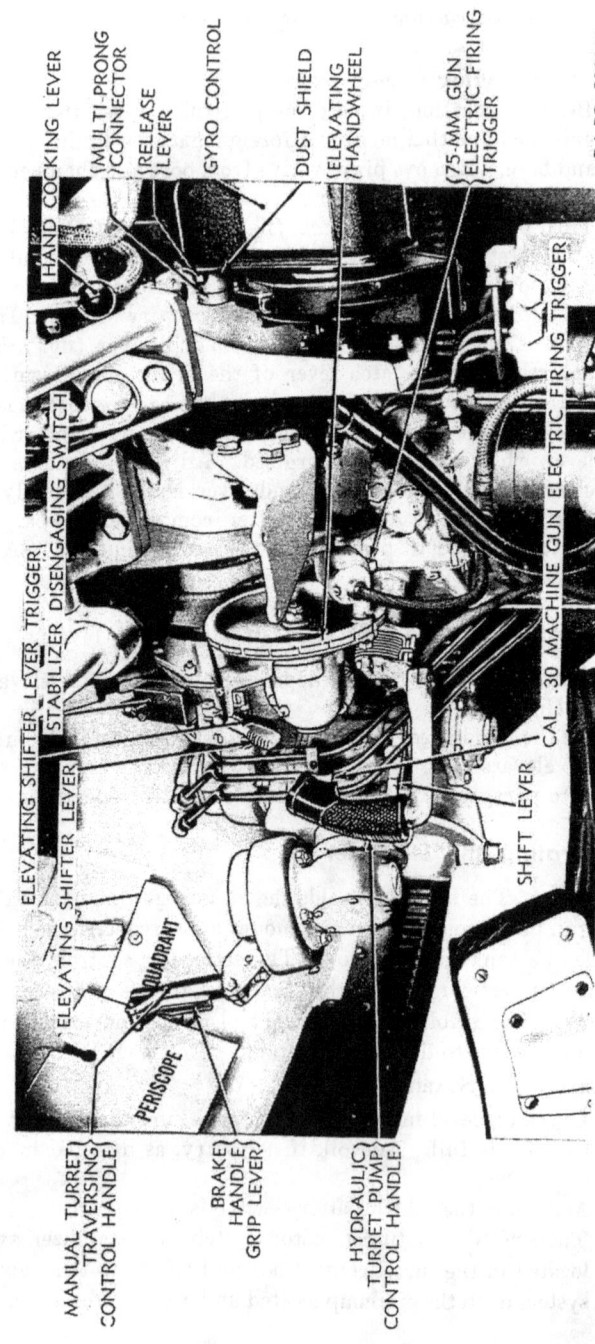

Figure 181. Elevating and firing controls.

warm to the touch. If the oil level drops, there is air in the system which must be removed. Refer to air removal instructions in paragraph 208 before proceeding.

(4) Place the gun in stabilizer control by moving the elevating shifter lever to the right. Depress the trigger in this lever to move it and, if necessary rotate the handwheel slightly back and forth to release the pinion gear from the elevating arc. The shifter lever operates a clutch mechanism which takes the elevating arc and pinion gear out of mesh and closes the disengaging switch contacts. The disengaging switch movable plunger must not protrude more than one-sixteenth inch when the elevating gears are in mesh and the switch open. The plunger protrudes about three-sixteenths inch when contacts are closed. The stabilizer is used only when the tank is in motion.

(5) Set the stiffness rheostat located in the control box to "0" and the recoil rheostat (fig. 19) to "5."

(6) Adjust stiffness rheostat as outlined in c below.

c. ADJUSTMENT OF STIFFNESS RHEOSTAT. The stiffness rheostat (fig. 19) regulates the action of the stabilizer. The correct setting is that which keeps the gun resistant to motion in a vertical plane from the aimed position. Adjustment is made by rotating the knob to the position which effectively reduces this motion. Starting at "0" slowly turn the knob clockwise. A vigorous vibration of the weapon indicates too great an adjustment. Turn the knob counterclockwise until vibration is eliminated. Failure of the gun to remain in its aimed position indicates insufficient adjustment. Resetting of rheostat to maintain satisfactory performance may be necessary due to change in battery voltage and oil viscosity.

d. ADJUSTMENT OF RECOIL RHEOSTAT. The recoil rheostat (fig. 22) in the turret control box can be adjusted so that the gun will vary only slightly from its aimed position during recoil. This adjustment is made by trial and error while the gun is being fired. Position number "5" is approximately normal setting. If the breech end of the gun drops during recoil, turn knob to a higher setting. Turn knob to a lower setting if breech rises.

e. CHECKING STABILIZER OPERATION.
 (1) Put stabilizer in operation (b above).
 (2) Move elevating shifter lever to right (b(4) above).
 (3) Turn the handwheel clockwise rapidly and observe depression of gun and elevation of breech; reverse motion of handwheel and observe how gun responds to the wheel.
 (4) Select a fairly rough terrain for a trial run.

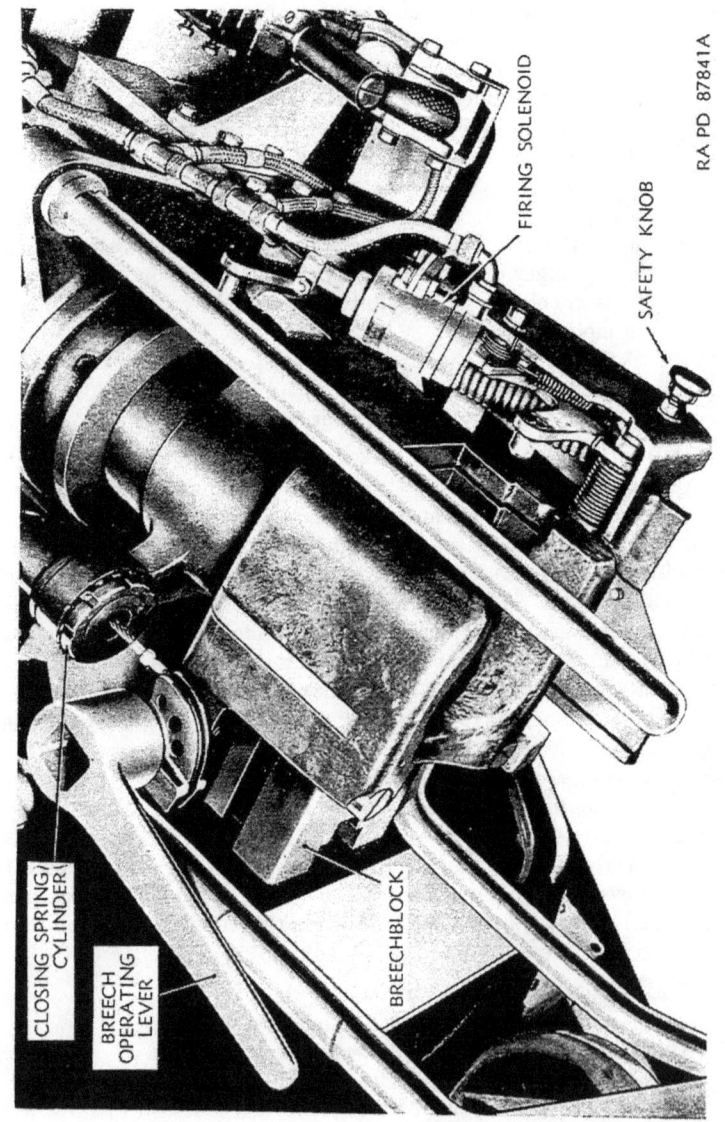

Figure 182. Breech in open position.

(5) Aim gun at a stationary target and operate tank at a steady speed directly toward target.

(6) Check and make sure that gun can be held on the target by adjusting the elevating handwheel. Make necessary adjustments to the stiffness rheostat (*c* above).

(7) If the stabilizer fails to hold the gun on target, test the electrical circuits and oil pump as outlined in paragraph 207.

221. Operating the Breech Mechanism

a. OPENING THE BREECH (fig. 182). Remove the breech operating lever from the bracket in the rear of the recoil guard and place it on the squared end of the cable terminal crank (fig. 182). Rotate the lever to the rear and left until the breechblock is held open by the extractors. Remove the lever and replace in its bracket before firing. When firing, the breech is automatically opened during counterrecoil of the gun.

b. LOADING THE GUN. Place the nose of the projectile into the breech recess and move the round forward with sufficient force to trip the extractors and close the breech automatically. The loader's hands should be moving to the right as the round is inserted, in order to clear the breechblock as it closes.

c. CLOSING THE BREECH. Release extractors from the breechblock by pushing them forward with the base of an empty cartridge case or block of wood, allowing the expanding closing spring to bring the breechblock to the closed position.

Caution: Do not use fingers to release the extractors.

Install breech operating lever in its bracket.

222. Firing the 75-MM Gun

a. ELECTRIC FIRING.

(1) For electric firing, move the gun master switch (fig. 19) in the turret control box to the "ON" position.

(2) Pull out on safety knob, releasing hand cocking lever.

(3) To fire, squeeze the electric firing trigger in the elevating handwheel handle (fig. 181). Release the trigger after firing.

b. MANUAL FIRING. For manual firing, depress the hand firing plunger knob located on the left side of the gun mount (fig. 183).

223. Gun Fails To Fire

a. If the gun fails to fire electrically, see that it is in battery and that the breech is closed; move the gun master switch to "OFF" position; then to "ON" position; and attempt to fire again. If the gun

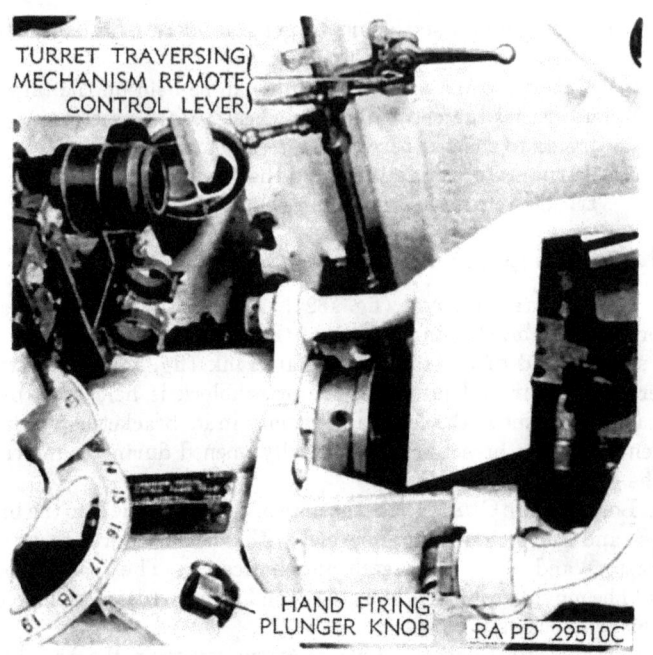

Figure 183. Hand firing and hydraulic traverse controls.

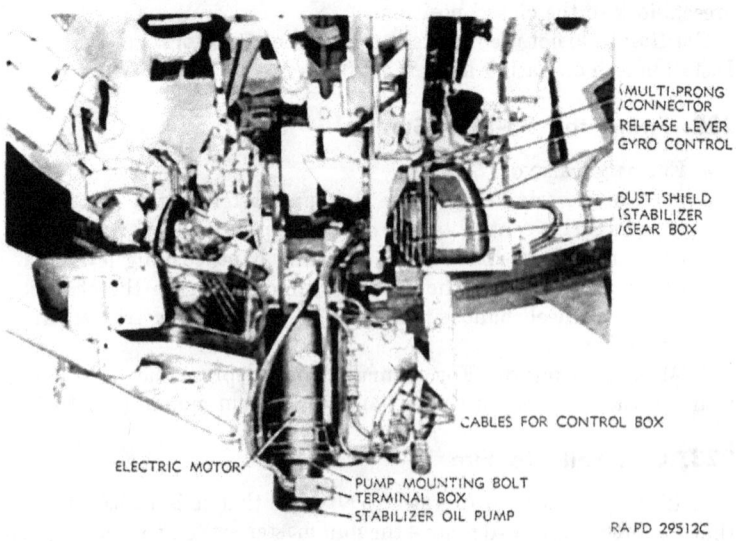

Figure 184. Under side of 75-mm gun.

still fails to fire, move the gun master switch to "OFF" position, recock by pulling the hand cocking lever (fig. 181) all the way to the rear, release safety and attempt to fire manually by depressing the hand firing plunger knob (fig. 183). If all attempts to fire fail and the gun tube is not hot, clear all personnel not necessary to the operation from the vicinity, wait 1 minute, open the breech, and unload (par. 225).

b. Inspect primer of round for indent. If the primer is deeply indented it indicates a defective round. Remove the defective round to a point 200 yards from the vicinity of firing. Reload and resume firing. If the primer is not indented or only slightly indented, the firing lock is not functioning properly. Remove the firing lock by turning it to its marked "OPEN" position and withdrawing it. Insert spare firing lock, turn it to its closed position, which is horizontal on M6 guns, and resume firing. Refer to TM 9–313 for other than immediate action remedies in case the gun fails to fire.

c. In the event that the gun has been subjected to continuous fire for a considerable length of time before misfire occurs, the gun tube will be very hot and this heat can cause the fuze, projectile filler, or propellant to explode or cook off. Allow the gun to cool, then unload and inspect the round as outlined in *b* above.

224. Extracting Empty Case

a. The breech opens automatically during counterrecoil and the cartridge case will be automatically extracted from the chamber and thrown to the rear of the gun.

b. If the case is not automatically extracted, assemble the rammer staff sections to rammer head and insert the rammer, head first, in the muzzle of the tube and push it carefully down the bore until it contacts the cartridge case. Exerting a steady pressure, push the case until it is free in the chamber and can be removed from the breech. Protect the hands with asbestos gloves when handling the empty case, as it may still be very hot.

c. Sometimes a partially extracted case can be pried loose from the chamber end by using the base of an empty cartridge case as a lever. Protect the hands as outlined in *b* above.

d. A ramming and extracting tool—41–T–3308–25 is supplied with each tank for battlefield emergency use. The general purpose of this tool is to act as a pry bar to remove and insert rounds or cases which do not yield to the breechblock extracting or seating action. In forcing a loaded round to its seat in the breech recess, the trunnions of the tool fit into, and fulcrum against, the breechblock guides in the breech ring. That part of the tool in contact with the cartridge case straddles the primer.

Caution: Do not contact the primer of live ammunition.

225. Unloading

a. UNFIRED ROUND. Whenever possible, live, fuzed rounds must be unloaded from the gun by firing.

(1) If it is not possible to remove the round by firing, station a man at the rear of the breech to catch the base of the round as it emerges to prevent it from slipping out and dropping to the floor. Open the breech slowly (par. 221).

Caution: Do not attempt to open the breech rapidly or the case may become separated from the projectile. Remove the round and return it to its rack.

(2) If the round cannot be removed by opening the breech or the case becomes separated from the projectile, it will be removed under the direct supervision of an officer.

(3) To remove a stuck round, open the breech as outlined in (1) above. Assemble the unloading rammer and insert it, head first, into the muzzle of the tube and push it carefully down the bore until it is seated on the ogive of the projectile. Exerting a steady pressure on the staff, push the round clear so that it may be removed from the breech end. If outlined procedure fails to remove the round, notify ordnance maintenance personnel.

(4) In combat, to avoid exposing personnel to enemy fire, the round may sometimes be pried out by using the base of an empty case as a lever or using ramming and extracting tool as outlined in paragraph 224.

b. STUCK PROJECTILE. To remove a stuck projectile, open the breech and fill the chamber with clean rags, waste, or material suitable to form a cushion and close the breech. Proceed as outlined in *a* (3) above. After the projectile is free in the chamber, open the breech, remove projectile, dispose of it in accordance with SR 385-310-1, and clean the bore and chamber.

226. Firing the Coaxially Mounted Cal. .30 Machine Gun

a. ELECTRIC FIRING.

(1) Load the gun. Refer to appropriate publications listed in the appendix.

(2) Move the gun master switch in turret control box (fig. 19) to the "ON" position.

(3) The indicator light should burn brightly when this switch is on and the firing solenoids are energized.

(4) To fire, depress the cal. .30 electric firing trigger located on the power traverse control handle (fig. 181). Release the trigger after attempting to fire.

b. Manual Firing. If the machine gun fails to fire electrically, it can be fired manually by squeezing the trigger located at the rear of the machine gun. This operation must be performed, however, by the loader, as the gun is mounted on the loader's side.

c. Cleaning After Firing. Clean the armament as outlined in lubrication order (par. 62).

227. Disassembly for Cleaning and Lubrication of Breech Mechanism

a. Disassembly.
 (1) Remove extractor plunger plug, spring, and plunger. Remove extractors (figs. 185 and 186).

  Note. The extractor plunger parts may be removed before or after the breechblock is removed.

 (2) Place the breech operating lever on the cable terminal crank and open breech until the ball fitting on the cable clears the closing spring housing nut. Place the breechblock closing spring locking washer 41–W–22 ahead of the ball fitting and close the breechblock until the spring tension is supported by the washer.
 (3) Remove the breech operating crank screw. Pull shaft completely out of breech ring. Slide the breechblock and breechblock crank completely out of the breech ring.
 (4) Hook the ball fitting, at the end of the closing spring, in the cable terminal crank. Place the breech operating lever on the crank and pull back on the lever until the breechblock closing spring locking washer can be removed. Let up on the lever until the closing spring piston starts to come out of the closing spring housing. Hold the piston with one hand and continue raising the operating lever until the ball fitting is forced out of the cable terminal crank. Remove the closing spring, piston, adapter, and cable from the housing (fig. 187).
 (5) Clean and lubricate in accordance with lubrication order.

b. Assembly.
 (1) If the closing spring housing has been removed, install the housing in the lug on the left side of the breech ring. Place the lock washer over the threaded end of the housing and screw on the lock nut.
 (2) Install the cable terminal crank and the breech operating crank in the lugs on the bottom of the breech ring. Install the two detent screws. Assemble the closing spring cable, adapter, piston, and spring in the closing spring housing. Place the breech operating lever on the cable terminal crank.

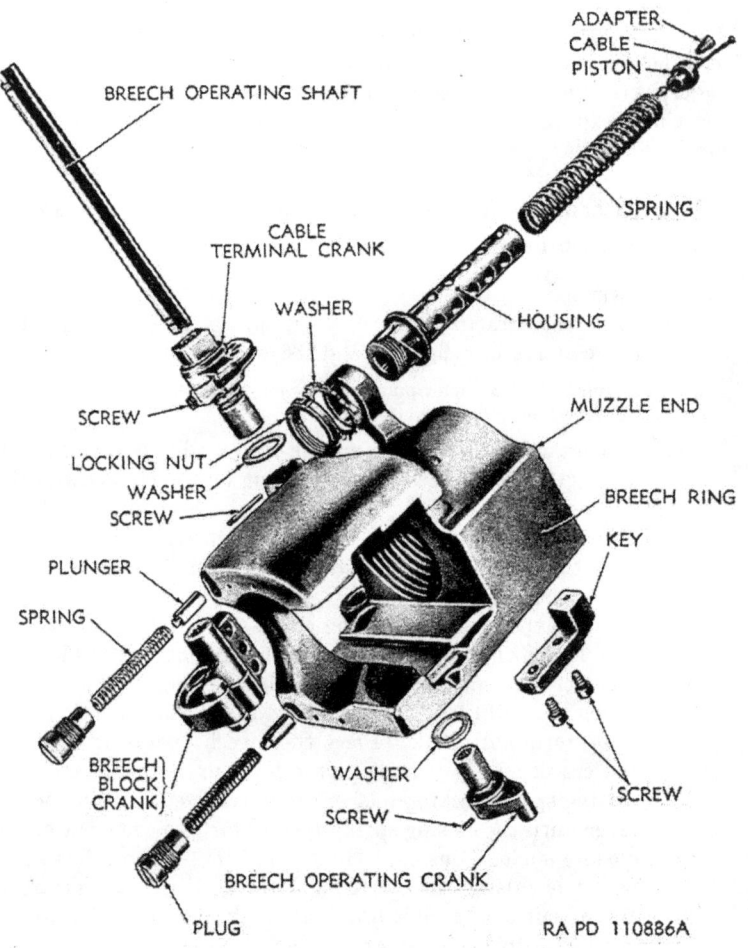

Figure 185. Breech mechanism—exploded view.

Place the closing spring parts in the closing spring housing. Compress the closing spring by pushing on the closing spring piston. Push the piston until the ball fitting on the end of the cable can be engaged with the cable terminal crank. Pull down on the operating lever until the breechblock closing spring locking washer can be dropped over the cable ahead of the ball fitting near the center of the cable. Install the closing crank cable screw.

(3) Install the breechblock crank in the breechblock with the trunnions of the crank engaging the T-slot in the breech-

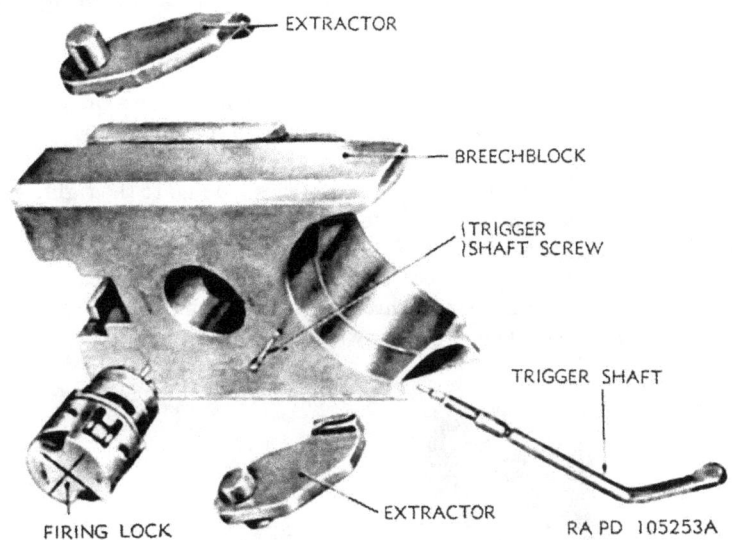

Figure 186. Breechblock—exploded view.

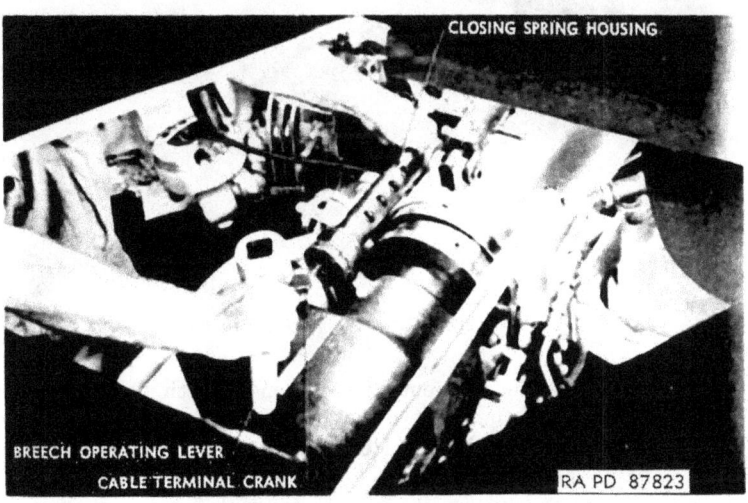

Figure 187. Removing closing spring.

block. Install the extractors in the breech ring. Slide the breechblock and crank in the breech ring, being careful not to dislodge the extractors. Line up the hub of the cable terminal crank and the breechblock crank. Aline the splines in the hubs and replace the breech operating shaft.

228. Disassembly for Cleaning and Lubrication of Firing Lock M15

a. DISASSEMBLY.

(1) The firing lock may be removed either before or after the breechblock has been removed. With the chamber empty and the breech closed, remove the trigger shaft screw (fig. 186). Withdraw the trigger shaft, downward and to the right, from the breechblock. Rotate the firing lock 60° in either direction and pull it out of the breechblock.

(2) Pry the trigger fork partially out of the firing case with a screwdriver inserted through the trigger shaft hole (fig. 188). Pull the trigger fork out with the fingers.

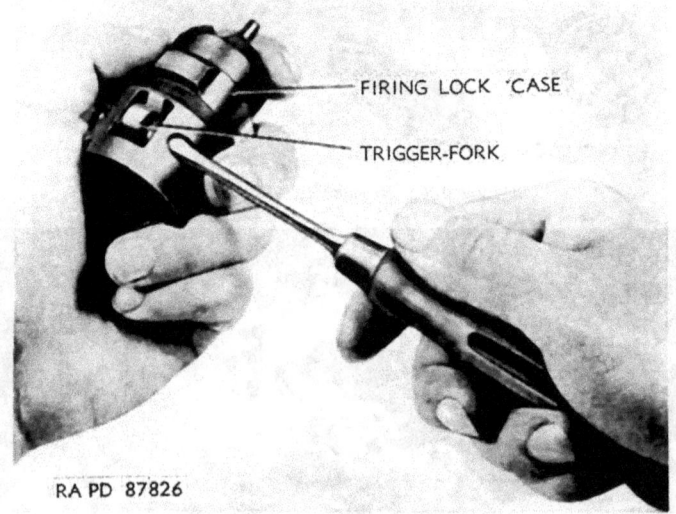

RA PD 87826

Figure 188. Removing trigger fork.

(3) Depress the sear out of engagement with the firing pin. Withdraw the firing pin and spring from the firing case. Remove the firing pin sleeve from the firing case. Remove the sear and sear spring.

(4) Clean and lubricate in accordance with lubrication order.

b. ASSEMBLY.

(1) Place the sear into the firing case and place the guide lug on the bottom of the sear into the sear spring.

(2) Assemble the firing pin spring and the firing pin sleeve on the firing pin. Insert a small screwdriver through the hole in the top of the firing lock case and depress the sear.

At the same time, insert the firing pin, spring, and sleeve in the firing lock case. Remove the screwdriver and push the firing pin into the case until the pin engages the sear.

(3) Insert the trigger fork in the case and push it in until it snaps into position.

229. Placing 75-MM Gun in Traveling Position

a. Pull hand cocking lever rearward, allowing safety plunger to lock it in the "SAFE" position.

b. Move the stabilizer switch, turret motor switch, and gun master switch in the turret control box to the "OFF" position (fig. 19).

c. Open breech and see that tube is clear. Close breech.

d. Traverse turret so that the gun faces to the front and is in line with the center of the vehicle. Engage the manual elevating control by moving the elevating shifter lever to the left and elevate or depress the gun sufficiently to engage the travel lock in the bracket under the turret roof (fig. 168). Move the elevating shifter lever to neutral position.

e. Remove the sighting and fire control equipment and install in carrying cases provided.

f. Install covers on all weapons.

g. Lock the turret traversing lock in traveling position as outlined in chapter 2, section IV.

Section II. SIGHTING AND FIRE CONTROL INSTRUMENTS

230. General

a. This section contains essential information necessary for the crew to properly identify, connect, and protect this equipment while being used or transported with the vehicle.

b. Complete detailed operating instructions for on-carriage sighting and fire control instruments as well as necessary instructions covering maintenance authorized to organizational personnel are contained in TM 9–313. Instructions covering off-carriage items are contained in separate technical manuals listed in the appendix.

231. Description

a. ARRANGEMENT. The arrangement of the on-carriage sighting and fire control instruments used in the tank is shown in table IV.

b. PERISCOPE M16, M10P, OR M4A1 ON PERISCOPE MOUNT M66. The periscope and periscope mount, on the left side of the combination gun mount, are used for direct fire with the 75-mm gun and coaxial cal. .30 machine gun. Periscope M16 (fig. 189) and M10P (fig.

Table IV. Arrangement of Sighting and Fire Control Instruments.

Used for Direct Fire with 75-mm Gun M6 or Coaxially Mounted Cal. .30 Machine Gun.	Used for Indirect Fire with 75-mm Gun M6 or Coaxially Mounted Cal. .30 Machine Gun.	Used for other Purposes.
MOUNT, telescope, M65 TELESCOPE, M83F or M71K and LIGHT, instrument, M33. MOUNT, periscope, M66 PERISCOPE, M16 or M10P or M4A1 (w/telescope M38A1 or M77G) and LIGHT instrument, M30.	QUADRANT, elevation, M9 and LIGHT, instrument, M30. INDICATOR, azimuth, M21 and LIGHT, instrument assembly, D78454.	PERISCOPE, M13 or M6 (for driving vehicle, par. 32). PERISCOPE, M15A1 LIGHT, instrument, M33 (for tank commander). SETTER, fuze, M27 or M14 (for setting fuzes).

190) each have a low power, wide field of view, with reticle pattern for observation and for close targets and a higher power narrower field of view with reticle pattern for more distant targets. The low power field of view of periscope M4A1 (fig. 191) is for observation only. If the head of the periscope should be damaged by a projectile, it may be readily replaced from within the tank by one of the spare heads provided for the purpose. Periscope mount M66 (fig. 192) is used directly with periscope M10P or M16 and an adapter is supplied for use with periscope M4A1. The mount is linked to the gun mount and moves in elevation with the gun and in azimuth with the turret. The mount holds the periscope in either viewing or retracted position.

c. TELESCOPE M83F AND M71K ON TELESCOPE MOUNT M65. The telescope and mount (figs. 167, 193, and 194), on the left side of the combination gun mount, are used for direct fire and, with the azimuth indicator, for indirect fire. The telescope has a smaller field of view, but greater magnification than the periscope and is mounted coaxially with the 75-mm gun. The reticle pattern is used primarily for direct fire. The mount moves with the gun in azimuth and elevation.

d. AZIMUTH INDICATOR M21. The azimuth indicator M21 (fig. 195) secured to the left side of the turret and geared to the turret ring gear is used in indirect fire. The pointers, driven through the gearing, indicate on coarse and fine scales the position of the gun with respect to the longitudinal axis of the tank and with respect to an aiming point. A gunner's aid dial is also provided for making corrections.

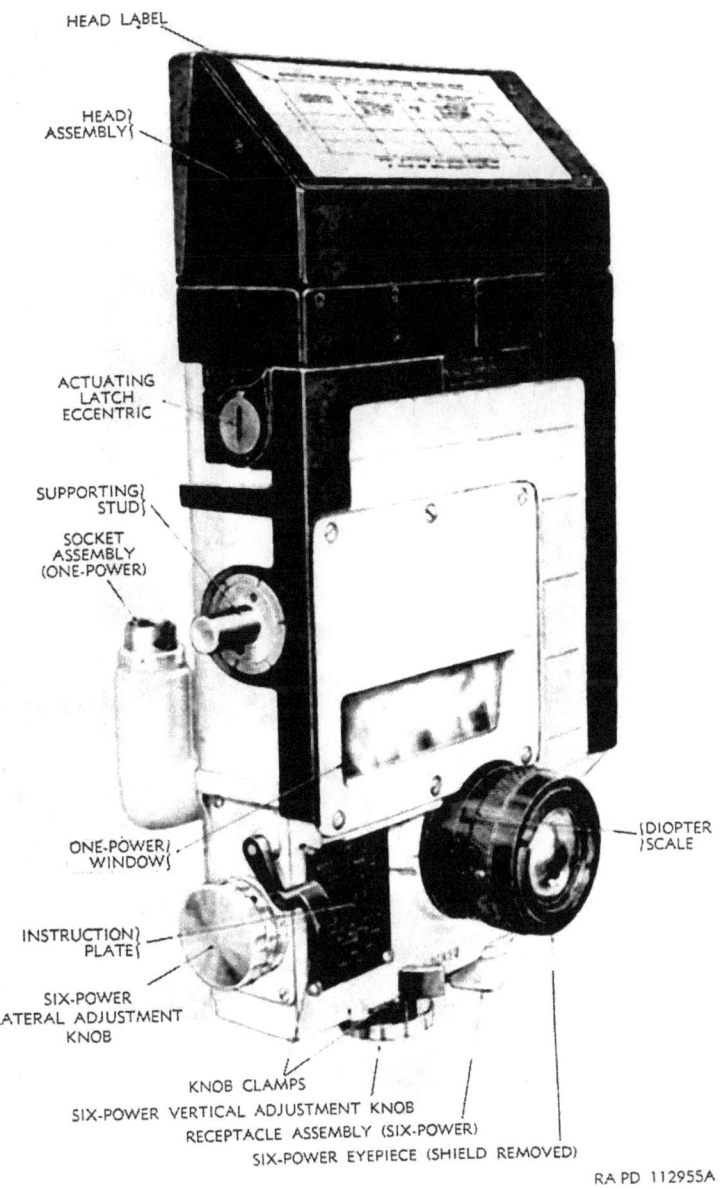

Figure 189. Periscope M16—rear view.

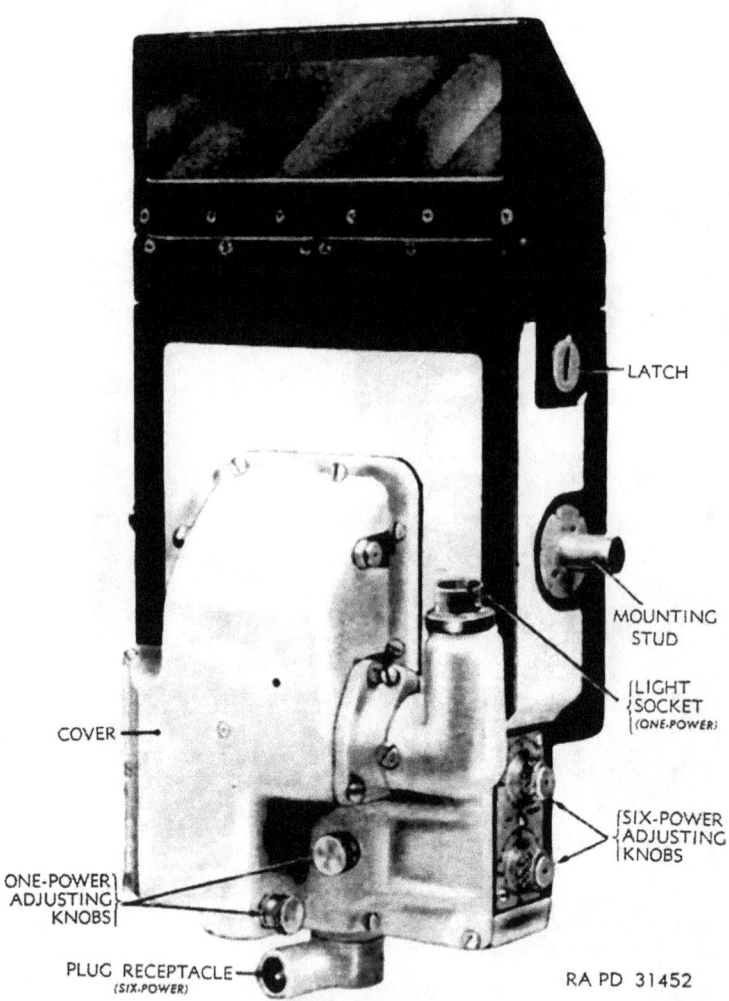

Figure 190. Periscope M10P—front view.

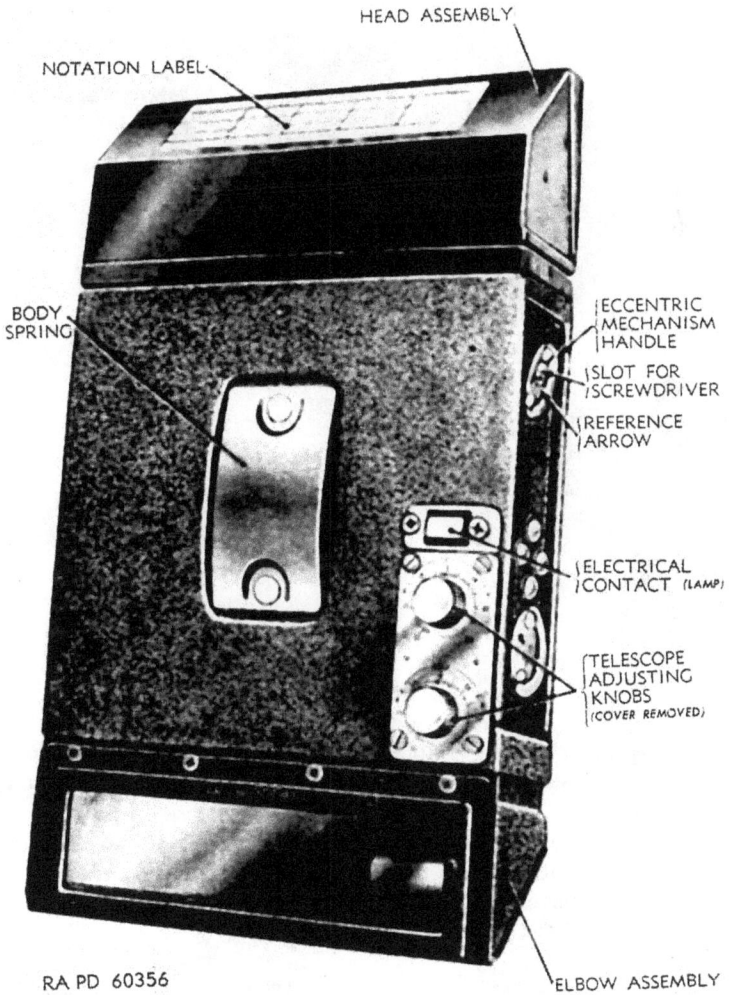

Figure 191. Periscope M4A1—rear view.

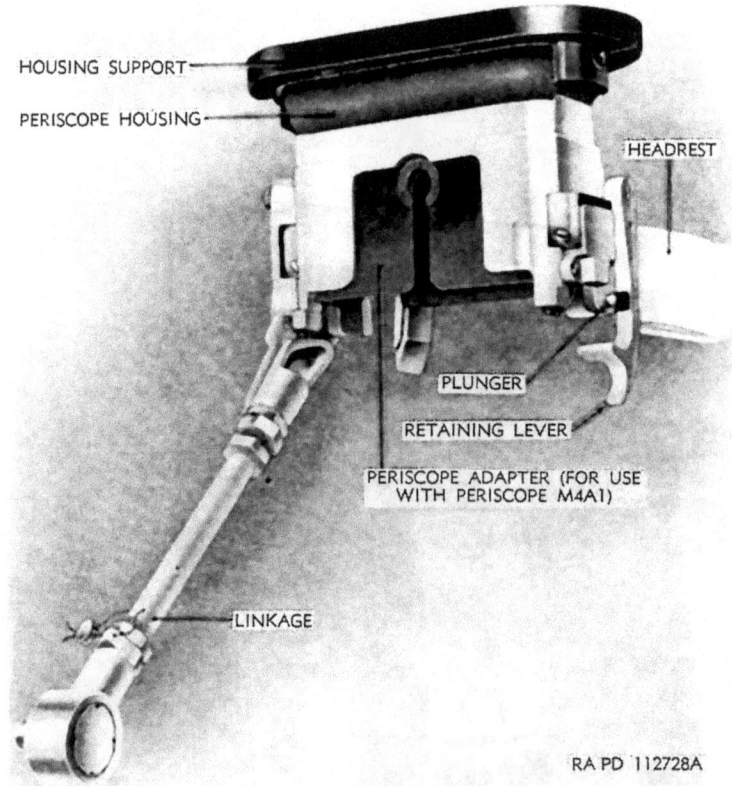

Figure 192. Periscope mount M66.

e. ELEVATION QUADRANT M9. The elevation quadrant (fig. 196), under the breechblock is used in indirect fire for laying the 75-mm gun in elevation.

f. AIMING DATA CHARTS. Aiming data charts (fig. 197), (in decalcomania or on metal plates) on the turret wall, facilitate use of the telescope and periscope reticles with ammunition other than that noted in the reticle pattern. The charts which may be furnished are—

ADC	Coaxial telescope	Periscope (w/telescope)	Periscope
75-H-2	M71G	M4A1 (M38A2)	
75-K-4	M71K	M4A1 (M77G)	M10P
75-L-3	M83F		

g. PERISCOPE M15A1. This binocular periscope (fig. 195) in the commander's cupola door is used for observation of fire and for designation of targets. It may be swung back or forward in its

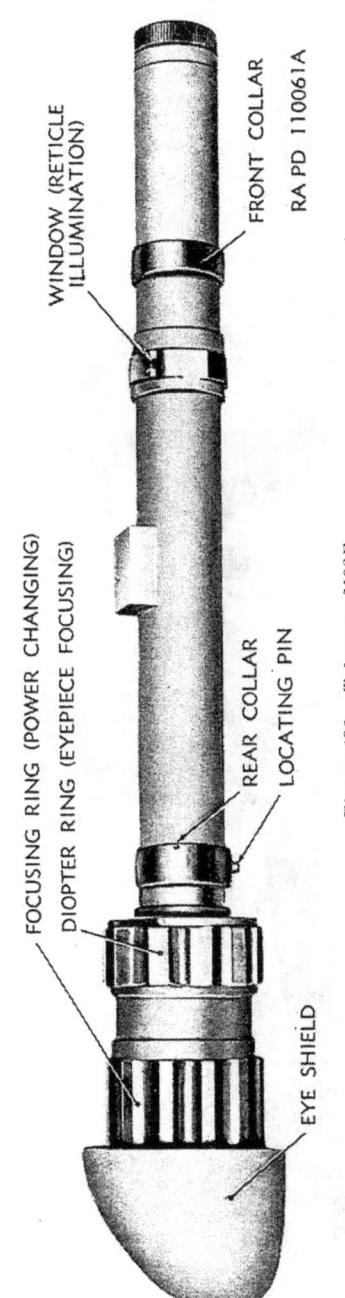

Figure 193. Telescope M83F

341

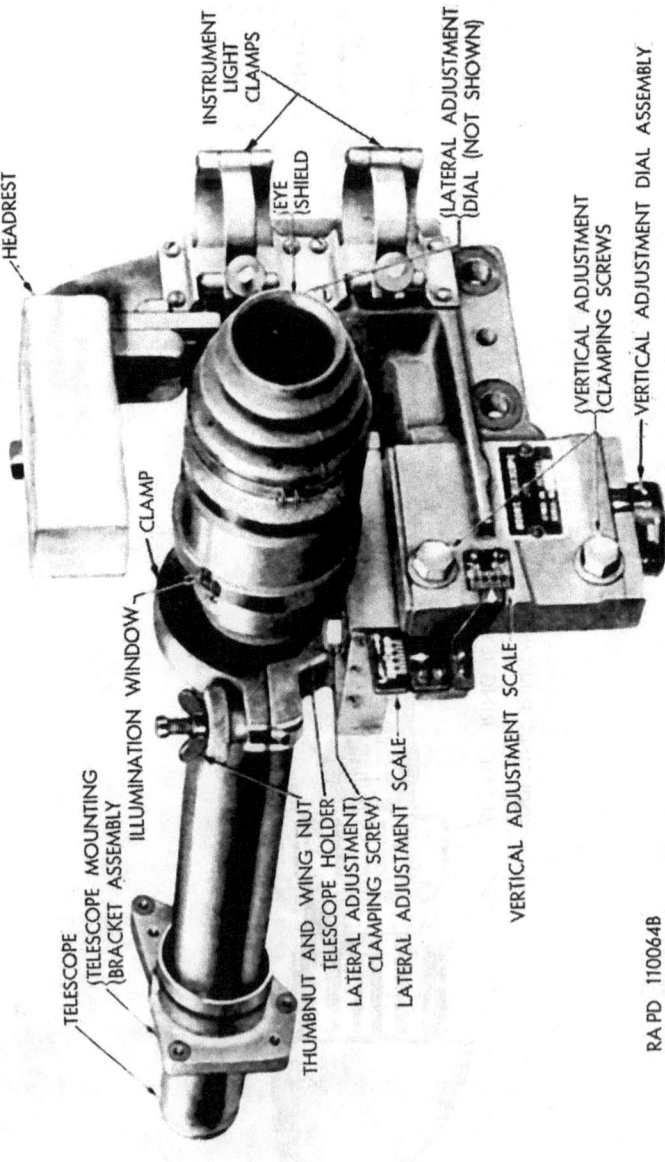

Figure 194. Telescope M71K and telescope mount M65.

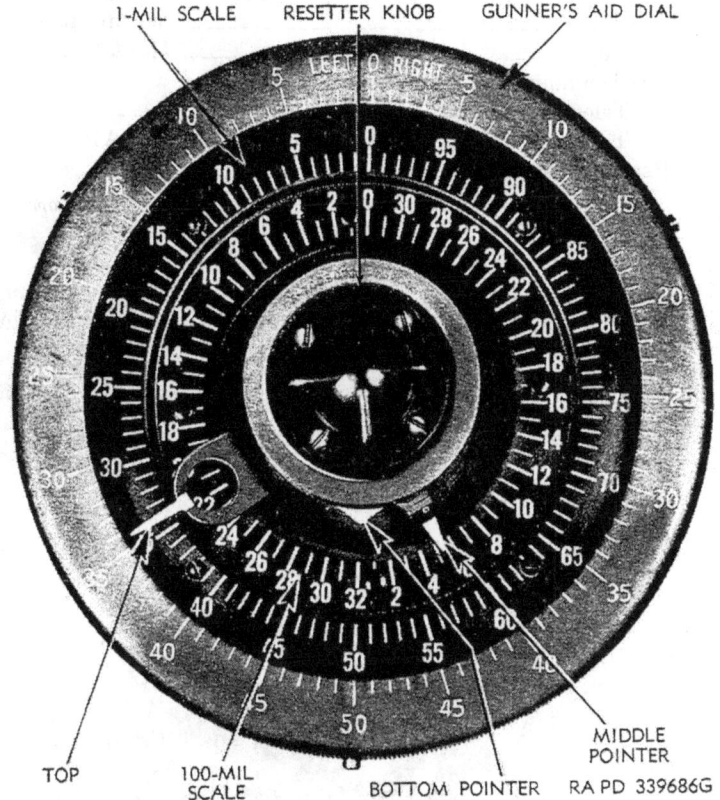

Figure 195. Azimuth indicator M21—scales and pointers.

holder to improve visibility on grades and may be rotated in azimuth with the cupola door for obtaining bearings on targets and aiming points. Where this periscope is not available, periscope M13 (fig. 15) or periscope M6 (fig. 14) may be furnished.

h. Fuze Setter M27 or M14 (Wrench Type). These fuze setters (fig. 199) are used for setting time fuzes for 75-mm ammunition.

232. Data

a. General. Data for the various on-carriage sighting and fire control instruments used on the tank are listed below.

b. Periscope M4A1.
    Magnification_____ 1.44X (telescope)
    Periscope offset_____ 8⅞ in.

343

Deflection:
    Horizontal _____ 30 mil R and L
    Vertical _____ 30 mil R and L
Telescope _____ M38A2 or M77G
Reticle ammunition graduations:
    Telescope M38A2 _____ 75–M61
    Telescope M77G _____ APC–T–M61

  c. PERISCOPE M10P AND M16.

	Periscope	Telescope
Magnification	1×	6×
Field of view:		
Horizontal	42 deg 10 min	10 deg 20 min
Vertical	8 deg 10 min	10 deg 20 min
Reticle ammunition graduations.	75–M61	75–M61
Reticle deflections	30 mil R and L	30 mil R and L
Range graduations	0–1,200 yd	0–4,200 yd
Weight		10 lb, 8 oz

d. PERISCOPE MOUNT M66. Moves in elevation with the gun and in azimuth with the turret.

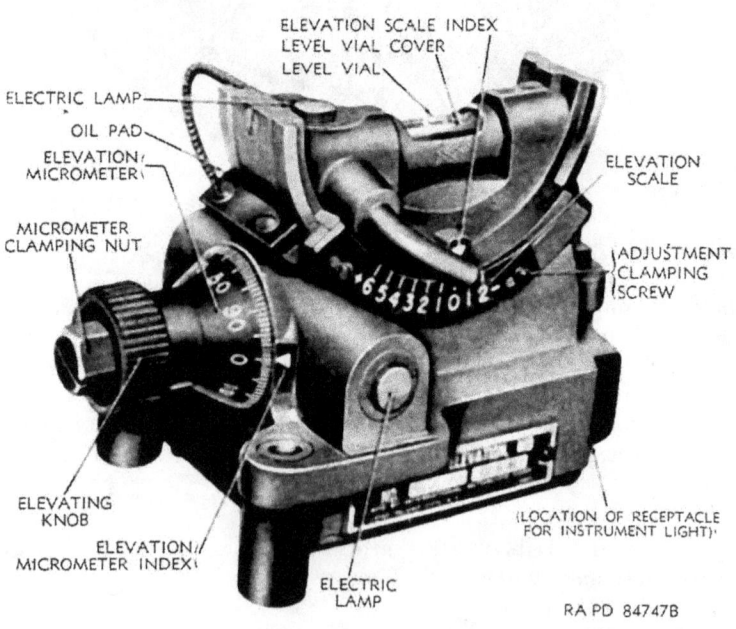

Figure 196. Elevation quadrant M9.

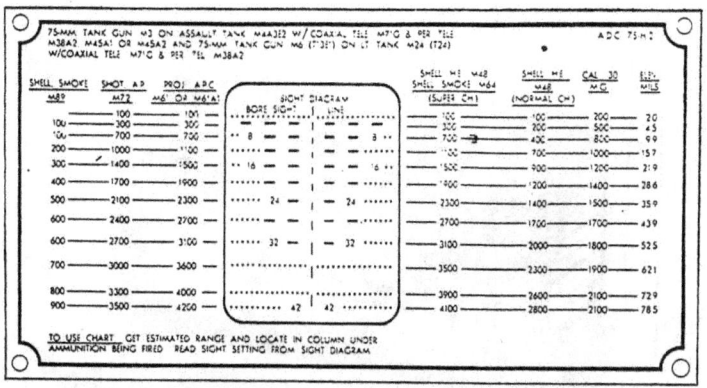

A—AIMING DATA CHART—ADC 75-H-2

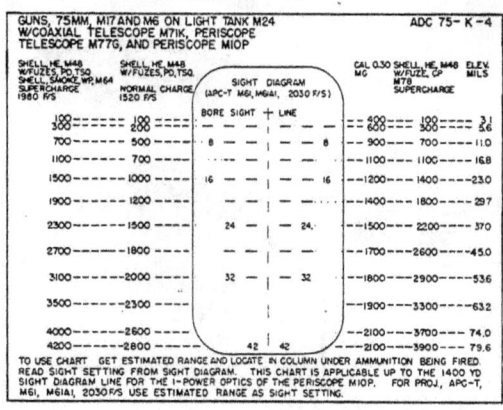

B—AIMING DATA CHART—ADC 75-K-4

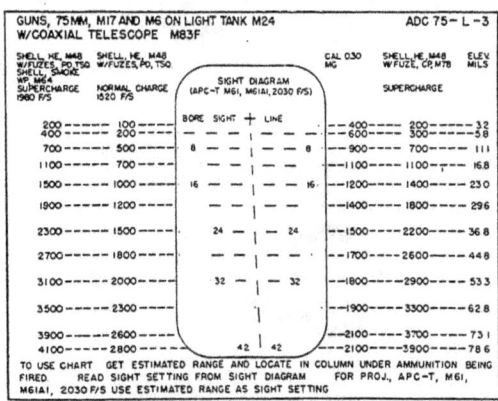

C—AIMING DATA CHART—ADC 75-L-3

RA PD 127798

Figure 197. Aiming data charts.

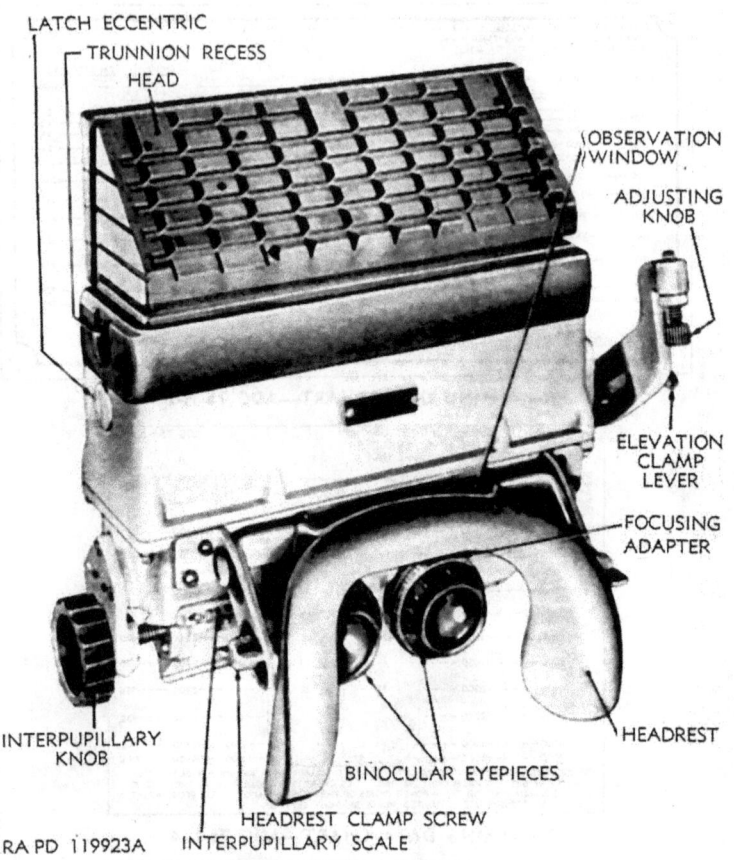

Figure 198. Periscope M15A1—rear view.

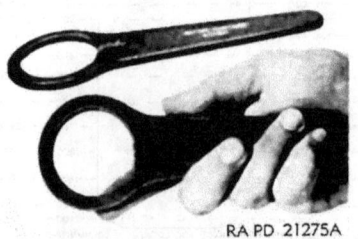

Figure 199. Fuze setter M14.

e. TELESCOPE M71K.
    Magnification -- 5X
    Field of view -- 13 deg
    Reticle ammunition graduations ----------------- APC-T M61
    Reticle identification ------------------------------- 75-M61
  f. TELESCOPE M83F.
    Magnification ---------------------------------- 4X or 8X
    Field of view:
      4X -- 7 deg 40 min
      8X -- 4 deg 15 min
    Reticle ammunition graduations ------------- 75 APC-T M61A1
  g. TELESCOPE MOUNT M65.
    Vertical scale ---------------------------------- EL 20-0-20 DEP
    Lateral scale ---------------------------------- R 20-0-20 L
  h. AZIMUTH INDICATOR M21.
    Fine scale -- 0-100 mils
    Coarse scale ---------------------------------- 0-3,200, 0-3,200 mils
    Gunner's aid dial ----------------------------- 0-50 mils R and L
  i. ELEVATION QUADRANT M9.
    Elevation scale -------------------------------- −200 to +600 mils
    Micrometer -- 0-100 mils
  j. PERISCOPE M15A1.
    Magnification ------------------------------------- 1X and 7X
    Field of view:
      1X --------------------------------- 60 deg hor, 13 deg 30 min vert
      7X --- 10 deg
    Interpupillary adjustment ----------------------- 58-70 mm
    Reticle pattern ----------------------------------- stadia scale

233. Controls

  a. TELESCOPE MOUNT M65. The telescope M83F or M71K (fig. 194) is clamped in position in the telescope mount. The telescope mount has lateral and vertical scales for boresighting adjustment.

  b. TELESCOPE M71K. The diopter ring (figs. 178 and 194) is used to bring the reticle and target into sharp focus.

  c. TELESCOPE M83F. This telescope (fig. 193) has the diopter ring, and also the focusing ring for changing the magnification from 4-8 power.

  d. PERISCOPE MOUNT M66. A retaining lever on the left-hand side of the periscope mount (fig. 192) is used for holding the periscopes M16, M10P, or M4A1 with adapter in either a viewing or a retracted position and to permit withdrawal of the periscope from the mount.

  e. PERISCOPE M16 OR M10P. The 6-power eyepiece (fig. 189) is rotated to bring the reticle and target image into sharp focus. Boresighting controls are provided.

f. PERISCOPE M4A1. There is no focusing eyepiece on the periscope M4A1. However, boresighting controls are provided.

g. PERISCOPE M15A1.

(1) An interpupillary knob at the lower left-hand portion of the periscope (fig. 198) is used to adjust the spacing of the eyepieces to suit the observer's eyes.

(2) A focusing adapter on the left eyepiece is used to bring the reticle and target image into sharp focus. A focusing adapter on the right eyepiece is used to bring the target image into sharp focus.

(3) An elevation clamp lever on the right-hand side of the periscope (fig. 198) is used to clamp the periscope at the desired elevation.

h. PERISCOPE M13 OR M6. A knob on the front face of the periscope (figs. 14 and 15) secures the periscope in either a viewing or a retracted position. The periscope is supported in a holder which is mounted in the driver's or assistant driver's door. These periscopes are used as vision devices for driving the vehicle (par. 32). Where periscope M15A1 is not available, the M6 or M13 may be employed.

i. ELEVATION QUADRANT M9.

(1) An elevating knob is used to set elevation or depression angles on the scale and micrometer.

(2) A level vial cover (fig. 196) protects the glass vial when not in use.

j. AZIMUTH INDICATOR M21. A resetter knob mounted in the central top portion of the azimuth indicator (fig. 195) is used to set the middle and top pointers.

234. Boresighting

a. PURPOSE. The purpose of boresighting is to establish the line of sight of the various sighting instruments so that each line of sight will intersect the projected axis of the gun bore at a distant aiming point. A method of boresighting on a distant point is described in this paragraph. For a method of boresighting on a target, refer to FM 23–100.

b. DISTANT AIMING POINT. The distant aiming point must be a sharp and distinct object, preferably in excess of the greatest range of employment and never less than the average range of employment or at approximately 1,500 yards if neither of these ranges is known.

c. EMPLACEMENT. Place the tank on ground as level as possible.

d. INSTALLATION OF BORE SIGHTS. Open the breechblock and insert the breech bore sight in the chamber. If the breech bore sight is not available, remove the firing lock (par. 223*b*) and, with the breechblock closed, use the firing pin hole in the bushing as a peep sight. Attach the muzzle bore sight, stretching the linen cord (or string) lightly across the score marks on the muzzle and hold in place by the strap or rubber bands.

c. PROCEDURE. Look through the breech bore sight or peep sight and accurately aline the intersection of the cross-strings of the muzzle bore sight on the distant aiming point. Look through the telescope or periscope to see if the cross of the reticle pattern falls on the aiming point (fig. 200). If it does not, shift the line of sight of the telescope

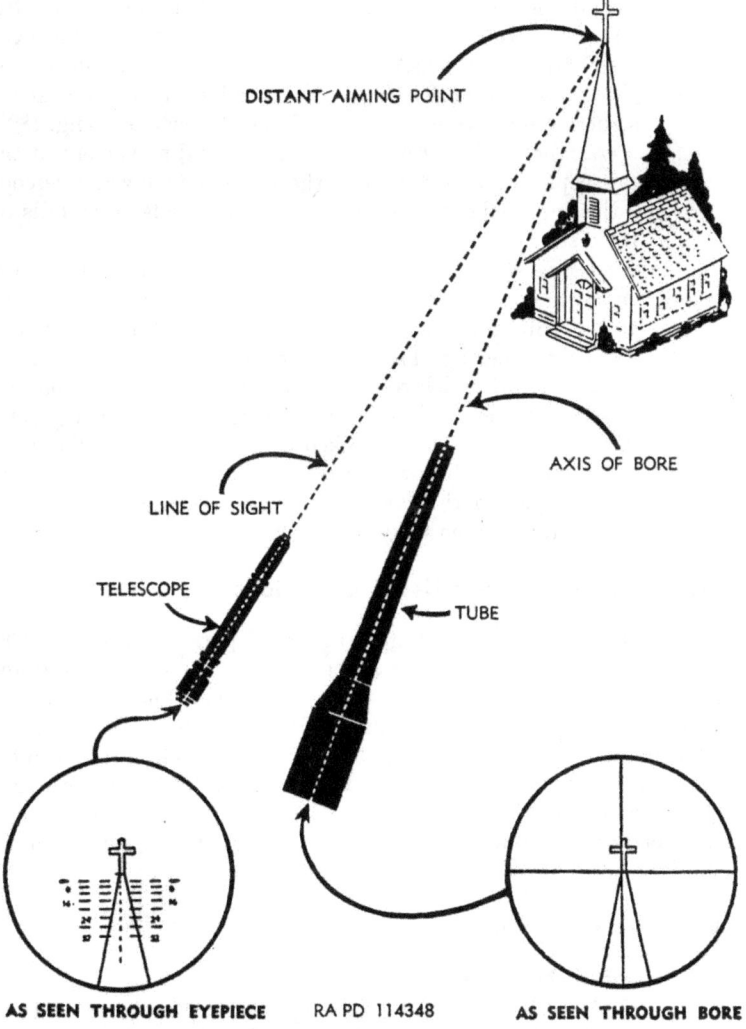

Figure 200. Gun boresighted on distant aiming point.

or periscope until coincidence is obtained (*f* below), being careful not to disturb the laying of the gun.

f. BORE SIGHT ADJUSTMENTS.

(1) *Telescope M83F or M71K.* For lateral adjustment of the line of sight, loosen the lateral adjustment clamping screws on the telescope mount M65 (fig. 194) and turn the lateral adjustment dial until the reticle cross falls on the distant aiming point; tighten the screws. For vertical adjustment of the line of sight, loosen the vertical adjustment clamping screws (fig. 194) and turn the vertical adjustment dial assembly until the cross falls on the distant aiming point; tighten the screws.

(2) *Periscope M16 or M10P.* Follow the instructions on the instruction plate which is attached to each periscope (fig. 189).

(3) *Periscope M4A1.* For lateral and vertical movement of the telescope line of sight, turn the upper and lower telescope adjusting knobs, respectively, until the reticle cross falls on the distant aiming point.

(4) *Periscope M15A1.* Swing the periscope down, using the depressing and elevating mechanism at the right of the periscope. Sight upon the periscope sighting vane (gunner's sighting vane, fig. 1). Move the periscope in the cupola door (part of which rotates) until the left eyepiece containing the reticle is alined with the periscope sighting vane. The cupola door is now locked against rotating. The periscope may then be swung up to cover the field of view and to make use of the range marks on the reticle and to differentiate between targets.

235. Operation Under Usual Conditions

Information in the following paragraphs (pars. 236 through 238) concerns the mechanical steps necessary to operate the sighting and fire control instruments of this tank under conditions of moderate temperature and humidity. Refer to FM 23–100 for information on determination of data for direct and indirect fire to be applied to the instruments of the tank for individual and group operation. For operation on uneven terrain requiring introduction of angle of site or cant corrections, refer to FM 23–100. For operation under unusual conditions, refer to TM 9–313.

236. Direct Fire Operations

a. TELESCOPE M83F OR M71K.

(1) Unlock the thumbnut and wing nut that secure the clamp to the telescope holder M65 (fig. 194). Drop the eye bolt

and open the clamp. Remove the telescope from its carrying case and insert the front end of the telescope in the telescope mounting bracket assembly. Aline the locating pin on the rear collar of the telescope with the slot in the telescope holder. Exert a slight forward pressure on the telescope to overcome the action of the spring loaded plunger and ease the telescope into position and then clamp. To remove the telescope, unclamp and withdraw the telescope. For night operation, attach the instrument light M33 and slide the lamp bracket into the dovetail slot over the reticle. Turn the rheostat knob until the reticle pattern is seen clearly through the eyepiece.

(2) Look through the eyepiece of the telescope M71K and turn the diopter scale to bring the reticle and target image into sharp focus. Focus the reticle pattern and target image of telescope M83F by turning the diopter ring; then turn the focusing ring to the desired magnification. Refocus the reticle pattern.

(3) The reticle pattern (fig. 201) aids the gunner in laying for range and deflection. The broken horizontal lines of the reticle pattern are graduated in hundreds of yards of range. Each dash or space comprising these lines represents a deflection of 5 mils. Angular drift has been incorporated in the broken vertical line of the reticle. The dashes and spaces represent intervals of 200 yards in range.

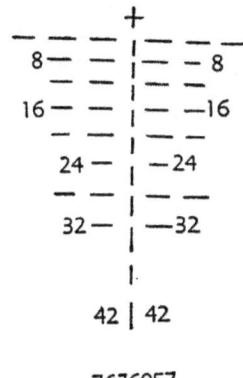

Figure 201. Reticle pattern as seen through the eyepiece, telescope M83F.

(4) With the gunner's eye against the eyeshield, bring the image of the target to the point on the reticle which represents the required range and deflection, by rotating the elevating handwheel and traversing the turret. If the required range falls within a 400-yard interval on the reticle, use the broken vertical line as a guide.

b. PERISCOPE M16 OR M10P OR M4A1.

(1) First see that the retaining lever of periscope mount M66 (fig. 192) is on the operator's side of the plunger. Then, insert the supporting studs of the periscope in the entering slots and push the periscope straight upward as far as it will go. Push the lever to the opposite side of the plunger and raise the periscope to the viewing position. Holding the periscope in the viewing position, pull the lever back past the plunger until it latches, in order to secure the periscope. To withdraw the periscope, first push the lever forward past the plunger and then pull the periscope straight down against the lever. The periscope is now in the retracted position. For complete removal, pull the lever back toward the operator past the plunger and pull the periscope straight down until clear of the mount.

(2) When the periscope M4A1 is to be installed in periscope mount M66, an adapter is used to seat the periscope in the mount. The clamping screw on the periscope secures the periscope in the adapter. The periscope and adapter are then seated in the periscope mount in the manner described for periscope M16 and M10P.

(3) Plug in the instrument light M30 for night operation of the 6-power optical system and turn the rheostat knob until the reticle pattern is clearly seen. Loosen the appropriate clamping screws and shift the head rest on the periscope mount inward or outward or sideways to suit the gunner's head; tighten the screws.

(4) Look through the eyepiece of periscope M16 or M10P or through the right-hand portion of the elbow of the M4A1 when sighting on distant targets. Bring the reticle pattern and target image into sharp focus (M16 and M10P only). Aline the reticle (the reticle pattern of the periscope M16 is similar to the reticle pattern of the periscope M10P shown in fig. 202) on the target as outlined in $a(4)$ above.

(5) Turn on the illumination (current from tank source) and look through the window when sighting on close targets. Aline the reticle (fig. 202) on the target as described in $a(4)$ above.

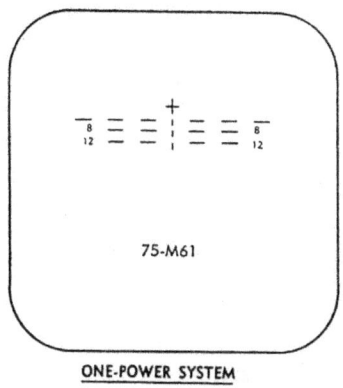
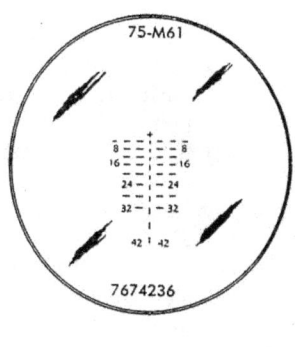

ONE-POWER SYSTEM   SIX-POWER SYSTEM

RA PD 110065A

Figure 202. Reticle patterns, periscope M10P.

(6) If the head of the periscope becomes damaged, turn the actuating latch eccentrics (fig. 189) to release the head from the body and lift off the head. Replace the head with one of the spare heads which are provided; tighten the actuating latch eccentrics. The periscope is withdrawn into the tank for the operation.

237. Indirect Fire Operations

a. LAYING GUN IN AZIMUTH.

(1) Look through the telescope and traverse the turret until the vertical broken line of the reticle falls on the aiming point. Allow for cant as described in FM 23–100.

(2) Turn the resetter knob of the azimuth indicator M21 (fig. 195) until the top pointer approaches the middle pointer, then press downward on the knob and turn; both pointers will move together when the knob is turned. Turn the pointers until they read zero on the 1-mil and 100-mil scales. For night operation, throw the toggle switch of the instrument light to illuminate the dials.

(3) Traverse the turret until the middle and top pointers read the required azimuth angle on their respective scales. The inner scale is graduated in 100-mil intervals and the middle scale is graduated in 1-mil intervals. If the azimuth angle is 550 mils, the middle pointer should read midway between the 500 and 600 graduations on the 100-mil scale and the top pointer should read 50 on the 1-mil scale. The gun is now laid in azimuth.

b. Making Deflection Corrections.
   (1) Rotate the gunner's aid dial (outer scale) (fig. 195) by hand until the zero graduation of its scale is opposite the top pointer.
   (2) Traverse the turret until the top pointer reads the desired deflection correction on the 1-mil scale.
c. Laying Gun in Elevation.
   (1) Rotate the level vial cover on the elevation quadrant M9 (fig. 196) to expose the graduations on the vial. Turn the elevating knob counterclockwise to set in plus (elevation) angles. Turn the knob until the sum of the scale and micrometer readings equals the desired angle. Allow for angle of site where applicable as described in FM 23–100. The scale is graduated in 100-mil intervals and the micrometer is graduated in 1-mil intervals. For night operation, attach the instrument light M30 and turn the rheostat knob until the desired illumination is obtained.
   (2) Turn the elevating knob clockwise to set in minus (depression) angles. Turn the knob until the scale reading plus the difference between the micrometer reading and 100 equals the desired angle. If the desired angle is minus 160 mils, turn the knob until the scale reads between -1 and -2. Continue turning the knob until the micrometer reads 100–60 or 40 mils.
   (3) Turn the elevating handwheel until the level bubble is centered. The gun is now laid in elevation.

238. Observation of Fire

a. Setting Up.
   (1) Install the periscope M15A1 (or M13 or M6, whichever is furnished) in its holder by rotating the two levers on the periscope holder to retract the trunnions and insert the periscope in the opening of the holder. Push the periscope straight upward until the trunnions engage the circular depression in each side of the periscope body as the levers are released, thereby securing the periscope in position. To remove the periscope from the holder, reverse the operation, pulling the periscope straight downward until free of the holder.
   (2) Loosen the appropriate clamping screws and shift the head rest of the periscope M15A1 inward or outward to suit the observer's head; tighten the screws.
   (3) Turn the interpupillary knob until the eyepieces are spaced to suit the observer's eyes.

(4) Look through the left eyepiece and turn the focusing adapter until the reticle pattern (fig. 203) and target image are in sharp focus. Turn the focusing adapter on the right eyepiece until the target image is in sharp focus.

(5) Slide the lamp bracket of the instrument light M33 into the dovetail slot under the left-hand eyepiece for night operation and turn the rheostat knob until the reticle pattern (fig. 203) is seen clearly.

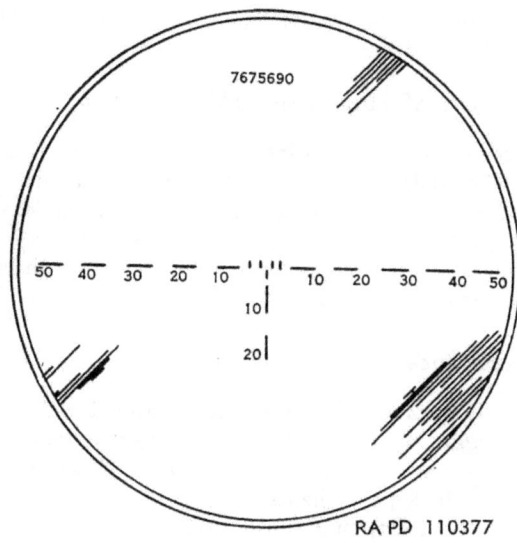

Figure 203. Reticle pattern as seen through the left eyepiece, periscope M15A1.

(6) Boresight as described in paragraph 234*f*(4).

b. OBSERVING. Push the elevation clamp lever on the periscope M15A1 (fig. 198) forward to release the elevation clamp. Look through the eyepieces and, with the hands, traverse and elevate or depress the periscope until the target appears at the center of the reticle pattern. Pull the elevation clamp lever backwards to clamp the periscope in the desired position. The position of the shell burst in relation to the target is noted on the reticle. Each horizontal line or space of the reticle pattern (fig. 203) represents a 5-mil deflection. The two short vertical lines either side of center are 1 mil high and spaced 1 mil and 3 mils either side of center. The broken vertical line is graduated in 2½-mil and 5-mil intervals.

c. REPLACING DAMAGED HEAD. If the head of the periscope becomes damaged, draw it into the tank, turn the latch eccentrics (fig. 198) to release the head from the body, and remove the head. Replace the head with one of the spare heads which are provided; tighten the latch eccentrics.

239. Observation of Terrain

Look through the observation window of the periscope M15A1 (or M13 or M6) or through the left-hand portion of the elbow of the periscope M4A1 to observe the general terrain from the interior of the tank. Grasp the sides of the periscope and rotate, elevate, or depress the holder and periscope until the desired panorama is brought into the field of view.

240. Operation of Fuze Setter M27 or M14

Remove the safety wire or wires from the fuze and then place the fuze setter (fig. 199) on the fuze with the tapered side of the hole fitting the fuze. Engage the key in the slot of the fuze and turn the handle clockwise (increasing direction) until the index mark on the fuze alines with the required time setting on the fuze scale. Use care in removing the fuze setter to avoid changing the setting.

241. Maintenance

a. CARE IN HANDLING. The instruments covered in this section are, in general, rugged and suited for the designed purpose. They will not, however, stand rough handling or abuse. When not in use, instruments must be kept in the carrying cases which are provided or covered and protected from dust and moisture.

b. BATTERIES. Batteries used in the instrument lights should habitually be removed whenever the instruments are not in use. To remove batteries, turn the cap on the end of the tube to disengage the two bayonet slots in the tube and pull off the cap. Extract the two type BA-30 dry cell batteries. When replacing batteries, be sure they go back in the tube in the same position as when removed. See that the bayonet pins in the cap engage the slots in the tube to insure a tight contact between battery terminals. Batteries showing signs of corrosion should be replaced. Examine tube for corrosive accumulation and clean out or, if far gone, turn in to ordnance maintenance personnel for repair.

242. Preparation for Traveling

a. Remove periscope M10P and stow in the bracket provided (fig. 181).

b. Remove periscope M15A1 and stow in bracket near turret overhang.

c. Remove telescope M83F or M71K from the telescope mount M65 and stow the telescope in its carrying case.

d. Make sure that rheostat or toggle switches on all instrument lights are turned off.

Section III. AMMUNITION

243. General

a. This section contains information for identification, preparation for firing, and precautions in use, of ammunition authorized for the 75-mm gun M6. For more complete information and illustrations, refer to TM 9–1901.

b. For information and instructions pertaining to cal. .30 and cal. .50 machine-gun ammunition, refer to TM 9–1990.

c. Ammunition should be protected from direct sunlight and other sources of heat. If it becomes necessary to leave ammunition in the open, raise it on dunnage at least 6 inches from the ground and cover it with a double thickness of tarpaulin leaving enough space for circulation of air. Suitable trenches should be dug to prevent water from running under the pile.

d. WP shell M64 should be stored so as to avoid excessive temperature, since the white phosphorus filler will melt at temperatures above 105° F. When this is not practicable, WP rounds should be stored on their bases so that, should the WP filler melt, it will resolidify in its normal position when the temperature decreases. Prematures have been caused by voids in the base end of WP shell and erratic performance may result from voids in the side.

244. Authorized Rounds

a. Ammunition authorized for use in the 75-mm gun M6 (figs. 204 and 205) is listed in table V. Standard nomenclature, used in the listing, completely identifies the ammunition except for lot number.

b. Ordinarily, issue of this ammunition is in proportion by types to meet tactical requirements so that substitution of fuzes in the field is not required. For special purposes, certain substitutions are authorized. The concrete-piercing fuze M78A1 is authorized as a separate item of issue for assembly in the field to the HE shell M48, supercharge round. Booster M25 is used in conjunction with the CP fuze M78A1.

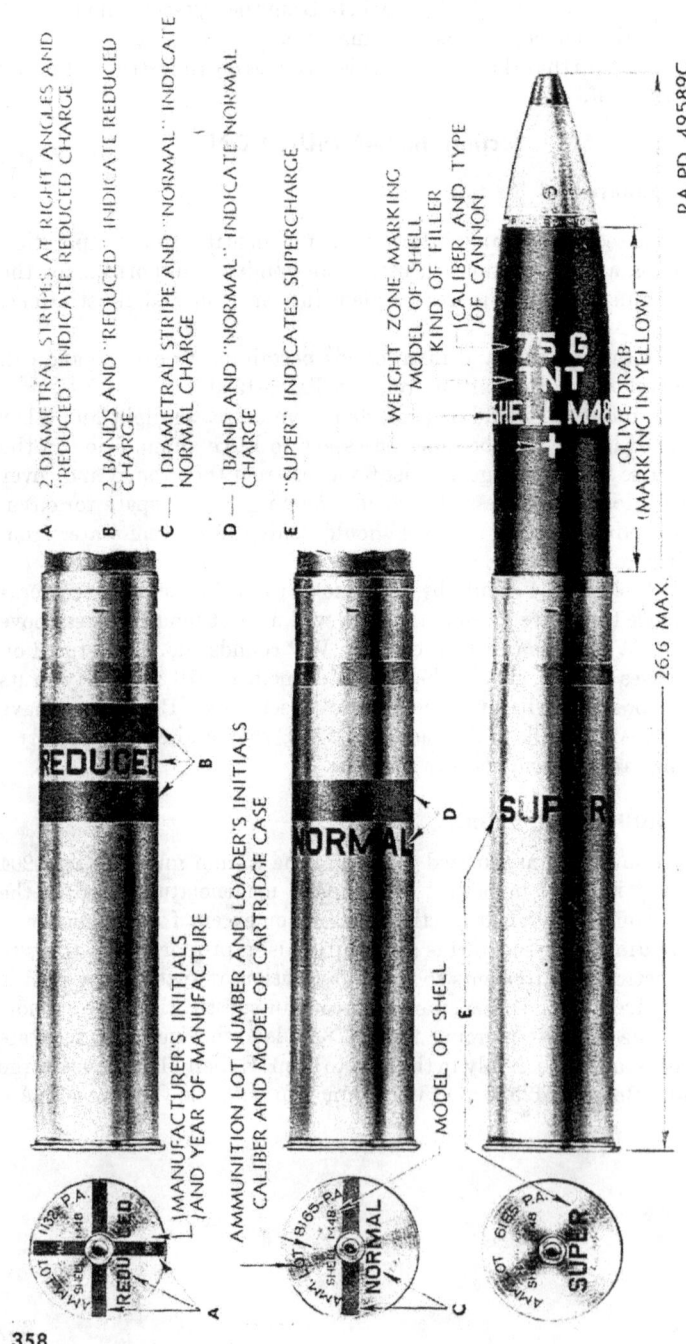

Figure 204. Markings for reduced, normal, and supercharge HE rounds.

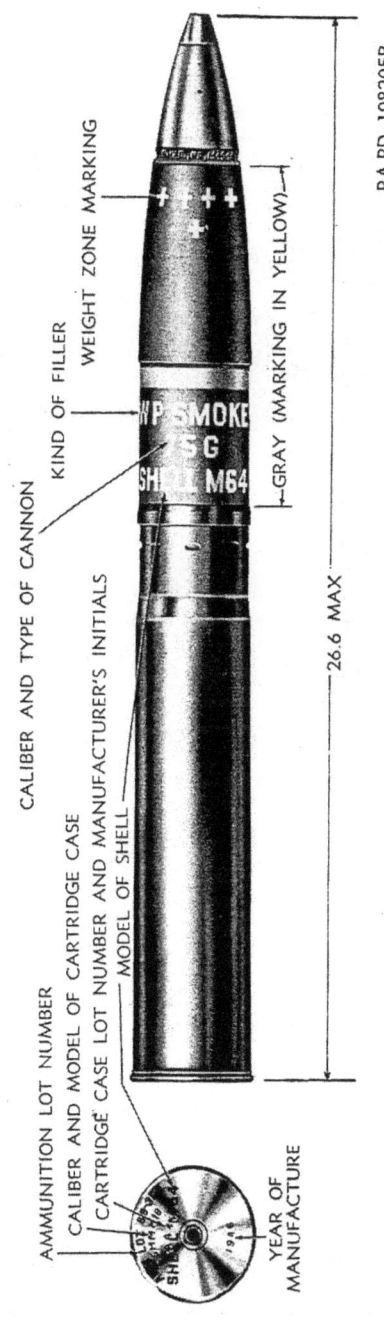

Figure 205. Shell, fixed, smoke, WP, M64, w/fuze, PD, M48A3, for 75-mm guns.

Table V. Authorized Rounds for 75-MM Gun M6

Standard Nomenclature	Complete Round		Projectile Weight as Fired (lb.)	Action of Fuze
	Weight (lb.)	Length (in.)		
SERVICE AMMUNITION				
PROJECTILE, fixed, APC-T, M61A1, for 75-mm guns.	19.91	26.29	14.96	Delay.
PROJECTILE, fixed, APC-T, M61, w/fuze, BD, M66A1, for 75-mm guns	19.92	26.29	14.96	Delay.
SHELL, fixed, HE, M48, normal charge, w/fuze, PD, M51A4 (M48A2), .05-sec delay, for 75-mm guns.[1]	18.74	26.63	14.70	SQ or 0.05-sec delay.
SHELL, fixed, HE, M48, normal charge, w/fuze, PD, M51A5, .05-sec delay, for 75-mm guns.[1]	18.74	26.63	14.70	SQ or 0.05-sec delay.
SHELL, fixed, HE, M48, normal charge, w/fuze, PD, M51A4 (M48A2), .15-sec delay, for 75-mm guns.[1]	18.74	26.63	14.70	SQ or 0.15-sec delay.
SHELL, fixed, HE, M48, normal charge, w/fuze, TSQ, M55A3 (M54), for 75-mm guns.[1]	18.74	26.63	14.70	Time (to 25 sec) and SQ.
SHELL, fixed, HE, M48, normal charge, w/fuze, MTSQ, M500, for 75-mm guns.[1]	18.74	26.63	14.70	Mechanical time (to 75 sec) and SQ.
SHELL, fixed, HE, M48, reduced charge, w/fuze, PD, M51A4 (M48A2), .05-sec delay, for 75-mm guns.[1]	18.18	26.63	14.70	SQ and 0.05-sec delay.
SHELL, fixed, HE, M48, reduced charge, w/fuze, PD, M51A5, .05-sec delay, for 75-mm guns.[1]	18.18	26.63	14.70	SQ or 0.05-sec delay.
SHELL, fixed, HE, M48, reduced charge, w/fuze, PD, M51A4 (M48A2), .15-sec delay, for 75-mm guns.[1]	18.18	26.63	14.70	SQ and 0.15-sec delay.
SHELL, fixed, HE, M48, reduced charge, w/fuze, TSQ, M55A3 (M54), for 75-mm guns.[1]	18.18	26.63	14.70	Time (to 25 sec) and SQ.
SHELL, fixed, HE, M48, reduced charge, w/fuze, MTSQ, M500, for 75-mm guns.[1]	18.18	26.63	14.70	Mechanical time (to 75 sec) and SQ.
SHELL, fixed, HE, M48, supercharge, w/fuze, PD, M51A4 (M48A2), .05-sec delay, for 75-mm guns.[1]	19.55	26.63	14.70	SQ and 0.05-sec delay.
SHELL, fixed, HE, M48, supercharge, w/fuze, PD, M51A5, .05-sec delay, for 75-mm guns.[1]	19.55	26.63	14.70	SQ and 0.05-sec delay.

See footnotes at end of table.

Table V. Authorized Rounds for 75-MM Gun M6—Continued

Standard Nomenclature	Complete Round		Projectile Weight as Fired (lb.)	Action of Fuze
	Weight (lb.)	Length (in.)		
SERVICE AMMUNITION—Continued				
SHELL, fixed, HE, M48, supercharge, w/fuze, TSQ, M55A3 (M54), for 75-mm guns.[1]	19.55	26.63	14.70	Time (to 25 sec) and SQ.
SHELL, fixed, HE, M48, supercharge, w/fuze, MTSQ, M500, for 75-mm guns.[1]	19.55	26.63	14.70	Mechanical time (to 75 sec) and SQ.
SHELL, fixed, smoke, WP, M64, w/fuze, PD, M48A2, .05-sec delay, for 75-mm guns.[2]	20.20	26.63	15.24	SQ and 0.05-sec delay.
SHELL, fixed, smoke, WP, M64, w/fuze, PD, M57, for 75-mm guns.	20.20	26.63	14.70	SQ.
SHELL, fixed, smoke, WP, M64, w/fuze, PD, M48A3, .05-sec delay, for 75-mm guns.[1]	20.20	26.63	14.70	SQ and 0.05-sec delay.
BLANK AMMUNITION				
AMMUNITION, blank, double pellet charge, for 75-mm guns and 75-mm howitzers.	3.07	7.25	--------	None.
AMMUNITION, blank, single pellet charge, for 75-mm guns and 75-mm howitzers.	2.68	7.25	--------	None.[4]
DRILL AMMUNITION				
CARTRIDGE, drill, M16, w/fuze, dummy, M59, for 75-mm guns.[3]	18.75	26.25	--------	Inert.
CARTRIDGE, training, M28 for 75-mm guns, and 75-mm howitzers.	·18.24	23.49	--------	None.

ABBREVIATIONS

APC-T	Armor-piercing-capped with tracer.	sec	Second.
BD	Base-detonating.	SQ	Superquick.
CP	Concrete-piercing.	TSQ	Time and superquick.
HE	High-explosive.	w/	With.
MTSQ	Mechanical time and superquick.	WP	White phosphorus.
PD	Point-detonating.		

FOOTNOTES

[1] Fuze, CP, M78A1, and Booster M25, authorized for use with all (supercharge, normal charge, and reduced charge) HE rounds, are issued separately for substitution of fuze and booster as required (par. 241).

[2] Includes Initiator, burster, M1 in M6 booster casing.

[3] Drill cartridges made of malleable iron are designated "M16B1" and are painted black.

[4] Training cartridges are designed to use a 10-gauge blank shotgun shell.

245. Preparation for Firing

a. GENERAL. After removal from packing materials, rounds of fixed ammunition for the 75-mm guns are ready for firing except for setting the point detonating fuzes to obtain the desired action or for substitution of fuzes. The CP fuze M78A1 and booster M25, which are separate items of issue, are assembled to HE projectiles in the field as described in *b*(2) below. Rounds and their components prepared for firing but not fired, will be returned to their original condition and packings and appropriately marked. Such rounds and components will be used first in subsequent firings in order that stocks of opened packings may be kept at a minimum.

Note. Upon removing a round with point detonating fuze from its fiber container, withdraw the "U" shaped packing stop from the fuze-wrench slots in the fuze. Serious damage to the gun tube may result if this stop is not removed.

b. SHELL.
 (1) *Shell with assembled fuzes.* Rounds to be fired with their original fuzes require only the adjustment of fuzes as described in *c* below.
 (2) *Shell with CP fuzes.* To prepare rounds for firing with CP fuzes proceed as follows:
  (*a*) Place round to be refuzed on its side. Protect the primer and cartridge case from being struck or damaged.
  (*b*) When present, loosen set screw in ogival surface of shell with screw driver end of the fuze wrench M18.
  (*c*) Using fuze wrench M18, remove the fuze (turn wrench in a counterclockwise direction as viewed from the fuzed end of the round).
  (*d*) Remove the booster, using booster end of fuze wrench M16.
  (*e*) Inspect threads of shell and CP fuze and its booster. Do not use components with damaged threads.
  (*f*) Remove the safety pin from booster M25 and assemble booster to shell, tightening it in place with fuze wrench M16. Discard boosters issued without safety pins.
  (*g*) Assemble the fuze M78A1 to shell and tighten securely with fuze wrench M16. There should be no space between the shoulder of the fuze and shell. Do not stake fuze to shell.
  (*h*) Although the booster set screw is not required, if originally present, it should be tightened sufficiently so that no part of the screw projects above the ogival surface of the shell.

c. FUZES.
 (1) PD fuze M57, BD fuze M66A1, and CP fuze M78A1 do not require any preparation for firing.
 (2) The PD fuzes M48A3 and M51A5 are prepared for delay action by turning the slot in the setting sleeve to point to

"DELAY." The fuze is set for "SQ" as shipped. If, after setting the fuze "DELAY," it is desired to fire it with "SQ" action or to return the fuze to its packing, it is necessary only to turn the setting sleeve back to its original position—the slot alining with "SQ."

(3) Prior to firing the TSQ fuze M55A3 (M54) or the MTSQ fuze M500 with either superquick or time action, withdraw the safety pull wire. If superquick action is required, the graduated time ring may be left as shipped—set at safe (s)— or set for a time greater than the expected time of flight. For time action, it is necessary only to set the graduated time ring to the desired time of flight. Use fuze setter M26 to set these fuzes accurately. In the event this fuze setter is not available, the M500 and M55A3 fuzes may be set accurately with fuze setters M23 (75 sec.) and M22 (25 sec.), respectively. As a last resort, the wrench type fuze setter M27 (or M14) may be employed. The fuze setter values may differ from the calibrations on the fuze. These fuze setter settings should be used in preference indicated for consistent performance.

246. Precautions in Firing

The following precautions should be closely observed in order to prevent injury to personnel or damage to matériel.

a. If the PD fuze M48A3, M51A5, or M57, the TSQ fuze M54 or M55A3, or the MTSQ fuze M500 is fired during extremely heavy rainfall, premature functioning may occur. The rainfall necessary to cause such malfunctioning is comparable to the exceedingly heavy downpours commonly occurring during summer thunderstorms. In the case of the M48A3 and M51A5 fuzes, such prematures may be prevented by setting the fuze for delay action, making the SQ action inoperative. However, no corresponding change can be made in the M54, M55A3, M57, and M500 fuzes, since their SQ action is always operative.

b. Do not remove safety devices from fuzes until just before use.

c. Before loading into the weapon, the ammunition should be free of foreign matter, sand, mud, moisture, or grease.

d. Make sure that the "U" shaped packing stop has been removed from the wrench slots of the point fuze before loading the round into the weapon.

247. Subcaliber Ammunition

Cartridge, ball, cal. .30, M2 is authorized as subcaliber ammunition for use in the coaxially mounted cal. .30 machine gun when that weapon is used for subcaliber purposes.

Section IV. COMMUNICATION SYSTEM

248. Description and Data

a. DESCRIPTION. The light tank M24 is equipped with either radio set SCR-508, SCR-528, SCR-608-B, or SCR-628-B, located in the turret overhang. Certain tanks are also equipped with radio set AN/VRC-3, suspended from the turret ceiling forward of the loader's position, and interphone equipment RC-99, with the amplifier located in the turret overhang. For complete information on operation and channel adjustment of the radios used in the vehicle, refer to appropriate publications listed in the appendix.

b. DATA.

    Radio _____ SCR-508, SCR-528, SCR-608-B, or SCR-628-B
    Radio (certain tanks) _____ AN/VRC-3
    Interphone (certain tanks) _____ RC-99
    Interphone _____ 5 stations
    Interphone extension _____ 1 station

>   *Note.* This is the external interphone mounted on the rear of each tank to be used by ground troops for communication with tank crew personnel.

    Others _____ signal flags, flares, and panel set

249. Radio Set SCR-508, SCR-528, SCR-608-B, or SCR-628-B

a. GENERAL. These radio sets provide radio communication between tanks and include an integral interphone system for communication between crew members in the tank.

b. DESCRIPTION.

(1) *Radio set SCR-508.* This radio set consists of radio transmitter BC-604, two radio receivers BC-603, chest CH-264 (for spare parts, tubes, headsets, etc.), mounting FT-237, phantom antenna A-62, interphone control boxes BC-606, one antenna system used for both transmitting and receiving, interphone cordage, and microphones and headsets. The radio transmitter and receiver are secured on mounting FT-237, which provides shock protection and interconnection of the units. Phantom antenna A-62 is installed on left front corner of mounting FT-237. The chest CH-264, containing spare tubes, parts, headsets, etc., is stowed in the tool box or other convenient location.

(2) *Radio set SCR-528.* This radio set is identical to radio set SCR-508, except that it has only one radio receiver BC-603. Chest CH-264 is placed on the mounting between the transmitter and receiver, rather than being stowed in the tool box

as in the case when radio set SCR-508 is installed. The chest is removed by loosening the thumb screws which secure it to mounting FT-237 and withdrawing it from mounting.

(3) *Radio set SCR-608-B.* This radio set is similar in appearance to radio set SCR-508 differing mainly in frequency range, using radio-transmitter BC-684-BM, and two radio receivers BC-683-BM. It uses mounting FT-237 and interphone components identical to SCR-508. Antenna A-83 and chest CN-96 are used, instead of the phantom antenna A-62 and chest CN-264 used with SCR-508, Antenna A-83 is installed on front left corner of mounting FT-237, near the antenna terminal. Chest CN-96 is stowed in the vehicle. Installation procedure for SCR-608-B is identical to that for SCR-508.

(4) *Radio set SCR-628-B.* This set is identical to radio set SCR-608-B, except that it has only one radio receiver BC-683-BM.

c. REMOVAL.

(1) *Receiver BC-603, or BC-683-BM.* Loosen the two knurled thumb screws until they are free and can be pushed below the receiver. Grasp the handle located below the tuning controls and pull the receiver away from the mounting FT-237.

(2) *Transmitter BC-604 or BC-684-BM.* Unscrew the two knurled thumb screws in the end of the transmitter which secure the transmitter to the mounting base plate. These are captive screws and are not removed from the transmitter. Grasp the transmitter and withdraw it away from the end of the mounting.

(3) *Mounting FT-237, interphone control boxes, interphone cordage, and power cable.*
   (a) Turn master battery switch off (fig. 209).
   (b) Remove cover plate of the radio terminal box.
   (c) Remove the interphone cordage cables from the terminal strip and carefully tag each wire with its proper terminal number.
   (d) Disconnect the power cord leads in the radio terminal box.
   (e) Unscrew the coupling nuts of the connectors securing the cables to the terminal box, and withdraw the cables from the terminal box.
   (f) Remove cable clips securing the cables to the vehicle.
   (g) Remove interphone control boxes BC-606-A to G by removing the cover of the box and removing the three No. 8 machine screws holding the box to its bracket. Interphone control box BC-606-H is removed by loosening the four hex

nuts on the sides of the box and withdrawing the box from its mounting. Install covers to prevent loss.

(*h*) Remove screws which fasten mounting FT-237 to vehicle. These screws are accessible through the rectangular opening in the bed plate.

(*i*) Disconnect antenna cable from mounting.

(*j*) Carefully coil the interphone cordage and power cable, place interphone control boxes on the mounting, and remove the equipment from the vehicle.

d. INSTALLATION.

(1) *Mounting FT-237, interphone control boxes, interphone cordage, and power cable.*

(*a*) Connect antenna lead to mounting.

(*b*) Fasten mounting FT-237 to vehicle by installing screw through the rectangular openings in the bed plate.

(*c*) Install interphone control boxes BC-606-A to G by removing the cover and installing the three mounting screws. Install cover. Install control box BC-606-H in its mounting after loosening the four nuts on the sides of the box. Tighten nuts to hold box in place.

(*d*) Install the cables in the radio terminal box and secure by fastening the coupling nuts to the terminal box.

(*e*) Install the cable clips which secure the cables to the vehicle.

(*f*) Connect the power cord cables to terminal box.

(*g*) Connect interphone cordage cables to the terminal strip on mounting FT-237. These cables were tagged when removed to facilitate assembly.

(*h*) Install cover plate on radio terminal box.

(2) *Transmitter BC-604 or BC-684-BM.*

(*a*) Place radio transmitter BC-604 or BC-684-BM on mounting FT-237. A rail on the base plate is provided to assist in alining the transmitter with the mounting.

(*b*) Press the transmitter against the rail and push the transmitter toward the closed end of the mounting. Make sure that the guide pins in the end of the mounting engage in the holes in the transmitter and the connector pins on the transmitter are mated into a receptacle on the mounting.

(*c*) Secure the transmitter by screwing the thumb screws at the right end of the transmitter into the taped holes provided in the mounting base plate. Tighten the screws securely.

(3) *Receiver BC-603 or BC-683-BM.*

(*a*) Place radio receiver BC-603 or BC-683-BM on mounting FT-237. Make sure that the guide pins on the wall of the mounting seat fully in the guide pin holes in the receiver and that the connector on the receiver are mated into the receptacle on the mounting.

(b) Tighten thumb screws to secure receiver in place.

(4) *New installation.* If a complete new installation of radio set SCR-508, SCR-528, SCR-608-B, or SCR-628-B is necessary, refer to ordnance maintenance personnel.

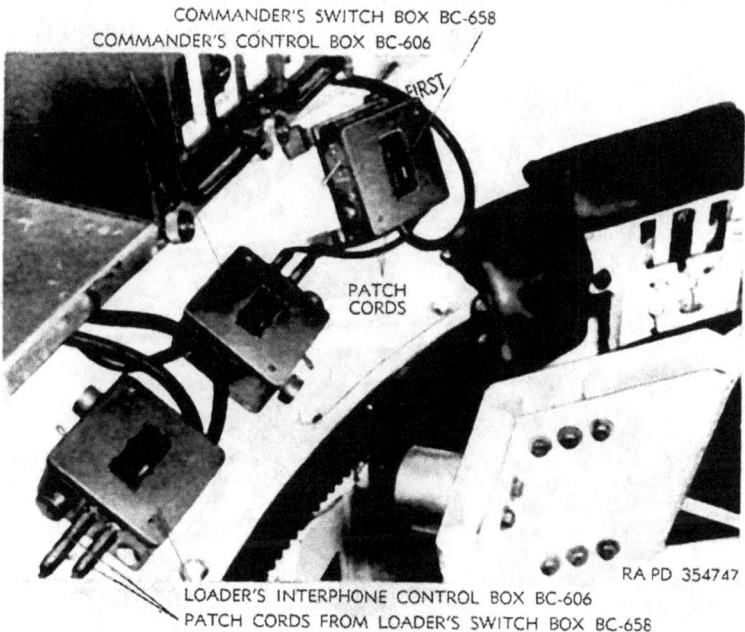

Figure 206. Radio-interphone control units at rear of turret.

250. Radio Set AN/VRC-3

a. GENERAL. Radio set AN/VRC-3 (fig. 207) is the vehicular adaptation of the standard infantry pack type radio communication set known as the "walkie-talkie." It is a two-way set, receiving and transmitting, consisting of a transmitter and receiver in one case. It is used for communication in combat areas between tanks and ground troops.

b. OPERATION. Radio set AN/VRC-3 can be operated either on a battery contained in a case secured to the set or by a vibrator power pack, which is contained in a case similar to the battery case and operates on 24-volt d-c power obtained at the vehicle radio terminal box. Two switch boxes BC-658 are provided to permit the commander and loader to switch their microphones and headsets from the AN/VRC-3 to the interphone system installed in the tank. The commander's switch box (fig. 206) is located on a bracket just forward of the turret overhang and the loader's switch box is located on the right wall of the turret. The switch boxes connect to the regular interphone boxes by means of patch cords plugged into the jacks on the control boxes. The loader's switch box has another pair of patch cords which connect to the AN/VRC-3 (fig. 207). A transformer is wired into the cord leading to the "Phones

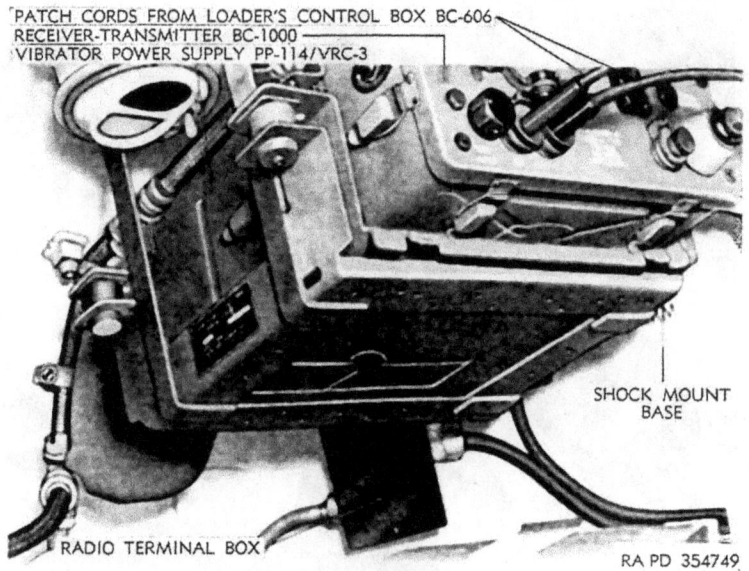

Figure 207. Radio set AN/VRC-3.

No. 1" jack on the radio set. The transformer matches the headset to the radio set output and must be used to obtain best clarity and distinctness from the radio set. The radio set will not operate when there is no plug inserted in the "Phones No. 1" jack, since there is a switch interlock actuated by insertion of the plug. The two switch boxes are connected together by a length of interphone cordage, so that the radio signals are heard in both boxes. The loader and tank commander plug their microphones and headsets into the switch boxes and can operate both the AN/VRC-3 and the interphone by throwing the "RADIO-INT" switch on the switch box to the proper position. The tank commander and the loader are the only two in the vehicle crew who operate the AN/VRC-3. The loader "listens-in" and, when the vehicle is called, switches over to interphone and notifies the commander who then switches over his control box to receive on this set. The vehicle commander is normally the only one who transmits on the AN/VRC–3 radio set, although the loader may transmit when required.

 c. REMOVAL.
  (1) If the vibrator power supply unit is installed (fig. 207), disconnect the power cord from the vehicle radio terminal box and remove cable clips.
  (2) Disconnect the plugs from the jacks on the panel of the radio set.
  (3) Disconnect the antenna lead-in cable. If an antenna matching transformer is used, the cable is disconnected by turning the coupling nut on the lead-in plug in a counterclockwise direction. When the antenna matching transformer is not used, the antenna lead-in cable is disconnected by unscrewing the $\frac{3}{8}$-inch hex head screw in the antenna socket.
  (4) Loosen the two wing nuts which secure the metal strap to the mounting and remove the strap.
  (5) Remove the radio set from the mounting.
  (6) Remove switch box BC–658–A or B by removing the box cover and removing the three screws which secure the box to its bracket. Switch box BC–658–C is removed by loosening the four hex nuts on the sides of the box.
  (7) Remove cable clips securing the interconnecting cordage and patch cords. Remove both boxes together with the cordage.

 d. INSTALLATION.
  (1) Install switch box BC–658–A or B by installing the three screws which secure the box to its bracket. Install switch box BC–658–C by installing the four hex nuts on the side of the box.
  (2) Secure the cordage and patch cords interconnecting the two boxes with cable clips.
  (3) Place radio set on mounting and place metal strap in position.
  (4) Install the two wing nuts which secure the metal strap to the mounting.
  (5) Connect the antenna lead-in cable. If an antenna matching transformer is used, the cable is connected by turning the coupling nut on the lead-in plug in a clockwise direction. When the antenna matching transformer is not used, the antenna lead-in cable is connected by screwing the $\frac{3}{8}$-inch hex-head screw in the antenna socket.
  (6) Install the plugs in the jacks on the panel of the radio sets.
  (7) If the vibrator power supply unit is to be installed, connect the power cord to the turret radio terminal box. Secure cord with cable clips.

e. NEW INSTALLATION. If a complete new installation of radio set AN/VRC-3 is required, refer to ordnance maintenance personnel.

251. Antennas

a. GENERAL. A separate whip type antenna is used for each radio set installed. Mast base AB-15/GR (fig. 208) consists of a flexible stem, feed-through porcelain insulator and provisions on the bottom for connecting the antenna lead-in. The mast sections are of flexible steel tubing. The location and components of each antenna are given in table VI.

Table VI. Location and Components of Antennas

Radio Set	Antenna Location	Mast Base Type No.	Mast Section Type Nos.	Lead-in
SCR-508, SCR-528, SCR-608-B, or SCR-628-B.	On side of turret, left rear.	AB-15/GR	One MS-116, one MS-117, and one MS-118.	8 ft of radio frequency cable WC-562.
AN/VRC-3	On top of turret, right front (fig. 4).	AB-15/GR	One MS-117 and one MS-118.	Cord CG-102/TRC-7 (7 ft long) with terminal box TM-217, or 32-in wire W-128.

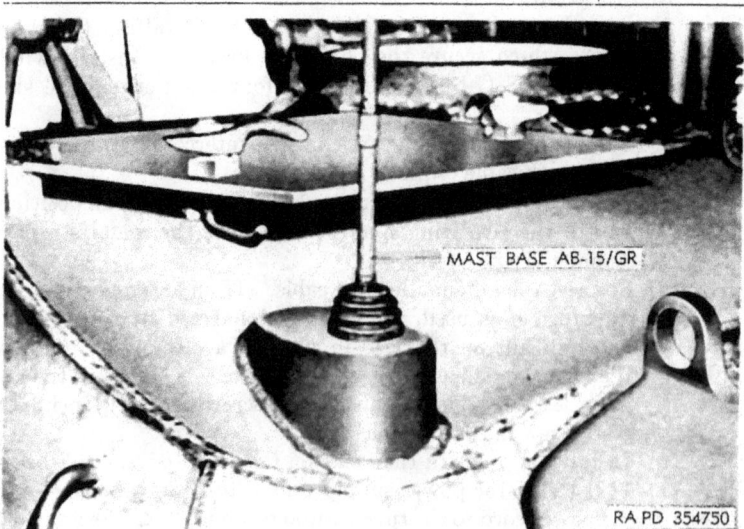

Figure 208. Mast base AB-15/GR.

b. PHANTOM ANTENNAS. Antenna A-62 (phantom), for radio set SCR-508 or SCR-528 and antenna A-83 for SCR-608-B or SCR-628-B are mounted on a bracket attached to the left front end of mounting FT-237.

Warning: The phantom antennas should be used when alining the transmitters of radio sets SCR-508, SCR-528, SCR-608-B, and SCR-628-B under combat conditions, in order to prevent radiation and resultant enemy detection. To use phantom antenna, follow instructions on case.

c. REMOVAL.
  (1) On SCR-508, SCR-528, SCR-608-B, or SCR-628-B mast base, remove top plate of mounting bracket on side of vehicle.
  (2) Unscrew bottom mast section from mast base.
  (3) Separate sections from each other and stow in roll BG-56.
  (4) Disconnect antenna lead-in.
  (5) Hold bottom insulator of mast base stationary and turn flexible stem counterclockwise, until upper and lower sections of mast base can be separated and removed.

d. INSTALLATION.
  (1) Install upper and lower sections of mast base by holding bottom insulator stationary and turning flexible stem clockwise.
  (2) Connect antenna lead-in.
  (3) Take out mast sections from roll BG-56 and connect them to each other.
  (4) Screw bottom mast section to mast base.
  (5) On SCR-508, SCR-528, SCR-608-B, or SCR-628-B mast base, install top plate of mounting bracket on side of vehicle.

252. Interphone System

a. GENERAL. The vehicle is equipped with a five-station interphone system, employing the audio stage of radio transmitter BC-604 or BC-684-BM for voice amplification. The system not only permits conversation between members of the crew, but also enables crew members to individually receive radio orders and messages direct. It consists of the necessary control boxes, headsets, microphones, and one spare hand-type microphone for testing, or for alternate use to obtain greater intelligibility. Each crew member is provided with control box mounted on a bracket at his station and a hook for hanging up his headset and microphone. Headset H-16/U or headset HS-30 and microphone T-45 are used together with chest set TD-4 or cords CD-307-A and CD-318. Both headset and microphone extension wires or cords have a plug at the end for plugging into the jacks in the control boxes. Plugs are of different shapes and sizes to permit quick correct plug-ins and to eliminate the possibility of wrong connections. Each interphone control box BC-606 is provided with a toggle-type switch. This switch is wired in the circuit at the commander's box only and is not used at crew members' positions. A spring holds the switch in the "IN" position on the commander's box. A volume control knob is provided on each box to adjust radio or interphone volume.

Figure 209. Interphone control units in driver's compartment.

 b. REMOVAL AND INSTALLATION. Refer to paragraph 249*c* and *d.*
 c. TESTING INTERPHONE SYSTEM.
  (1) Assemble headsets and microphones with their extension cords. Insert the plugs on cords into jacks on interphone control boxes BC-606 (fig. 209). Turn the volume control on each box to maximum, then turn the control back slightly from maximum.
  (2) Turn the "RADIO-INT" switch on radio transmitter BC-604 to the "RADIO" position. Turn the "OFF-ON" switch on the front panel of radio transmitter BC-604 and radio receiver BC-603 to the "ON" position. Turn the "RADIO & INT-INT ONLY" switch on radio receiver BC-603 to "RADIO & INT." Tune to a strong signal from another radio set located in the vicinity. Adjust the volume of the

radio receiver BC-603 so that the signal received in the headsets is at a comfortable level.

(3) Pressing the button of any of the microphone switches should start the dynamotor of the radio transmitter. Interphone communication should be checked between one box and all other boxes, in turn, by speaking into the microphones at a normal voice level.

(4) The commander's control box (fig. 206) is the only box wired for radio transmission in addition to interphone transmission. To talk on the interphone, the commander presses his microphone switch. To operate the radio transmitter, first press one of the channel selector push buttons on the transmitter turning panel, hold the radio-interphone switch in the radio position, then press the button of the microphone switch before speaking into the microphone. When the radio-interphone switch on the interphone control box is released, the commander's microphone and headset are automatically switched to the interphone system.

(5) At the crew member's positions, pressing the button of the microphone switches should cut off radio transmissions and permit communication to the commander over the interphone system. Releasing the switch reopens the radio circuit.

(6) The commander must set the switch on his switch box BC-658 to "RADIO" to use the AN/VRC-3 or to "INT" to use the SCR-508 and SCR-528.

d. DESCRIPTION OF INTERPHONE EQUIPMENT RC-99. This interphone system consists of an interphone amplifier BC-667, a mounting bracket and five interphone control boxes BC-606 type, interphone cordage CO-213 and microphone, and headsets. The interphone amplifier BC-667 is secured to the mounting which in turn is secured to the radio shelf in the turret overhang (fig. 210). The interphone control boxes are installed in same locations as for radio set SCR-508. The removal and installation procedure for interphone equipment RC-99 is similar to that of the interphone equipment of radio set SCR-508 (par. 249*c* and *d*). Interphone amplifier BC-667 is detached by removing eight screws near the edge of the front panel and removing the four screws holding the cabinet to the mounting bracket in the vehicle.

253. Interphone Extension Kit RC-298

a. GENERAL. Interphone extension kit RC-298 is installed on each tank to permit communication with ground troops outside the tank. This kit consists of external interphone box BC-1362 (fig. 209) and switch box BC-1361 (fig. 6) interconnected with cordage CO-213 encased in flexible conduit. The switch box is mounted at the rear of the

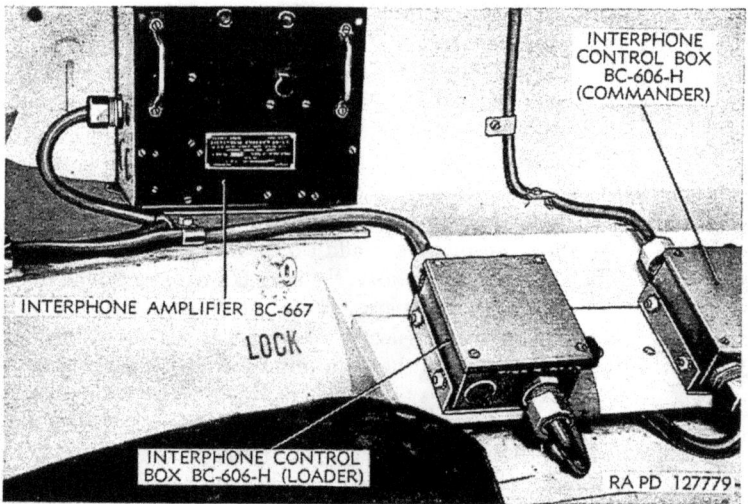

Figure 210. Interphone amplifier BC–667 and interphone control box BC–606–H in turret overhang.

assistant driver's position and makes connection to the vehicle interphone system by a length of cordage CO–213 connected into the vehicle terminal box at the left of the assistant driver's seat. The external interphone box is mounted on a bracket secured by the tail light housing bolts at the right rear of the vehicle.

  b. REMOVAL.

    (1) *External interphone box BC–1362* (fig. 211).

      (*a*) Open door of external interphone box.

      (*b*) Remove screws holding bracket on which volume control is mounted.

      (*c*) Unsolder cordage wires from terminal block and carefully tag each conductor with its proper terminal number to facilitate installation.

      (*d*) Remove coupling nut of cordage connector on bottom of box and pull cordage from box.

      (*e*) Install volume control bracket.

      (*f*) Dismount external interphone box from bracket on vehicle by removing three mounting screws through lugs on outside of box.

    (2) *Switch box BC–1361* (fig. 211).

      (*a*) Remove screws from cover of vehicle radio terminal box in driving compartment.

      (*b*) Remove nuts from terminal block and lift off terminals of cordage leading to switch box without disturbing other terminals on block and install nuts.

(c) Carefully tag cordage with proper terminal numbers for reinstallation.
(d) Remove coupling nut from connector and withdraw cordage from terminal box, with connector coupling nut and grommet attached.
(e) Remove clamps securing cordage to vehicle.
(f) Remove screws from switch box cover and lift off cover.
(g) Unsolder wires of the cordage leading into right battery compartment from terminal block inside switch box, carefully tagging each conductor with its proper terminal number for reinstallation.
(h) Dismount switch box from bracket by removing screws through lugs on box.

(3) *Cordage and flexible conduit.*
(a) Unscrew nut of connector securing short length of conduit to the fitting in tail light housing and withdraw the cordage from the conduit.
(b) Remove right gasoline tank and battery compartment cover plates.
(c) Detach clamps securing interphone conduit.
(d) Detach clamps securing conduit and cordage in crew compartment.

Figure 211. External interphone box BC-1362.

(e) Withdraw cordage from conduit into crew compartment and remove.

(f) Withdraw conduit through grommet in bulkhead into battery compartment and remove.

c. INSTALLATION.

(1) *Cordage and flexible conduit.*

(a) Pull conduit through grommet in bulkhead into battery compartment.

(b) Pull cordage out of crew compartment and through conduit.

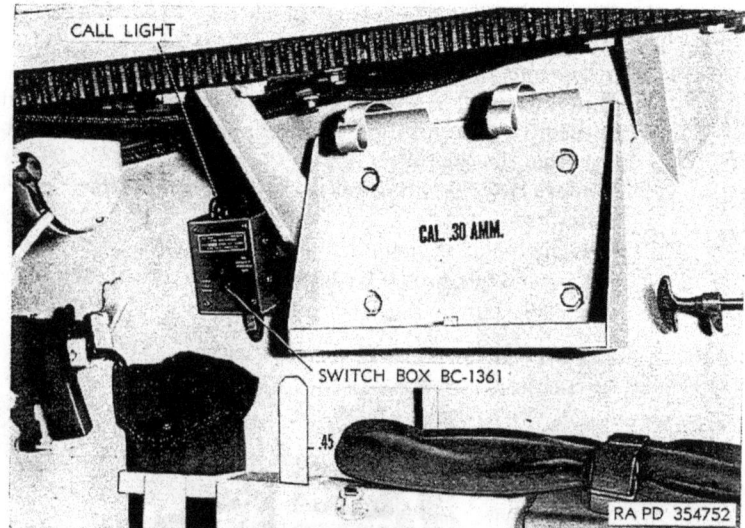

Figure 212. Switch box BC-1361.

(c) Secure clamps to conduit and cordage in crew compartment.

(d) Secure clamps to interphone conduit.

(e) Install battery compartment cover plates and right gasoline tank.

(f) Pull cordage through conduit and screw nut of connector to secure short length of conduit to the fitting in tail light housing.

(2) *Switch box BC-1361* (fig. 212).

(a) Place switch box on bracket and secure with screws.

(b) Solder wires of cordage leading into right battery compartment after making connections called for by tags on conductors.

(c) Install cover on switch box.
(d) Secure cordage to vehicle with clamps.
(e) Insert cordage into terminal box and install coupling nut on connector.
(f) Remove nuts from terminal block and connect cordage leading to switch box as called for by tags on cordage.
(g) Secure cover of radio terminal with screws.

(3) *External interphone box BC-1362* (fig. 211).
  (a) Install external interphone box on bracket by securing it with three mounting screws.
  (b) Remove volume control bracket.
  (c) Insert cordage into bottom of box and secure coupling nut of cordage connector.
  (d) Connect each connector as called for by tags and solder cordage wires to terminal block.
  (e) Install volume control bracket.
  (f) Close door of external interphone box.

254. Inspections

a. GENERAL. To insure the continued and efficient operation of the radios and interphone system, frequent and careful inspections must be made of the various units and wiring.

b. RADIO MOUNTING. Inspect radio mounting screws to see that they are tight and that shock mountings are in good condition. Rock units to determine if they bump against anything. Be sure all connections are clean, dry, and tight and that wiring is in good condition. Make sure units are securely fastened to mounting.

c. ANTENNA.
  (1) *Mast sections.* Inspect antenna mast sections. Be sure they are screwed securely together and are not damaged or excessively bent.
  (2) *Flexible stem.* Inspect flexible upright stem of mast base. Be sure it maintains a vertical position and is not damaged to prevent flexibility.
  (3) *Porcelain insulators.* See that both upper and lower insulators are present, free of paint, grease, or other foreign matter, and not cracked or damaged.
  (4) *Leads to sets.* Inspect leads to radios. Be sure nothing interferes with them or that there is nothing present which could damage the wires. Be sure connections are clean and tight.

d. WIRES AND CONNECTIONS. Inspect interphone and power cables which connect the radio sets or interphone amplifier. Make sure they are not damaged and that they are properly secured in clips. Inspect all connections for tightness. Make sure headset and microphone cords are not twisted or knotted to prevent free movement. Inspect plugs to see that they are not damaged or bent.

e. INTERPHONE CONTROL BOXES, MICROPHONES, AND HEADSETS. Inspect control boxes for damage, looseness, and correct operation. Check headsets and microphones periodically to insure proper operating condition, by listening to each of the headset receivers independently while someone is speaking into the interphone system. Both receivers should be approximately the same strength. If the entire headset response is believed to be weak, it may be compared with that of another headset known to be good.

Caution: Exercise care in the operation of the interphone system to prevent damage to the headsets. Continued chattering of the headsets by excessive volume output will damage them if it happens over a long period of time.

Hang headsets and microphones on hooks provided in vehicle, when not in use. When not used for an extended period of time, remove from vehicle and store in a dry place.

f. COVERS. Be sure all covers for the protection of the radios are present in the vehicle. Install covers when equipment is not in use. See that zippers and fasteners are in good condition and operate properly.

255. Precautions

a. ANTENNA. Tie antenna back securely, when vehicle is in motion and radio is not in use, to prevent damage to antenna. Make sure antenna is vertical and not touching anything when radio is in use. Disjoint antenna as necessary when aboard decked landing craft.

b. RADIO.

(1) Keep radios covered, when not in use, to prevent entrance of dust and moisture. Keep all hatch doors closed and securely fastened.

(2) Turn off all radio switches when sets are not in use. Do not turn off 24-volt battery switch with radio switch on.

(3) Do not store equipment on or behind radios where it can prevent action of shock mountings or damage connections.

(4) Do not start engine with radio switch on.

(5) Do not attempt to operate radio transmitter BC–604 unless one of the push buttons on the transmitter control panel is depressed.

(6) Do not operate radio receiver BC–603 or receiver of BC–1000 with squelch switch on, unless squelch control is carefully adjusted in accordance with instructions in the technical manual included with each radio set.

(7) Carefully read the operating instructions in the manuals of each unit of equipment before attempting to operate the communication system.

c. BATTERIES AND GENERATING SYSTEM.

(1) Make sure batteries are properly charged at all times to ensure satisfactory operation of the radios. Low batteries will cause sets to be noisy, which will result in poor reception.

(2) See that all battery terminals are in good condition, clean, and tight.

(3) Test operation of generator and regulator (par. 69). Excessive charge will damage radio and make it noisy.

(4) Do not race engine while radio is in operation.

d. INTERPHONE EQUIPMENT. This equipment is sensitive and delicate. Do not drop or misuse it. Keep microphones and headsets on hooks when not in use. Keep cords clear of moving equipment, parts, and doors.

CHAPTER 5

SHIPMENT AND LIMITED STORAGE AND DESTRUCTION TO PREVENT ENEMY USE

Section I. SHIPMENT AND LIMITED STORAGE

256. Domestic Shipping Instructions

a. PREPARATION FOR SHIPMENT IN ZONE OF INTERIOR. When shipping the light tank M24 interstate or within the zone of interior, except directly to port of embarkation, the officer in charge of preparing the shipment will be responsible for furnishing vehicles to the carriers for transport in a serviceable condition, properly cleaned, preserved, painted, lubricated, etc., as prescribed in SB 9-4.

Note. For loading and blocking instructions of vehicles on freight cars, refer to paragraphs 258 and 259.

b. PREPARATION FOR SHIPMENT TO PORTS.
   (1) *Inspection.* All used vehicles destined for oversea use will be inspected prior to shipment in accordance with TB ORD 385.
   (2) *Processing for shipment to ports.* All vehicles destined to ports of embarkation for oversea shipment will be further processed in accordance with SB 9-4.

   Note. Ports of embarkation will supplement any necessary or previously omitted processing upon receipt of vehicle.

c. REMOVAL OF PRESERVATIVES FOR SHIPMENT. Personnel withdrawing vehicles from a limited storage status for domestic shipment *must not remove preservatives*, other than to insure that the matériel is complete and serviceable. If it has been determined that preservatives have been removed, they must be restored prior to domestic shipment. The removal of preservatives is the responsibility of depots, ports, or field installations (posts, camps, and stations) receiving the shipments.

d. ARMY SHIPPING DOCUMENTS. Prepare all Army shipping documents accompanying freight in accordance with TM 38-705.

e. ARTILLERY GUN BOOK. During transfer or shipment, place the artillery gun book in the gun book envelope and secure to the top of the breech mechanism with water-resistant, nonhygroscopic, adhesive tape. Under one of the wrappings of tape, insert one end of a tab reading "GUN BOOK HERE."

257. Limited Storage Instructions

a. GENERAL.
  (1) Vehicles received already processed for domestic shipment is indicated on DA AGO Form 9-3, need not be reprocessed unless the inspection performed on receipt of vehicles reveals corrosion, deterioration, etc.
  (2) Completely process vehicle if the processing data recorded on the tag indicates that vehicle has been rendered ineffective by operation, freight shipping damage, or upon receipt of vehicles directly from manufacturing facilities.
  (3) Vehicles to be prepared for limited storage must be given a limited technical inspection and processed as prescribed in SB 9-63. The results and classification of vehicle will be entered on DA AGO Form 461-5.

b. RECEIVING INSPECTIONS.
  (1) Report of vehicles received in a damaged condition or improperly prepared for shipment will be reported on DD Form 6 in accordance with SR 745-45-5.
  (2) When vehicles are inactivated, they are to be placed in a limited storage status for periods not to exceed 90 days. Stand-by storage for periods in excess of 90 days will normally be handled by ordnance maintenance personnel only.
  (3) Immediately upon receipt of vehicles, inspect and service as prescribed in chapter 2, section I. Perform a systematic inspection and replace or repair all missing or broken parts. If repairs are beyond the scope of the unit and the vehicle will be out-of-service for an appreciable length of time, place vehicles in a limited storage status and attach a tag to the vehicles specifying the repairs needed. The report of these conditions will be submitted by the unit commander for action by an ordnance maintenance unit.

c. INSPECTIONS DURING STORAGE. Perform a visual inspection periodically to determine general condition. If corrosion is found on any part, remove the rust spots, clean, paint, and treat with the prescribed preservatives.

Note. Touch-up painting will be in accordance with TM 9-2851.

d. REMOVAL FROM LIMITED STORAGE.
  (1) If the vehicles are not shipped or issued upon expiration of the limited storage period, vehicles may either be processed

for another limited storage period or be further treated for stand-by storage (vehicles inactivated for periods in excess of 90 days up to 3 years) by ordnance maintenance personnel.

(2) If vehicles to be shipped will reach their destination within the scope of the limited storage period, they need not be reprocessed upon removal from storage unless inspection reveals it to be necessary according to anticipated in-transit weather conditions.

> *Note.* All used vehicles that are to be reissued to troops within the continental limits of the United States will be inspected prior to shipment or issue in accordance with TB ORD 385.

(3) Deprocess vehicles when it has been ascertained that they are to be placed into immediate service. Remove all rust-preventive compounds and thoroughly lubricate as prescribed in chapter 3, section II. Inspect and service vehicles as prescribed in chapter 2, section I.

(4) Repair and/or replace all items tagged in accordance with *b*(3) above.

e. STORAGE SITE. The preferred type of storage for vehicles is under cover in open sheds or warehouses whenever possible. Where it is found necessary to store vehicles outdoors, they must be protected against the elements as prescribed in SB 9–47.

258. Loading the Vehicles for Rail Shipment

a. PREPARATION.

(1) When vehicles are shipped by rail, every precaution must be taken to see that they are properly loaded and securely fastened and blocked to the floor of car. All "on vehicle matériel" (OVM) will be thoroughly cleaned, preserved, packed (boxed or crated), labeled, and securely stowed in or on the vehicle during transit.

(2) Prepare all vehicles for rail shipment in accordance with paragraph 252*a*. In addition take the following precautions:

(*a*) Disconnect the batteries to prevent their discharge by vandalism or accident. This is accomplished by disconnecting the positive leads, taping the ends, and tying them back away from the batteries.

> *Note.* Before removing any cables, turn master relay switch to "OFF" position.

(*b*) Apply the parking brakes and place the transmission in neutral position after the vehicle has been finally spotted on the flatcar.

> *Note.* The vehicles must be loaded on the car in such a manner as to prevent the car from carrying an unbalanced load.

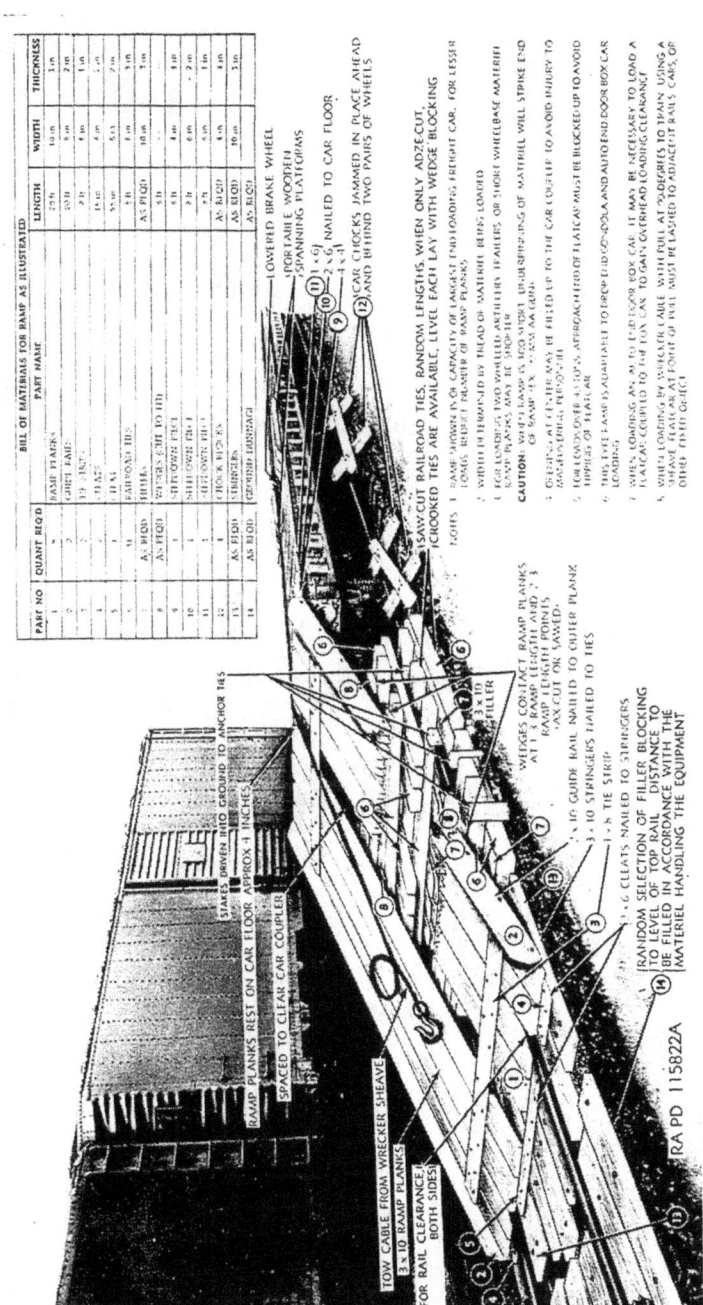

Figure 213. Construction of improvised loading ramp.

Caution: If matériel is equipped with steel tool boxes, all padlocks and keys will be removed from the matériel in order to prevent pilferage while matériel is in transit. Lids of steel tool boxes will be secured by wiring the hasp to prevent damage during shipment. Padlocks and keys will be preserved with preservative engine oil (grade 1) and wrapped in greaseproof-barrier material for domestic shipment. For oversea shipment the item will be sealed in a waterproof-greaseproof wrapping or bag. Locate all wrapped padlocks and keys in the shipping container with the accessories.

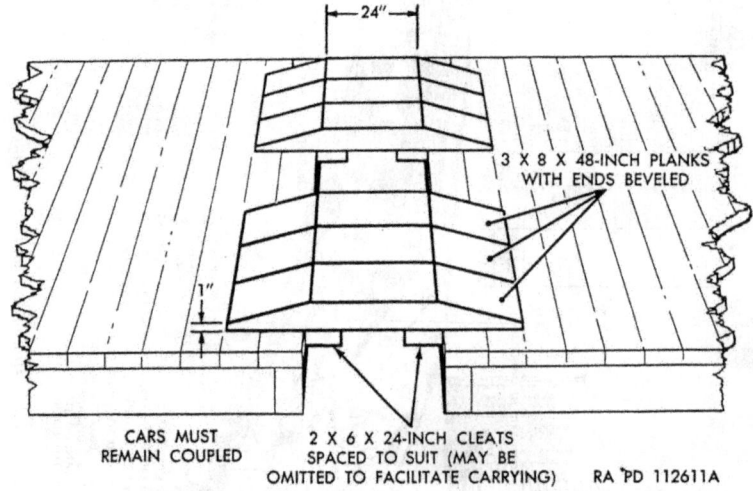

Figure 214. Construction of improvised spanning platforms.

 b. TYPES OF CARS. Instructions contained herein pertain to the loading of vehicles on flatcars (cars with wooden floors laid over sills and without sides or ends, but equipped with stake pockets).

 c. METHOD OF LOADING VEHICLES.

  (1) Vehicles will be loaded and unloaded with the use of hoisting equipment when available. When suitable hoisting equipment is not available for loading or unloading vehicles, an end ramp must be used in cases where the vehicle is not on a level with the flatcar deck. Vehicles on a warehouse platform or loading dock can be pivoted over spanning platforms aboard a flatcar spotted adjacent to the platform, then again pivoted into lateral position on the flatcar.

  (2) When vehicles must be loaded from ground level, a ramp may be improvised (*d* below) by borrowing railroad ties normally

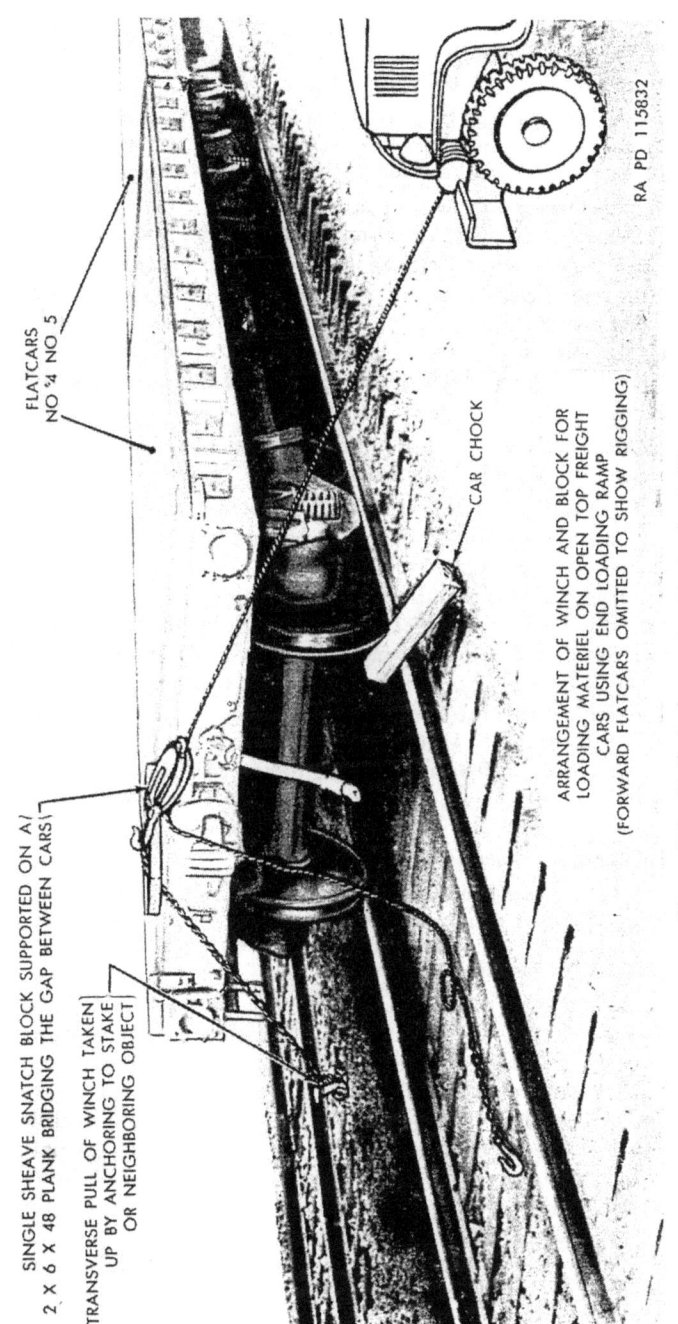

Figure 215. Method of powering the towing cable.

found stacked in railroad yards and by procuring necessary planking. An efficient end ramp is shown in place in figure 213.

Note. Railroad ties alone, stacked without deck planking and not securely anchored, provide a very unstable ramp and should not be attempted except under conditions of extreme emergency.

(3) Vehicles which can be loaded under their own power will be driven on to improvised apron at base of ramp and then be carefully guided up ramp to their positions on the flatcar.

(4) To load vehicles which cannot be operated due to processing, tow onto the improvised apron at base of ramp and unhitch. Using a cable laid along the center line of the flatcar, attached to vehicle, the vehicle is pivoted to point towards the ramp.

Caution: Follow up forward movement of the vehicle by chocking behind tracks on the ramp.

(5) For powering the towing cable, a 6-ton wrecker using rear winch is substituted for the vehicle illustrated in figure 215. If the capacity of the wrecker rear winch is insufficient for the load, additional pulleys may be employed. The wrecker is spotted at *right angles* to the train, located at about the third or fourth flatcar to facilitate signaling and because of cable length limits. A single-sheave snatch block located between cars on the train center line will provide the necessary *lateral* pull. Vehicles passing this point can be towed by a vehicle on the ground with personnel guiding its passage. A long tow cable from the towing vehicle will lessen the tendency of the vehicle to stray from the center line of the train.

Note. The snatch block fastening chain must be lashed to an adjacent fixed object or stake to offset the cross pull of the powered winch (fig. 215). Snatch block movement is allowed for low front winches and high rear (wrecker) winches.

(6) After the first vehicle is loaded on the flatcar, additional vehicles may be similarly loaded by passing the towing cable beneath the loaded vehicle. When a train of flatcars is being loaded, steel or wooden spanning platforms or bridges are used to cover the gaps between cars. Flatcar brake wheels must first be lowered to floor level to permit passage. A pair of improvised spanning platforms is shown in place in figure 214. These spanning platforms are moved along the train by hand as the vehicle advances.

(7) The above method of train loading requires careful advance planning as to the order of loading, so that vehicles are ar-

ranged on each flatcar under prescribed methods and combinations.

d. LOADING RAMP.

(1) A ramp for end-loading of vehicles on flatcars may be improvised when no permanent ramps or hoisting facilities are available. An ideal ramp suitable for the loading of most ordnance items is shown in figure 212.

Caution: Personnel guiding the vehicle up the ramp must exercise care when working close to the edges of the ramp planking.

(2) The flatcar bearing the ramp must be securely blocked against rolling, particularly when the car brakes are not applied as in train loading. Successive cars must remain coupled and be additionally chocked at several points along the train when ground towing of vehicles aboard the train is being effected.

(3) Whenever the flatcars are not on an isolated track or blocked siding, each end approach to the train must be placarded (blue flag or light) to advise that men are at work and that the siding may not be entered beyond those points.

(4) Upon completion of the loading operation, the ramp planks and bridging devices should be loaded on the train for use in unloading operations. Random sizes of timbers used in building the approach apron up to rail level should be included. All materials should be securely fastened to the car floors after vehicles are blocked in place and entered upon the bill of lading (B/L). Railroad ties borrowed for the operation need not be forwarded to the unloading point unless specifically required and only with the consent of the owner.

e. LOADING RULES. For general loading rules pertaining to rail shipment of ordnance vehicles, refer to TB 9–OSSC–G.

Warning: The height and width of vehicles when prepared for rail transportation must not exceed the limitations indicated by the loading table as prescribed in section II, AR 700–105. Whenever possible, local transportation officers must be consulted about the limitations of the particular railroad lines to be used for the movement to avoid delays, danger, or damage to equipment.

259. Blocking the Vehicles for Rail Shipment

a. GENERAL. All blocking instructions specified herein are minimum and are in accordance with Association of American Railroads "Rules Governing the Loading of Commodities on Open Top Cars." Additional blocking may be added as required at the discretion of the officer in charge. Double-headed nails may be used if available, except in the lower piece of two-piece cleats. All item reference

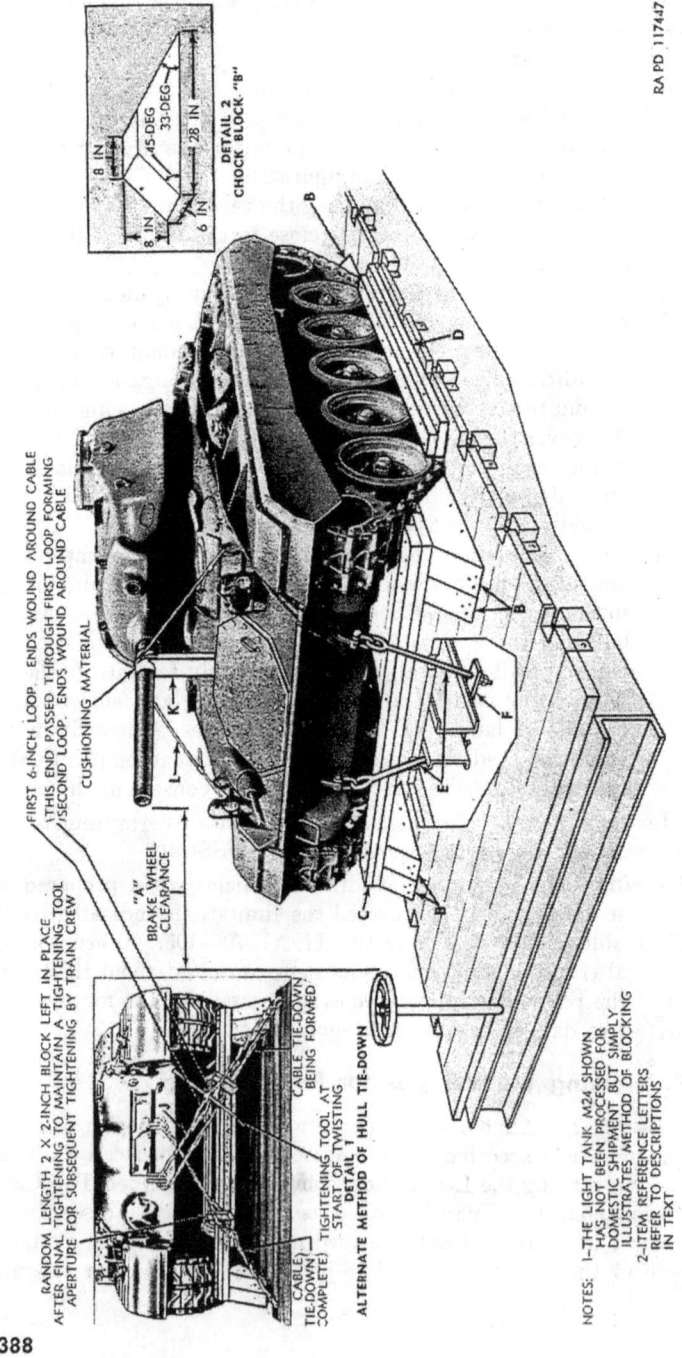

Figure 216. Method of blocking light tank M24 for rail shipment.

letters given below refer to the details and locations as shown in the figures.

Note. Any loading methods or instructions developed by any source which appear in conflict with this publication or existing loading rules of the carriers, must be submitted to the Chief of Ordnance, Washington 25, D. C., for approval.

b. BRAKE WHEEL CLEARANCE "A." Load vehicles on cars with a minimum clearance of at least 4 inches below and 6 inches above, behind, and to each side of the brake wheel (fig. 216). Increase clearance as much as is consistent with proper location of load.

Note. When loading the light tank M24 on flatcars, less than 10 feet in width, track inside cleats and spacers must be used (*f* and *g* below). Locate and nail track inside cleats to car floor prior to loading. This is accomplished by measuring the width between inside of tracks and applying chalk marks on car floor to establish exact location of cleats.

c. CHOCK BLOCKS "B," FIGURE 4, DETAIL 2 (6 x 8 x 28-INCH, EIGHT REQUIRED). Locate the 33-degree surface of two chock blocks on car floor against the front of each track. Locate the 45-degree surface of two chock blocks against the rear of each track. Nail heel of chock blocks to car floor with three fortypenny nails. Toenail sides of each chock block to car floor with two fortypenny nails.

d. CROSS CLEATS "C" (2 x 6-INCH AND 2 x 4-INCH, LENGTH TO SUIT, TWO EACH REQUIRED). Locate the two 2 x 6-inch cleats on top of chock blocks against track. Nail to each chock block with three thirtypenny nails. Place the 2 x 4-inch cleats on top of the 2 x 6-inch cleats with one edge against the track and nail to lower cleats into each chock block with two fortypenny nails.

e. TRACK OUTSIDE CLEATS "D" (2 x 6-INCH AND 2 x 4-INCH, LENGTH TO SUIT, TWO EACH REQUIRED). Locate a 2 x 6-inch cleat against outside of each track on car floor. Nail cleats to car floor with thirtypenny nails staggered along their length. Place a 2 x 4-inch cleat against track on top of the 2 x 6-inch cleats. Nail to lower cleats and car floor with fortypenny nails staggered along their length.

Note. Paragraphs *f* and *g* below will be used when loading on cars less than 10 feet in width (fig. 216 detail 1).

f. TRACK INSIDE CLEATS "G" (2 x 6-INCH AND 2 x 4-INCH, LENGTH TO SUIT, TWO EACH REQUIRED.) Locate the track inside cleats on car floor prior to loading vehicle. Nail the 2 x 6-inch cleats and the 2 x 4-inch cleats with the outside edges located so that after the vehicle is loaded, the outside edges will lie flush against the inside edges of track.

g. TRACK INSIDE SPACERS "H" (2 x 4-INCH, LENGTH TO SUIT, TEN REQUIRED). Locate five track inside spacers on car floor equidistant along the length and between cleats "G" with ends butting inside of lower cleats "G." Nail to car floor with thirtypenny nails staggered along their length. Place a spacer on each of the five spacers with ends butting inside of upper cleats "G." Nail to lower spacers and car floor with fortypenny nails staggered along their length.

h. HULL TIE-DOWN, FIGURE 216, DETAIL 1 (EYEBOLTS "E", 1-INCH DIAMETER, LENGTH TO SUIT, FOUR REQUIRED; CLAMPING BAR "F," 1 x 3-INCH, LENGTH TO SUIT, TWO REQUIRED). Locate and drill holes through car floor, front and rear of tank, where eyebolts will penetrate to engage in clamping bar "F." Insert threaded ends of eyebolts through holes in ends of clamping bars and secure eye to clevises inserted in towing lugs. Locate clamping bars "F" under center sill of flatcar. Insert threaded ends of eyebolts through clamping bar and secure with standard washers and 1-inch hex and jam nuts.

i. GUN SUPPORT "K" (2 x 8 INCH, LENGTH TO SUIT). Cut one end of support to conform with the contour of gun tube. Bevel other end to conform with contour of tanks. Wrap the portion of gun tube, which will contact support, with cushioning material before support is positioned. Locate support between turret gun and tank and secure with a strip of $1\frac{1}{4}$ x 0.035-inch hot-rolled steel banding placed over cushioning material and nailed to each side of support with sixpenny nails. Pass an end of another length of $1\frac{1}{4}$ x 0.035-inch hot-rolled steel "L" through one of the lifting rings on top of tank and seal band about 12 inches from lifting ring. Extend other end across top of gun tube on top of cushioning material through the lifting ring on the opposite side of tank. Seal other end about 12 inches from ring.

Note. Turret gun must be in traveling position, and turret lock handwheel and elevating mechanism handwheel must be wired to prevent rotating.

j. HULL TIE-DOWN "J," ALTERNATE METHOD, FIGURE 216, DETAIL 1 (SIX STRANDS NO. 8 GAGE BLACK ANNEALED WIRE). Twist-tie the six strands of wire together to form a cable. Pass one end of cable through a towing lug on front right side of hull. Extend end beyond half the distance to a stake pocket left side forward of tank. Pass other end of cable through the stake pocket on left side of tank and form a 6-inch loop in end, twisting each strand of wire *tightly* around cable. Make certain that the loop is positioned above the span of the cables. Insert the free end of cable through the loop hand tight and form another 6-inch loop in the free end. Insert a tightening tool in one of the loops. Place a random length 2 x 2-inch block between cables. Twist-tie cables with tightening tool just enough to take up slack. Keep random block intact for subsequent tightening by train crew when required. Repeat operation for other side of hull, cross-wiring tank to opposite side of flatcar. Strap rear end of tank as described above.

Section II. DESTRUCTION OF MATERIEL TO PREVENT ENEMY USE

260. General

a. Destruction of the vehicle and equipment when subject to capture or abandonment in the combat zone, will be undertaken by the using arm only when, in the judgment of the unit commander, such action is necessary in accordance with orders of, or policy established by, the army commander.

b. The information which follows is for guidance only. Certain of the procedures outlined require the use of explosives and incendiary grenades which normally may not be authorized items for the vehicle. The issue of these and related materials and the conditions under which destruction will be effected are command decisions in each case, according to the tactical situation. Of the several means of destruction, those most generally applicable are—

Mechanical	Requires ax, pick mattock, sledge, crowbar or similar implement.
Burning	Requires gasoline, oil, incendiary grenades, or other inflammables.
Demolition	Requires suitable explosives or ammunition.
Gun fire	Includes artillery, machine guns, hand and rifle grenades, and antitank rockets.

In general, destruction of essential parts, followed by burning, will usually be sufficient to render the matériel useless. However, selection of the particular method of destruction requires imagination and resourcefulness in the utilization of the facilities at hand under the existing conditions. Time is usually critical.

c. If destruction to prevent enemy use is resorted to, the matériel must be so badly damaged that it cannot be restored to a usable condition in the combat zone either by repair or cannibalization. Adequate destruction requires that all parts essential to the operation of the matériel, including essential spare parts, be destroyed or damaged beyond repair. However, when lack of time and personnel prevents destruction of all parts, priority is given to the destruction of those parts most difficult to replace. Equally important, the same essential parts must be destroyed on all like matériel so that the enemy cannot construct one complete operating unit from several damaged ones.

d. If destruction is directed, due consideration should be given to—
    (1) Selection of a point of destruction that will cause greatest obstruction to enemy movement and also prevent hazard to friendly troops from fragments or ricocheting projectiles which may occur incidental to the destruction.
    (2) Observance of appropriate safety precautions.

261. Evacuation of Sighting and Fire Control Equipment and Machine Guns

a. SIGHTING AND FIRE CONTROL EQUIPMENT. All items of sighting and fire control equipment, including such items as periscopes, telescopes, binoculars, and azimuth indicators, are costly, difficult to replace, yet relatively light, hence, whenever practicable, they should be conserved and evacuated rather than destroyed. In the event of subsequent abandonment, the equipment will be completely destroyed and all optical elements and mounts, smashed; firing tables, trajectory charts, and inflammable items, burned.

b. MACHINE GUNS. If machine guns are removed for protection of troops during withdrawal and are subsequently abandoned, they will be destroyed as indicated in FM 23–55 for cal. .30 and FM 23–65 for cal. .50 machine guns.

262. Method No. 1—Destruction by Burning

a. SIGHTING AND FIRE CONTROL EQUIPMENT. Evacuate or smash with sledge carried as pioneer equipment.

b. COMMUNICATIONS EQUIPMENT. Smash. Burn any circuit or wiring diagrams and performance data.

c. CAL. .30 AND CAL. .50 MACHINE GUNS. Evacuate or smash.

d. 75-MM GUN M6 AND COMBINATION GUN MOUNT M64. If time permits, open drain plug on recoil mechanism and allow recoil fluid to drain. With gun at near zero elevation, jam a fuzed HE round (fuze set SQ) in the muzzle. Load the gun with a fuzed HE round (fuze set SQ). Jam breech half open to facilitate ignition from fire in tank if it is positive that a very hot fire can be started and if time permits to continue with destruction of the vehicle. Otherwise, close the breech and fire the weapon from cover, using a lanyard about 100 feet long. The danger area is approximately 200 yards. Elapsed time: about 3 minutes.

e. TANK.
   (1) Remove and empty portable fire extinguishers. Discharge fixed fire extinguisher system.
   (2) Drain fuel tanks and collect fuel for use as described in (6) below. Puncture tanks and allow to leak to ground.
   (3) If combustible is limited in quantity, smash spark plugs, distributor, carburetors, and levers.
   (4) Smash turret control switch box. Smash stabilizer control box and sever oil lines. If time permits, smash engine block, crankcase, and transmission.
   (5) Remove protection from ammunition. Leave loose rounds on floor or on gun mount.

(6) With all doors and hatches open to admit air for combustion, pour gasoline and oil in and over entire vehicle. Slosh gun mount.

(7) With all personnel clear, ignite and take cover. Elapsed time for crew: about 10 minutes.

263. Method No. 2—Destruction by Demolition

Note. For the successful execution of this method, all personnel concerned should be thoroughly familiar with provisions of FM 5–25. Training and careful planning are essential.

 a. SIGHTING AND FIRE CONTROL EQUIPMENT. Evacuate or smash with sledge carried as pioneer equipment.

 b. COMMUNICATIONS EQUIPMENT. Smash. Burn any circuit or wiring diagrams and performance data.

 c. CAL. .30 AND CAL. .50 MACHINE GUNS. Evacuate or smash.

 d. 75-MM GUN M6 AND COMBINATION GUN MOUNT M64. Insert a 2-pound charge of explosive in the muzzle end of the tube and a 4-pound charge in the chamber, after preparing the charges for simultaneous detonation by the use of detonating cord (primacord). Place a 2-pound charge on top of the elevating mechanism, adjacent to the recoil mechanism. Connect this charge to those in the tube for simultaneous detonation. Insert an object such as a hammer handle in the breech opening. Close the breechblock upon this. The obstruction will prevent full closing and hence severing of the detonating cord. Plug the muzzle tightly with mud, rags, etc. to a distance of approximately 9 inches. If time permits, continue with destruction of the vehicle; if not, destroy the armament first (*f* below). Elapsed time: about 3 minutes.

 e. TANK.

  (1) Remove and drain the portable fire extinguishers. Discharge the fixed fire extinguisher system.

  (2) Drain the fuel tanks or puncture them as near the bottom as possible.

  (3) Smash vital elements indicated in method No. 1.

  (4) Prepare and place a 3-pound charge on each of the two clutch housings. The charge should touch the crankcase. Place a 2-pound charge between each engine and adjacent fuel tank. Connect all with primacord.

  (5) For tracks, if sufficient time and materials are available, prepare and place a 2-pound charge at the center of each track assembly. Connect the two charges with detonating cord.

 f. DETONATION. All charges should be connected for simultaneous detonation. Dual priming should be provided to minimize the possibility of a misfire. For priming, either a nonelectric blasting cap crimped to a length of safety fuse or an electric blasting cap and firing wire may be used. The safety fuse may be ignited by a fuse lighter or a match. The electric blasting cap requires a blasting machine or equivalent source of electricity. Prepare and detonate the charges as prescribed in FM 5-25. The danger area is approximately 200 yards. Elapsed time: about 10 minutes.

264. Method No. 3—Destruction by Gunfire

a. This method cannot be relied upon to destroy the same parts in different tanks nor produce the same degree of destruction unless an intense fire results from the gunfire.

b. Remove and empty portable fire extinguishers and discharge the fixed fire extinguisher system.

c. Drain the fuel tanks or puncture them as low on the fuel tank as possible.

d. Smash all vital elements as outlined in method No. 1.

e. Fire on the tank, using adjacent gun motor carriages, tanks, antitank weapons, other artillery, or grenades. Aim at the engine compartment, suspension, and armament in the order named. If an intense fire is started, the vehicle may be considered destroyed. Elapsed time: about 6 minutes.

f. Unless evacuated, destroy the last remaining tank or weapon by the best means possible.

APPENDIX

REFERENCES

1. Publication Indexes

The following publication indexes and lists of current issue should be consulted frequently for latest changes or revisions of references given in this appendix and for new publications relating to matériel covered in this manual:

Index of Administrative Publications	SR 310-20-5
Index of Army Motion Pictures and Film Strips.	SR 110-1-1
Index of Army Training Publications (Field Manuals, Training Circulars, Firing Tables and Charts, Army Training Programs, Mobilization Training Programs, Graphic Training Aids, Joint Army-Navy-Air Force Publications, and Combined Communications Board Publications.)	SR 310-20-3
Index of Blank Forms and Army Personnel Classification Tests.	SR 310-20-6
Index of Technical Manuals, Technical Regulations, Technical Bulletins, Supply Bulletins, Lubrication Orders, Modification Work Order, Tables of Organization and Equipment, Reduction Tables, Tables of Allowances, Tables of Organizations, Tables of Equipment.	SR 310-20-4
Introduction and Index	ORD 1
Military Training Aids	FM 21-8

2. Supply Catalogs

The following catalogs of the Department of the Army Supply Catalog pertain to this matériel:

a. AMMUNITION.

Ammunition, Blank, for Pack, Light and Medium Field, Tank, and Antitank Artillery.	ORD 11 SNL R-5

Ammunition, Fixed and Semifixed, Including Subcaliber, for Pack, Light and Medium Field, Aircraft, Tank, and Antitank Artillery, Including Complete Round Data.	ORD 11 SNL R-1
Ammunition, Instruction Matériel for Pack, Light and Medium Field, Aircraft, Tank, and Antitank Artillery.	ORD 11 SNL R-6
Ammunition, Rifle, Carbine, and Automatic Gun.	ORD 11 SNL T-1
Service Fuzes and Primers for Pack, Light and Medium Field, Aircraft, Tank, and Antitank Artillery.	ORD 11 SNL R-3
b. ARMAMENT.	
Gun, 75-mm, M6; and Mount, Combination Gun, M64, (T90).	ORD (*) SNL C-66
Gun, Machine, Cal. .30, Browning, M1919A4, Fixed and Flexible; M1919A5, Fixed; M1919A6; and Ground Mounts.	ORD (*) SNL A-6
Gun, Machine, Cal. .50, Browning, M2, Heavy Barrel, Fixed and Flexible; and Ground Mounts.	ORD (*) SNL A-39
Mount, Machine Gun, AA, Cal. .50 (D80030)	ORD (*) SNL A-55, Sec. 38
Stabilizers (All types)	ORD (*) SNL C-56
c. DESTRUCTION TO PREVENT ENEMY USE.	
Land Mines and Fuzes, Demolition Material, and Ammunition for Simulated Artillery and Grenade Fire.	ORD 11 SNL R-7
d. MAINTENANCE AND REPAIR.	
Cleaners, Preservatives, Lubricants, Recoil Fluids, Special Oils, and Related Maintenance Materials.	ORD 3 SNL K-1
Items of Soldering, Metallizing, Brazing, and Welding Materials; Gases and Related Items.	ORD 3 SNL K-2
Lubricating Equipment, Accessories, and Related Dispensers.	ORD (*) SNL K-3
Tool-Sets (Common), Specialists' and Organizational.	ORD 6 SNL G-27, Sec. 2
e. SIGHTING AND FIRE CONTROL EQUIPMENT.	
Mount, Periscope, M66	ORD (*) SNL F-312
Mount, Telescope, M65	ORD (*) SNL F-296

See footnote at end of table.

Periscope M4A1, M6, M13, M15, M16, M10P   ORD (*) SNL F-235
Quadrant, Elevation, M9_____   ORD (*) SNL F-281
Setter, Fuze, M14, M27_____   ORD (*) SNL F-245
Telescope, M71K, M83F_____   ORD (*) SNL F-235
  f. VEHICLE.
Tank, Light, M24_____   ORD (*) SNL G-200

  (*) See ORD 1, Introduction and Index, for published catalogs of the ordnance section of the Department of the Army Supply Catalog.

3. Forms

The following forms are applicable to this matériel:
Standard Form 91, Operator's Report of Motor Vehicle Accident.
Standard Form 91A, Transcript of Operator's Report of Motor Vehicle Accident.
Standard Form 93, Report of Investigating Officer.
Standard Form 94, Statement of Witness.
DA Form 30b, Report of Claims Officer.
DA AGO Form 9-3, Processing Record for Storage Shipment.
DA AGO Form 9-4, Vehicular Storage and Servicing Record.
DA AGO Form 9-69, Spot Check Inspection Report for all Full-Track and Tank-Like Wheeled Vehicles.
DA AGO Form 9-74, Motor Vehicle Operator's Permit.
DA AGO Form 9-75, Daily Dispatching Record of Motor Vehicles.
DA AGO Form 348, Driver's Qualification Record.
DA AGO Form 460, Preventive Maintenance Roster.
DA AGO Form 461-5, Limited Technical Inspection.
DA AGO Form 462, Work Sheet for Full-Track and Tank-Like Wheeled Vehicles—Preventive Maintenance Service and Technical Inspection.
DA AGO Form 468, Unsatisfactory Equipment Report.
DA AGO Form 478, MWO and Major Unit Assembly Replacement Records and Organizational Equipment File.
DA AGO Form 811, Work Request and Job Order.
DA AGO Form 811-1, Work Request and Hand Receipt.
DD Form 6, Report of Damaged or Improper Shipment.
DD Form 317, Preventive Maintenance Service Due (Sticker).
OO Form 5825, Artillery Gun Book.

4. Other Publications

The following publications contain information pertinent to this matériel and associated equipment:
  a. AMMUNITION.
Ammunition, General_____ TM 9-1900
Artillery Ammunition_____ TM 9-1901

Ballistic Data, Performance of Ammunition	TM 9-1907
Qualification in Arms and Ammunition Training Allowances.	AR 775-10
Regulations for Firing Ammunition for Training, Target Practice, and Combat.	SR 385-310-1
Report of Accident Experience	SR 385-10-40
Small Arms Ammunition	TM 9-1990

b. ARMAMENT.

75-mm Gun M6 and Combination Gun Mount M64	TM 9-313
Browning Machine Guns, Cal. .30, M1917A1, M1919A4, and M1919A6.	FM 23-55
Browning Machine Gun, Cal. .50, HB, M2	FM 23-65
Crew Drill and Service of the Piece, and Stowage Light-Tank M24.	FM 17-75
Tank and Tank Destroyer Gunnery	FM 23-100

c. CAMOUFLAGE.

Camouflage	TM 5-267
Camouflage, Basic Principles	FM 5-20
Camouflage of Vehicles	FM 5-20B

d. COMMUNICATIONS.

Installation of Radio and Interphone Equipment in Tank, Light, M24.	TM 11-2754
Radio Set AN/VRC-3	TM 11-637
Radio Set SCR-506-A	TM 11-630
Radio Sets SCR-508-A, C, D, AM, CM, DM; SCR-528-A, C, D, AM, CM, DM; and AN/VRC-5.	TM 11-600

e. DECONTAMINATION.

Decontamination	TM 3-220
Decontamination of Armored Force Vehicles	FM 17-59
Defense Against Chemical Attack	FM 21-40

f. DESTRUCTION TO PREVENT ENEMY USE.

Explosives and Demolitions	FM 5-25

g. GENERAL.

Cooling Systems: Vehicles and Powered Ground Equipment.	TM 9-2858
Driver Selection, Training and Supervision, Half-Track and Full-Track Vehicles.	TM 21-301
Instruction Guide: Operation and Maintenance of Ordnance Matériel in Extreme Cold (0° to —65° F.).	TM 9-2855
Manual for the Full-Track Vehicle Driver	TM 21-306
Motor Vehicles	AR 700-105

Mountain Operations	FM 70-10
Operations in Snow and Extreme Cold	FM 70-15
Precautions in Handling Gasoline	AR 850-20
Principles of Automotive Vehicles	TM 9-2700
Spark Plugs	TB ORD 313
Supplies and Equipment—General: Unsatisfactory Equipment Report.	SR 700-45-5
Storage Batteries—Lead-Acid Type	TM 9-2857

h. MAINTENANCE AND REPAIR.

Basic Maintenance Manual	TM 38-650
Cleaning, Preserving, Sealing, and Related Materials Issued for Ordnance Matériel.	TM 9-850
Cold Weather Lubrication: Operation and Maintenance of Artillery and Sighting Fire Control Matériel.	TB ORD 193
Hand, Measuring, and Power Tools	TM 10-590
Lubrication Order—Light Tank M24	LO 9-729
Maintenance and Care of Hand Tools	TM 9-867
Motor Vehicle Inspection and Preventive Maintenance Services.	TM 37-2810
Painting Instructions for Field Use	TM 9-2851
Preparation of Ordnance Matériel for Deep-Water Fording.	TM 9-2853
Tracklaying Vehicles: Tracks Currently Applicable	TB ORD 391

i. SIGHTING AND FIRE CONTROL EQUIPMENT.

Auxiliary Sighting and Fire Control Equipment	TM 9-575
Instruction Guide: Elementary Optics and Applications to Fire Control Instruments.	TM 9-2601
12-Inch Graphical Firing Tables	TM 9-524

j. SHIPMENT AND LIMITED STORAGE.

Army Marking Directive	TM 38-414
Army Shipping Document	TM 38-705
Instruction Guide: Ordnance Packaging and Shipping (Posts, Camps, and Stations).	TM 9-2854
Ordnance Storage and Shipment Chart—Group G	TB 9-OSSC-G
Preparation of Unboxed Ordnance Matériel for Shipment.	SB 9-4
Protection of Ordnance Matériel in Open Storage	TB ORD 379
Shipment of Supplies and Equipment: Report of Damaged or Improper Shipment.	SR 745-45-5

Standards for Oversea Shipment and Domestic Issue of Ordnance Matériel Other Than Ammunition and Army Aircraft.   TB ORD 385

Storage, Inspection, and Issue of Unboxed Serviceable Motor Vehicles; Preparation of Unserviceable Vehicles for Storage; and Deprocessing of Matériel Prior to Operation.   SB 9-63

INDEX

	Paragraph	Page
Accelerator pedal and hand throttle	18	16
Accessories	8	7
Accidents	2	2
Air cleaner (crankcase)	90	114
Air cleaners:		
Installation and removal	115	164
Maintenance	8	7
Ammeter:		
Inspection	8	7
Installation and removal	134	203
Main battery	34	24
Ammunition:		
Authorized rounds (table V)	244	357
General	243	356
Precautions in firing	246	363
Preparation for firing	245	362
Subcaliber ammunition	247	363
Ammunition boxes	54, 189	41, 286
Antennas:		
General	251	370
Location and components (table VI)	251	370
Antiaircraft machine gun mount	199	299
Arm bumper spring bracket	181	281
Armament	5, 69, 215	5, 69, 319
Azimuth indicator M21	231, 233	335, 347
Batteries:		
Clean	8, 68	7, 65
Maintenance	212	317
Removal	241	356
Trouble shooting	77	97
Batteries and lighting system:		
Description and data	126	184
General	127	193
Installation and removal:		
Circuit breakers	130	198
Conduits and cables	131	199
Lights	129	195
Master battery switch	128	194
Bearings	173, 176, 179	262, 270, 277
Belts:		
Adjustment and replacement	88	110
Inspection	8	7
Blocking for rail shipment	259	387
Boresighting	234	348

	Paragraph	Page
Brake band adjustment	162	239
Brake shoes	163	240
Brakes:		
Operation	22	18
Testing	9	11
Breathers	8	7
Breech mechanism:		
Cleaning	227	331
Lubrication	227	331
Bulkhead doors	185	283
Bulkhead ventilating doors	30	22
Buttons, circuit breaker reset	33	24
Caution plates. (*See* Plates.)		
Charging stabilizer with oil	208	307
Charging system:		
Data	123	179
Description:		
Generator regulators	125	183
Generators	124	182
Trouble shooting	76	96
Choke	8, 19	7, 17
Circuit breaker reset buttons	33	24
Cleaning	67	58
Clutch	23	18
Coil	103	142
Cold weather (extreme)	211	317
Combination gun mount M64:		
Filling and draining recoil mechanism	203	302
Maintenance	202	301
Communication system:		
Data	248	364
Description	248	364
Inspection	254	377
Precautions	255	378
Compensating wheel	173	262
Compensating wheel support arm assembly	174	265
Conduits and cables	131	199
Controlled differential. (*See* Differential.)		
Controls:		
Description:		
Fuel pump	14	15
Shifting	24	18
Spark	20	17
Turret traversing	42	29
Ventilating blower	29	22
General	10	12
Description and data	117	167
Installation and removal:		
Fan	121	174
Hose and connections	119	172
Radiator	121	174
Radiator thermostat	120	172
Water pump	122	178

	Paragraph	Page
Controls—Continued		
Maintenance	118	169
Trouble shooting	75	95
Crankcase air cleaner	90	114
Cylinder head gaskets	89	111
Destruction of matériel to prevent enemy use:		
Burning (Method No. 1)	262	392
Demolition (Method No. 2)	263	393
General	260	391
Gunfire (Method No. 3)	264	394
Sighting and fire control equipment	261	392
Differential:		
Adjustment of steering and brake band adjustment	162	239
Installation and removal:		
Oil cooler	165	244
Oil pump and screen	164	242
Steering and brake shoes	163	240
Replacement	166	245
Differential, controlled:		
Description	161	238
Trouble shooting	82	105
Distributors	102	138
Door:		
Bulkhead	185	283
Driver's	27, 184	19, 283
Escape	31	22
Replacements	194	289
Turret	41	26
Ventilating, in bulkhead	30	22
Drain valves	191	287
Driver or operator preventive maintenance services (table II)	68	65
Driver's doors	27, 184	19, 283
Driving instructions	46	34
Electrical circuit number list	101	136
Electrical outlet	40	26
Electrical testing of stabilizer	207	304
Elevating handwheel difficult to operate	210	314
Elevation quadrant M9	231, 233	35, 347
Emergency ignition switch	12, 136	13, 204
Engine:		
Description	85	107
Installation	98	130
Installation and removal:		
Belt adjustment and replacement	88	110
Cylinder head gaskets	89	111
Intake and exhaust manifolds	94	120
Mountings	95	123
Oil pan and gaskets	92	118
Oil pump strainer	93	119
Operation in vehicle	87	108
Removal	97	125

	Paragraph	Page
Engine—Continued		
Servicing:		
Crankcase air cleaner	90	114
Oil filter	91	115
Starting	44	30
Stopping	48	35
Trouble shooting	71	87
Tune-up	86	108
Engine temperature gages	36, 142	26, 206
Escape door	31	22
Exhaust system, pipes, and mufflers	116	167
Fan, removal	121	174
Fender, inspection	8	7
Final drives:		
Description and data	167	249
Installation and removal:		
Hubs and sprockets	168	250
Propeller shafts	159	236
Replacement	169	251
Fire extinguishers:		
Inspection	8	7
Installation	200	299
Maintenance	200	299
Removal	201	300
Types:		
Fixed	50	37
Portable	51	38
Firing lock M15	228	334
Fording:		
Maintenance after	213	318
Operation	57, 68	45, 65
Forms	2	2
Fuel	8	7
Fuel pump controls	14	15
Fuel pumps	113	162
Fuel switches	139	205
Fuel system:		
Description and data	111	151
Installation and removal:		
Carburetors	112	155
Fuel pumps	113	162
Tank and lines	114	163
Trouble shooting	74	94
Fuel tanks and lines	114	163
Fuze setter:		
M13	233	347
M14	231, 240	335, 356
M27	231, 240	335, 356
Gages:		
Engine temperature	36, 142	26, 206
Inspection	9	11

	Paragraph	Page
Generating and charging system:		
Data	123	179
Description:		
Generator regulators	125	183
Generators	124	182
Trouble shooting	76	96
Generator regulators	125	183
Generators	124	182
Gun, 75-mm, M6:		
Authorized rounds of ammunition	244	357
Breech mechanism operation	221	327
Description	216	319
Elevation operation	219	323
Extracting empty case	224	329
Fails to fire	223	327
Preparation for firing	218	322
Stabilizer operation	220	323
Tabulated data	217	321
Traveling position	229	335
Unloading	225	330
Gun fluctuates	210	314
Gun mount M64. (*See* Combination gun mount M64.)		
Guns, elevating and traversing	9	11
Gyro control and gear box	209	310
Handwheel difficult to operate	210	314
Hoses and connections	119	172
Hot weather (extreme), maintenance in	212	317
Hubs	168	250
Hull:		
Ammunition boxes	189	286
Care in extreme heat	212	317
Description:		
Drain valves	191	287
General	3	2
Periscopes	192	287
Propeller shaft tunnel	188	286
Subfloor	190	286
Description and data	182	282
General:		
Protective pads	186	284
Sealing hull parts	183	283
Inspection	8	7
Installation and removal:		
Bulkhead doors	185	283
Driver's doors	184	283
Seats	187	285
Hull armor	5	5
Hunting (rapid gun vibration)	210	314
Hydramatic transmission:		
Description and data	148	213
Installation and removal:		
Torus replacement	152	221
Transmission replacement	151	217
Valve body	153	222

	Paragraph	Page
Hydramatic transmission—Continued		
Manual control linkage adjustment	149	214
Test and band adjustment	150	215
Trouble shooting	79	103
Hydrostatic lock	44	30
Ignition switch	13, 135	13, 204
Ignition system:		
Data	100	136
Description	99	133
Electrical circuit number list	101	136
Installation and removal:		
Coil	103	142
Distributor	102	138
Spark plugs	106	146
Timing	104	143
Trouble shooting	72	91
Wiring	105	145
Ignition timing	104	143
Indicator, azimuth, M21	231, 233	335, 347
Instruction panel	133	202
Instruction panel, instruments, switches, and sending units, description.	132	200
Instruction plates. (See Plates.)		
Instruments:		
General	10	12
Inspection	9	11
Interphone box BC-1362, external	253	373
Interphone control boxes:		
BC-604-A to G	249	364
BC-606-H	249	364
Interphone extension kit RC-298	253	373
Interphone system	252	371
Lamps. (See Lights.)		
Latches	194	289
Leaks	8, 9	7, 11
Lever, steering brake	21	17
Lever, transfer unit control	155	225
Light, warning signal	35	24
Lighting switches	16, 138	15, 205
Lighting system, trouble shooting	77	97
Lights, installation and removal	129	195
Limited storage instructions	257	381
Lines, fuel	114	163
Loading the vehicles for rail shipment	258	382
Lock, hydrostatic	44	30
Lock, turret	198	298
Lubrication:		
Breech mechanism	227	331
Complete	8	7
Unusual conditions	63	52
Usual conditions	64	52
Lubrication order	62	52
Machine gun, cal. .30, firing	226	330

	Paragraph	Page
Machine gun mount, antiaircraft	199	299
Machine gun mount, cal. .30 (bow)	204	302
Machine gun mount, cal. .30 (combination gun mount M64)	205	304
Main propeller shafts	158	234
Maintenance operations	69	69
Manifolds, intake and exhaust	94	120
Manual control linkage adjustment	149	214
Master battery switch	11, 128	12, 194
Mount, antiaircraft machine gun	199	299
Mount, gun, M64. (See Combination gun mount M64.)		
Mount, machine gun, cal. .30 (bow)	204	302
Mount, machine gun, cal. .30 (combination gun mount M64)	205	304
Mounting, FT-237	249	364
Mountings, engine. (See Engines.)		
Muffler	116	167
Name plates. (See Plates.)		
Neutral pedal	25	19
Observation of terrain	239	356
Oil	8	7
Oil cooler	165	244
Oil filter	91	115
Oil pan and gaskets	92	118
Oil pump:		
Screen	164	242
Stabilizer	207, 209	304, 310
Strainer	93	119
Oil seals	173, 176	262, 270
Operation:		
Extreme-cold weather	53, 54	40, 41
Extreme-hot weather	55	44
Fording	57	45
Unusual conditions	52	38
Unusual terrain	56	44
Usual conditions	43	30
Organizational mechanic or maintenance crew preventive maintenance services (quarterly, monthly) (table III)	69	69
Outlet, electrical	40	26
Painting	65	53
Panel light access plugs	39	26
Pedal, neutral	25	19
Periscope:		
M4A1	231-234, 236	335, 350
M6	32, 233, 238, 239	23, 347, 354, 356
M10P	231-234, 236, 242	335, 350, 356
M13	32, 238, 239	23, 347, 356
M15A1	231-234, 238, 239, 242	335, 354, 356
M16	231-234, 236	335, 350
Periscope mount M66	23, 232, 233	18, 343, 347
Periscopes	192	287

	Paragraph	Page
Pipes, exhaust	116	167
Placing vehicle (tank) in motion	45	32
Plates (caution, instruction, name)	4	4
Plugs, panel light access	39	26
Portable fire extinguishers	51	38
Power train, description of	3	2
Prevent enemy use, destruction to:		
Burning	262	392
Demolition	263	393
General	260	391
Gunfire	264	394
Sighting and fire control equipment	261	392
Preventive maintenance services:		
Driver's or operator's (table II)	68	65
General	66	53
Organizational mechanic or maintenance crew (table III)	69	65
Propeller shaft tunnel	188	286
Propeller shafts, trouble shooting	81	105
Propeller shafts and support bearings:		
Description and data	157	231
Installation and removal:		
Final drive propeller shafts	159	236
Main propeller shafts	158	234
Propeller shaft support bearing	160	237
Publications, where stowed	9	11
Quadrant, elevation M9	231, 233	335, 347
Radiator	121	174
Radiator thermostat	120	172
Radio interference suppression:		
Early model vehicles	146	210
Late model vehicles	147	211
Purpose	145	210
Trouble shooting	78	101
Radio set:		
AN/VRC-3	250	367
SCR-508, SCR-528, SCR-608B, SCR-628B	249	364
Rail shipment:		
Blocking	259	387
Loading	258	382
Receiver, BC-603, BC-683-BM	249	362
Records	2	2
Relay	110	150
Reports	2	2
Road test	69	69
Run-in procedures	9	11
Scope	1	1
Screen	164	242
Seals	179	277
Seat adjustment	28	21
Seats	187, 195	285, 291
Sending unit	144	207

	Paragraph	Page
Servicing matériel	7, 8	7
Shifting controls	24	18
Shipping instructions:		
Blocking	258	382
Domestic	256	380
Loading	259	387
Shock absorbers	181	281
Sighting and fire control conditions (direct fire)	236	350
Sighting and fire control instruments:		
Arrangement (table IV)	230	335
Description	230	335
General	230	335
Indirect fire	237	353
Maintenance	241	356
Observation of fire	238	354
Operation, ordinary conditions	235	350
Preparation for traveling	242	356
Signal switches	144	207
Siren (or horn):		
Inspection	8	7
Installation	140	205
Removal	140	205
Trouble shooting	77	97
Siren switch	17	16
Solenoid (cranking motor)	109	150
Spark control	20	17
Spark plugs	106	146
Special tools and equipment (table I)	61	47
Speedometer:		
Installation and removal	143	206
Location	37	26
Spot light	26	19
Sprockets	168	250
Stabilizer:		
Gyro control	209	310
Isolation of functional defects	210	314
Maintenance of	209	310
Purging the system	208	307
Stabilizer electric circuit and operation of oil pump	207	304
Starter	108	149
Starter switch	15, 137	15, 204
Starting system:		
Data	107	147
Description	107	147
Installation and removal:		
Cranking motor solenoid	109	150
Relay	110	150
Starter	108	149
Trouble shooting	73	92
Starting the engines	44	30
Steering and brake band adjustment	162	239
Steering and brake shoes	163	240
Steering brake levers	21	17

	Paragraph	Page
Steering linkage	8	7
Stopping the engines	48	35
Stopping the vehicle	47	34
Subcaliber ammunition	247	363
Subfloor	190	286
Support bearing	160	237
Suspension. (*See* Tracks and suspension.)		
Switch box BC-1361	253	373
Switches:		
Emergency ignition	12, 136	13, 204
Fuel	139	205
Ignition	13, 135	13, 204
Lighting	16, 138	15, 205
Master battery	11, 128	12, 194
Signal	144	207
Siren (or horn)	17, 140	16, 205
Starter	15, 137	15, 204
Table I—Special tools and equipment	61	47
Table II—Driver's or operator's preventive maintenance services	68	65
Table III—Organizational mechanic or maintenance crew preventive maintenance services (quarterly, monthly)	69	69
Table IV—Arrangement of sighting and fire control instruments	231	335
Table V—Authorized rounds for 75-mm gun M6	244	357
Table VI—Location and components of antennas	251	370
Tabulated data	5, 217	5, 321
Tachometer:		
Inspection	8	7
Installation	143	206
Location	38	26
Removal	143	206
Tank:		
Placing in motion	45	32
Stopping	47	34
Towing	49	35
Telescope, M71K, M83F	231–234, 236, 242	335, 350, 356
Telescope mount M65	231–233	335
Temperature, test	9	11
Test and band adjustment	150	215
Throttle, pedal and hand	18	16
Tools and equipment	58–61, 69	47, 69
Tools and equipment (special) (table I)	61	47
Torsion bars	177	274
Torus replacement	152	221
Towing connections	8	7
Towing the vehicle (tank)	49	35
Track support rollers	179	277
Track wheels	176	270
Tracks and suspension:		
Adjustment of	171	256
Description and data	170	253

	Paragraph	Page
Tracks and suspension—Continued		
Installation and removal:		
Arm bumper spring bracket	181	281
Compensating link	175	269
Compensating wheel support and assembly	174	265
Shock absorbers	180	280
Torsion bars	177	274
Track support rollers, bearings, and seals	179	277
Track wheel support arm	178	277
Interchangeability and removal:		
Compensating wheel, bearings, and oil seals	173	262
Track wheels, bearings, and oil seals	176	270
Replacement of track	172	257
Trouble shooting	83	105
Tracks, tension	9	11
Transfer unit:		
Description and data	154	224
Inspection	9	11
Installation and removal of control levers	155	225
Replacement	156	229
Trouble shooting	80	104
Transmission	9	11
Transmission replacement	151	217
Transmitter, BC-604, BC-684-BM	249	364
Traversing mechanism	197	293
Trouble shooting	70	87
Tunnel, propeller shaft	188	286
Turret:		
Antiaircraft machine gun mount	199	299
Armor	5	5
Description of	3, 193	2, 287
Doors	41, 194	26, 289
Installation and removal:		
Doors and latches	194	289
Lock	198	298
Seats	195	291
Vision devices	196	292
Replacement of traversing mechanism	197	293
Traversing controls	42	29
Trouble shooting	84	106
Unusual conditions, driver's or operator's preventive maintenance services	68	65
Unusual conditions for operation	52-56, 68, 69	38, 65, 69
Unusual terrain	214	318
Usual conditions, driver's or operator's preventive maintenance services	68	65
Valve body	153	222
Vehicle. (*See* Tank.)		
Ventilating blower controls	29	22
Ventilating doors in bulkhead	30	22
Ventilator, trouble shooting	77	97

	Paragraph	Page
Vents	8	7
Vision devices	8, 196, 233	7, 292, 347
Warning signal lights	35	24
Water pump, removal	122	178
Wheel. (*See* Compensating wheel and track wheel.)		
Wiring, removal and installation	105	145

○

©2013 Periscope Film LLC
All Rights Reserved
ISBN#978-1-937684-33-4
www.PeriscopeFilm.com